W9-BSD-663

How do you choose your first telescope or build one from first principles? What can the deep sky offer you season-by-season? How do you get started in astrophotography? And progress to CCD imaging? *The Guide to Amateur Astronomy* answers the questions of the novice and the experienced amateur astronomer in one easy-to-use and comprehensive account. Throughout the emphasis is on practical methods to get you started and then develop your skills; with lavish illustrations to show you just what is possible.

This second edition of the highly successful *Guide* has been fully revised and updated. It now takes you from basic 'piggyback' astrophotography, through the use of a cold camera to state-of-the-art CCD imaging; from studies of the planets to the most distant objects in the Universe.

From guidelines for the care and adjustment of your telescope through to lists of the spectral classification of stars, amateur astronomy societies and clubs, all the information you need for your voyage of discovery and revelation is provided in this self-contained, helpful handbook.

THE GUIDE TO AMATEUR ASTRONOMY

THE GUIDE TO AMATEUR ASTRONOMY

Jack Newton and Philip Teece

Second edition

Dedication

Again, for Alice and for Wendy

Published by the Press Syndicate of the University of Cambridge
The Pitt Building, Trumpington Street, Cambridge CB2 1RP
40 West 20th Street, New York, NY 10011-4211, USA
10 Stamford Road, Oakleigh, Melbourne 3166, Australia

First published 1988
Reprinted 1990
Second edition 1995

Printed in Great Britain at Butler & Tanner Ltd, Frome, Somerset

A catalogue record for this book is available from the British Library

Library of Congress cataloging in publication data

Newton, Jack, 1942–
The guide to amateur astronomy / Jack Newton and Philip Teece –
2nd edn.
p. cm.
Includes bibliographical references and index.
ISBN 0 521 44492 6
1. Astronomy — Amateurs' manuals. I. Teece, Philip. II. Title.
QB64.N49 1994
522—dc20 93-40354
CIP

ISBN 0 521 44492 6 hardback

BT

Contents

Foreword
by Dr Helen Sawyer Hogg

Usually the amateur astronomer begins the study of the heavens with wonderment at the beauties to be seen with the naked eye on a clear, moonless night away from the illumination of a city. The starclouds of the Milky Way and easy-to-recognize constellations draw attention: the Big Dipper or Great Bear, visible all year in the north for northern hemisphere observers, the brilliant winter constellation of Orion, or in the southern hemisphere, the Southern Cross. Then an observer is attracted to other celestial objects of smaller apparent size, visible to the unaided eye: the Moon and the bright planets; the Pleiades, the star-cluster in the constellation Taurus, known as the Seven Sisters; the fuzzy patch known as the Great Nebula in Orion; and another misty patch in the constellation Andromeda. This latter has come to be known in this century as the most distant object the unaided human eye can see – the Andromeda Galaxy at a distance of 2.2 million light-years.

By this time, if not before, most amateurs are eager to have optical aids, binoculars or telescopes, perhaps cameras, and they need information on the materials they should acquire and how to use them. Jack Newton and Philip Teece in *The Guide to Amateur Astronomy* have provided exactly the information that large numbers of amateurs are scurrying around to obtain. Newton and Teece form a pair of remarkably dedicated and widely experienced amateurs whose knowledge can be relied on to lead others easily along the various types of pathways to the stars.

Furthermore, the authors draw our attention to the fact that astronomy is not all strenuous night-time observing. There is also joy to be found in associating with kindred spirits, as in the annual meetings of (for example) the Royal Astronomical Society of Canada, held in cities coast to coast, and the local meetings of its Centres.

The moods of these two activities are somewhat like the two moods described by Milton in his companion poems *L'Allegro* and *Il Penseroso*. The one mood is vivacious and gregarious, the other thoughtful and generally solitary. Of *Penseroso*, Milton wrote '... Or let my lamp at midnight hour/ Be seen in some high lonely tower/

Where I may oft outwatch the Bear'. If we think of the lamp as turned on to adjust the telescope or to change plates, we picture the attendant at work. And we share the activities with *Allegro* when 'Towered cities please us then,/ And the busy hum of men'.

I think the happiest astronomer is the one who engages in both activities, studying celestial objects by night or the Sun by day, and then exchanging views at astronomical meetings and learning more from well-informed speakers. The study of the heavens is certainly one of the most uplifting pursuits available to human beings. Determination and self control rank very high in the pursuit of observational astronomy, either professional or amateur, with weather vagaries and strenuous night-time hours.

The New England poet and philosopher Ralph Waldo Emerson wrote, 'If the stars should appear one night in one thousand years ...'. It is a stunning concept. How could amateur and professional astronomers ever plan for such a night? Fortunately we don't have to. We are given night after night to learn more about the wonderful Universe in which we live. Each of us might finish Emerson's sentence. Emerson finished it by writing 'how would men believe and adore!'

Helen Sawyer Hogg
David Dunlap Observatory
University of Toronto

Acknowledgements

Many people have contributed material and ideas to this book. In addition to the individual acknowledgements that appear throughout, in the contexts where they belong, we would like to add special thanks to our copy-editor Alison Litherland. Her contribution has removed many embarrassing infelicities, improved the readability of the *Guide*, and added insights that we have gratefully included in our work.

All photographs by Jack Newton except where otherwise captioned.

Part I First discoveries: the adventure begins

1.1 Introduction: the accessible Universe

A schoolboy once stood in a cornfield at dusk, watching excitedly for the earliest stars to emerge in the evening twilight. His brand new telescope, fixed on top of a fencepost, waited beside him to receive its first starlight.

The boy had christened his little mailorder telescope 'The Strawberry Spyglass', because its eighteen-dollar price had cost him weeks of hard work picking strawberries in his father's market garden. It was a simple instrument. With a main lens of only two inches (5 cm) diameter and a homebuilt mounting that incorporated a stone grinding wheel and an angled structure of wooden struts, the Strawberry Spyglass was, by most amateur astronomers' standards, very crude indeed. Yet for the boy, whose name was Leslie Peltier, the diminutive two-inch telescope was destined to become his spaceship for an unusually venturesome excursion across the depths of the cosmos.

His first glance at the surface of the Moon was a revelatory experience for Peltier. In his own words: 'I descended into craters by the score ... and down the blinding wall of Aristarchus. One night I walked across the strange and violent gash of the Alpine Valley and ... rested briefly in the long shadows of Pico and Piton.'

Soon his telescope had transported him far beyond the Moon. He explored the satellites and cloud belts of Jupiter, the intricacies of Saturn's divided rings and eventually the remote worlds of Uranus and Neptune. Even there, at the solar system's outer fringes, he was only on the threshold of far bolder odysseys into the broader Universe, 1500 light-years to the Orion Nebula, 20 000 light-years to the giant Hercules star cluster, more than two million light-years to the great galactic system in Andromeda.

These adventures with his first small telescope were only a beginning. Leslie Peltier's subsequent lifetime of cosmic explorations resulted in his discovery of a dozen comets, and his contribution to science of a hundred thousand observations of variable and peculiar stars. Shapley of Harvard once called him the world's greatest nonprofessional astronomer.

And the point of the story ... ?

The point is that, for even the uninitiated amateur, the Universe is *highly accessible*. Leslie Peltier was only one especially notable example of a general truth: that the key to the Universe is not a huge or costly telescope (or an inventory of high-tech accessories). Ultimately, it is not equipment of any kind. The key is simply your own enthusiasm, and your imagination.

Perhaps the most delightful aspect of the Universe is the ease with which you can explore its remote places, even as a casual observer. Eventually, you may undertake a more serious investment of thought and effort as an amateur astronomer. Then your adventure will carry you even further into a realm that is both eerily beautiful and scientifically fascinating.

The amateur's Universe

Even without optical aid, and in the sadly light-polluted environment of town or suburb, something beyond the Earth can be discovered. The lunar landscape reveals surprising details when studied carefully with the naked eye, the planets are easily found and followed, the brightest constellation can be recognized.

Away from city lights the unaided eye becomes a truly powerful astronomical instrument, discerning the bright star-clouds and dusty rifts of our Milky Way Galaxy and focusing on individual star clusters and pale nebulous patches that are the night sky's mysterious 'deep-sky objects'.

How far into deep space can you penetrate, using no equipment except your eye alone? The famous 'Double Cluster', which can be glimpsed as a misty glow in the northern sector of the constellation Perseus, lies an impressive 7500 light-years distant. Messier 24, a distinctive star-cloud in Sagittarius, is so remote that its light that you see tonight began the earthward journey 16 000 years ago. Yet, having crossed even that awesome gulf of space and time, your eye has reached only the threshold of a much deeper leap into the void. In Andromeda, as we have mentioned earlier, the pale oval swatch of light of which you may catch a glimpse on autumn evenings is a separate galaxy, 2.2 million light-years removed from our own.

The addition of very modest optical equipment – binoculars or a small telescope – allows a quantum leap into depths and details of the cosmos never seen by eye alone. A little bit of magnification shows the planets as spherical worlds with perhaps a hint of atmospheric or surface detail; the increased light-grasp of a good lens reveals many stars as colourful binary systems, resolves the myriad individual members of a globular cluster, permits you to view the complex inner depths of an interstellar nebula.

The range of your ventures in time and space will be vastly increased by a simple optical aid: the common $2\frac{1}{2}$-inch (6 cm) beginner's telescope enables you to glimpse star systems that lie twenty times more distant than the Andromeda Galaxy.

At this very modest level of astronomical equipment real scientific work in astronomy can begin, as well. The great amateur fields of variable star monitoring and solar recordkeeping are readily available to the keen owner of a small telescope, and some of our century's most dramatic nova discoveries have been made by astute observers using ordinary binoculars.

Someday your interest may lead you to the purchase or construction of a much larger and more sophisticated telescope. With optics in the eight-inch (20 cm) to twenty-inch (50 cm) diameter range you may find yourself committed to a very serious study of the Universe, with the capability to do highly detailed planetary observation, photometry of variable stars and astrophotography of deep-sky objects that will rival that achieved by the professional observatories. To the most dedicated amateur the cosmos may become a place not only for recreational exploration, but for intense and creative work.

The 'work' aspect suggests perhaps a danger to be avoided. Beware of letting the joy become lost amid the work. Unlike the professional astronomer whose funded projects are geared to specific scientific results, the amateur is a free agent whose ultimate motive is his or her own satisfaction. The truly successful amateur explorer of the Universe may be the casual lawnchair-and-binocular astronomer for whom it is all great fun, rather than the hundred-thousand-dollar observatory owner in whom the original childlike wonder has been lost.

The authors as amateurs

It is in the spirit described above that we have been privileged to enjoy the delight that we have found so readily accessible in the night sky. Each of us has indulged a passion for the planets, stars, nebulae and remote galaxies as an amateur astronomer for more than a quarter century.

Over the decades our interests have ranged from the most basic lunar and planetary observation to asteroid tracking, variable star work, solar eclipse chasing and deep-sky photography. We have built, used, sold and purchased a mountain of equipment, from small department-store refractors to giant reflecting telescopes weighing a ton. Yet fancy equipment and special observing projects have always remained secondary to the original simple goal – the thrill of acquiring a growing personal intimacy with the night sky.

The co-authors, incidentally, represent two rather different and complementary traditions of amateur astronomy. Jack Newton is an astrophotographer who is usually to be found perched at the guiding eyepiece of his 60 cm reflector, painstakingly time-exposing a distant galaxy with his homebuilt cold camera or acquiring a CCD image that will eventually be processed to reveal deep-sky objects that lie beyond the range of most amateur telescopes. Philip Teece

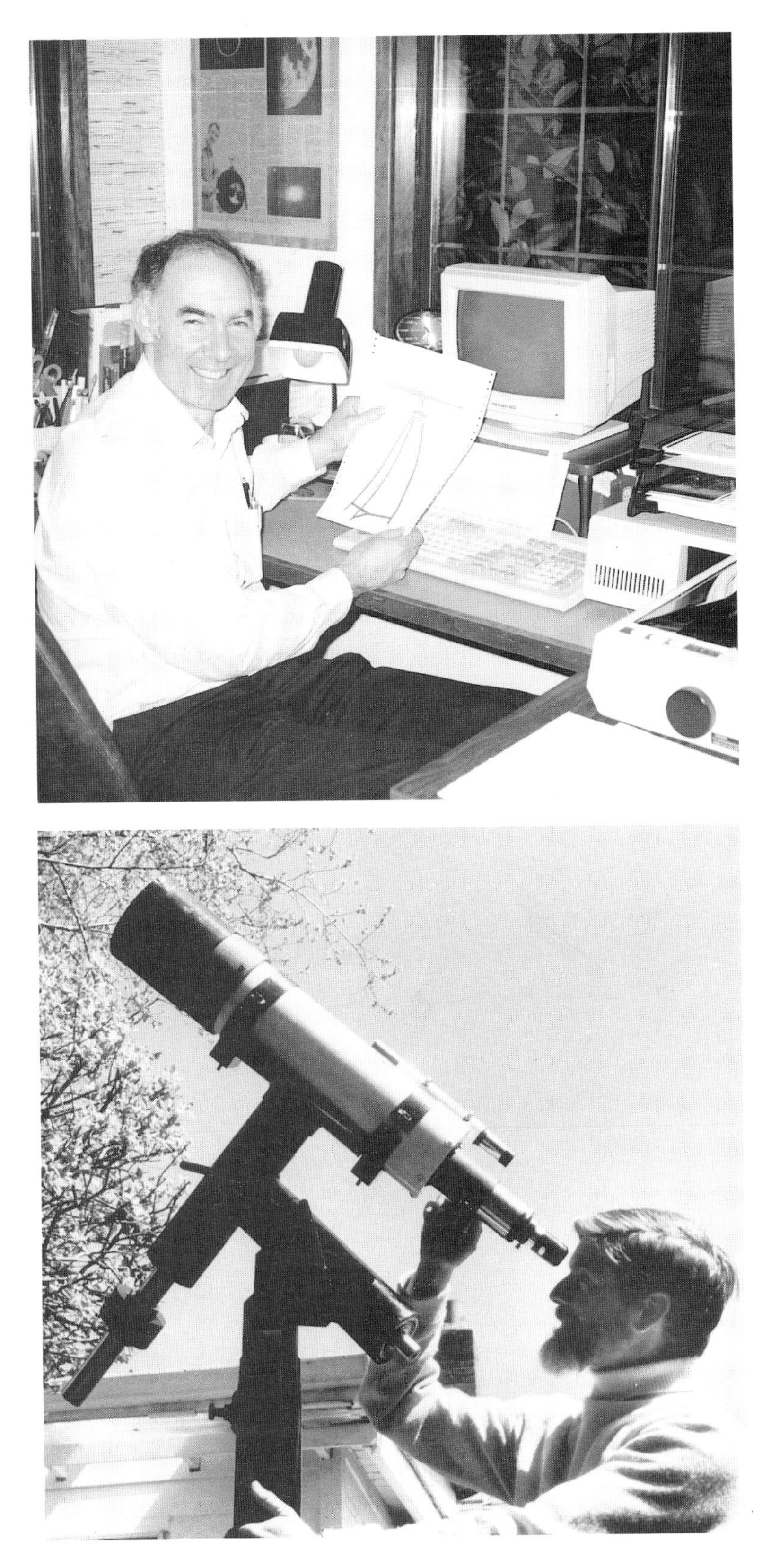

Plate 1. Jack Newton.

Plate 2. Philip Teece.

is a devotee of visual observation, often using a fine old 12 cm refractor to sketch elusive planetary features, or to locate and follow a tenth- or eleventh-magnitude asteroid.

It is our hope that we can share, on the pages that follow, every aspect of the various astronomical pursuits that we have so richly enjoyed. More than that, we offer also the observing secrets of many specialists whom we are grateful to number among our friends – amateur and professional experts on telescope construction, variable star observing, backyard photometry, comet hunting and much more.

If you are a beginner, however, don't be intimidated by the specialists' esoteric pursuits and complex gadgetry. For a view of many deep-sky objects you need only a dark sky and your unaided eye. Indeed, many of us who have spent half a lifetime with large telescopes and intricate cameras still return frequently to the profound satisfaction of merely scanning the heavens with binoculars or the naked eye.

We have stressed the ease which access can be gained to the Universe. Yet perhaps to the amateur astronomer the ultimate attraction in the remote star clouds and galaxies of deep space is their untouchability. We inhabit (we have in fact created) a planet that has become a commercial resource, to be consumed and despoiled. Yet our telescopes continue to be magic carpets by which we can still fly free, to wonders of the Universe that can never be used, manipulated or marketed. Their most reassuring attribute is that they lie in regions of space and in a framework of time forever beyond our reach.

1.2 Finding your way

A person who feels the first stirring of astronomical interest or curiosity is on the brink of something greater than he or she may imagine. Life offers few pleasures so intense as that of gaining familiarity with the night sky.

If you are at this introductory stage, beware of spoiling it by jumping into telescopic astronomy before you have enjoyed the rich (and necessary) delight of preliminary naked-eye exploration.

We recall a star party a few years ago at which a noted professional astrophysicist was present, in the midst of the little knot of amateurs with their varied small telescopes. His conversation fascinated us; he talked of the complex physics of stellar interiors and the bizarre orbital mechanics of multiple-star systems.

Yet when a telescope became free and he was invited to use it for a firsthand look at the night's best nebulae and galaxies, he declined with regret. His background, he explained, was purely theoretical; at no time in his life had he learned to find his way about the night sky without the aid of computer-controlled setting circles. The glory of the constellations overhead – the Lyre's neat little quadrangle, the widespread wings of Aquila, huge Cygnus with its central axis of Milky Way star clouds – held no welcoming familiarity for him.

Constellations: celestial road maps

Acquiring an easy, intimate familiarity with the night sky is the amateur astronomer's most important task. It is also his or her greatest joy.

At every level of amateur observing, a good thorough knowledge of the patterns of stars overhead is the observer's most important tool. The usual items of astronomical equipment – binoculars, cameras, telescopes – are all very secondary to this richly satisfying knowledge.

When amateur stargazers convene for an evening of telescopic observing, you can readily identify the most experienced observers

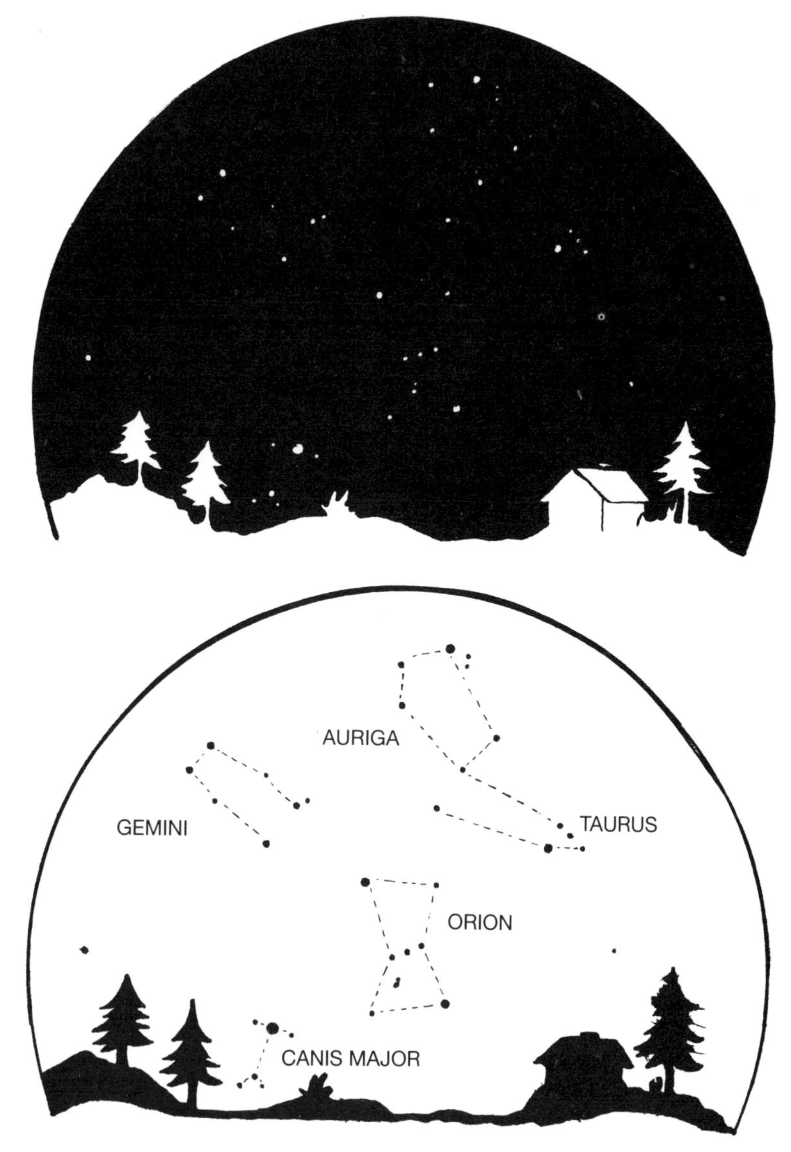

Figure 1.1. Winter night sky: a segment above the southern horizon. These prominent star patterns will be an easy first step in your programme of constellation recognition.

among them. It is not the cost or sophistication of their instruments that marks the group's really knowledgeable members; it is the impressive ease with which they locate a wide array of star clusters, nebulae and faint galaxies among the constellations overhead. Their intimacy with the heavens' subtle features is not a static achievement, but a continuing development; their personal catalogue of familiar deep-sky objects continues to grow throughout their lives. You too will want to develop your own mental catalogue of the night sky.

Where should you begin?

The location of every cluster, every notable double star, every dim, intriguing galaxy is a position within the star pattern formed by a specific constellation. Thus, your first step is to recognize the major constellations.

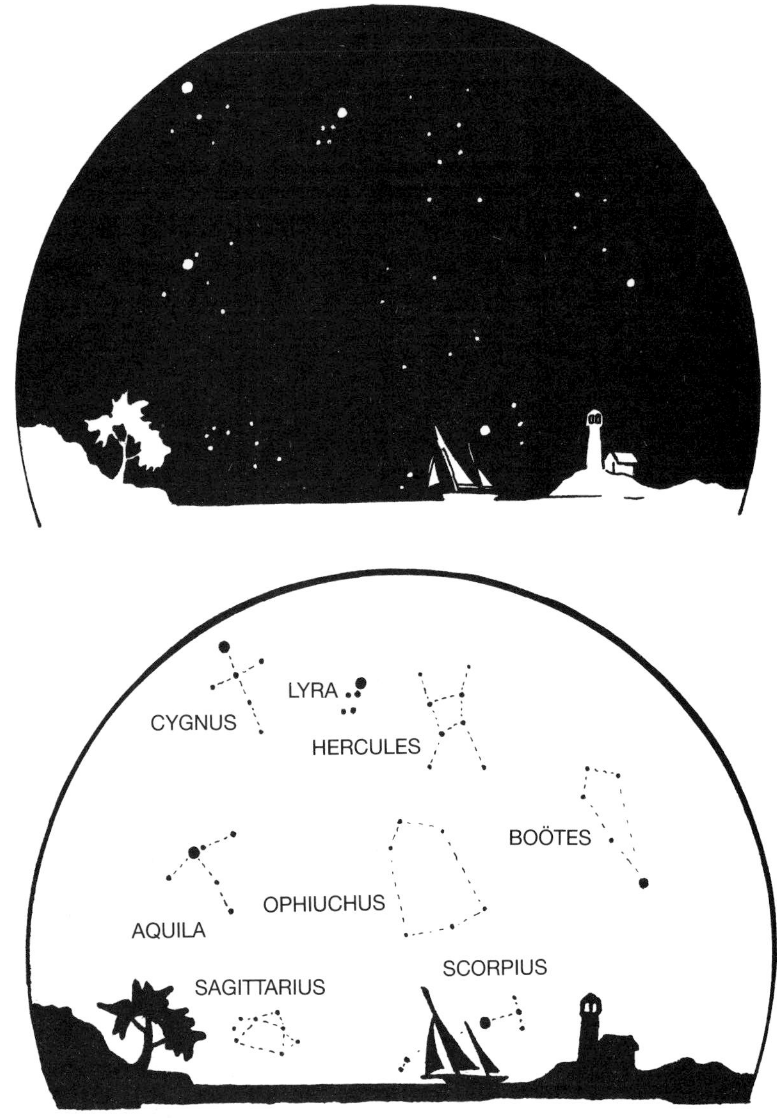

Figure 1.2. Summer night sky: the view facing south. Use the 'Summer Triangle' (the three principal stars of Cygnus, Lyra, Aquila) to establish your bearings.

If you are totally unfamiliar with the constellations, begin by attempting to recognize one bright, easy star pattern. At most seasons the Big Dipper, or Plough, is usually the beginner's first success. In winter Orion is an even easier first constellation, with its huge outer rectangle of bright stars and its distinctive transverse line of three central stars – the 'belt'.

Study the pattern first on a simple star map, like those in Figures 1.1 and 1.2. A planisphere, whose circular sky chart can be rotated to show the positions of constellations at each season, is an even better choice for initial study.

Next, look up at the sky, in a place reasonably far removed from bright lights, which complicate your task by hiding the stars in a haze of illumination. When you are facing the segment of the sky (southward, eastward, overhead, etc.) in which the star map or planisphere shows the constellation to be, attempt to match the pattern that you have studied on paper with an actual array of stars above.

The stumbling block at first will be the *scale* of the actual Dipper or Orion, or whatever else. Remember that the tiny, compact grouping of stars on the chart will be found in a greatly stretched and enlarged form in the sky. It is this initial problem of discovering the true scale that usually holds up a beginner's first recognition. When the scale of one prominent star pattern has become familiar, then it becomes an increasingly easy task to relate the small patterns on the chart to real asterisms in the night sky.

The stargazer's signposts

Each constellation that has been successfully found becomes a guidepost to others nearby.

At this point the reader may want to go outdoors under a dark night sky and begin the adventure of constellation-hopping, following guideposts illustrated in the small chart segments shown here.

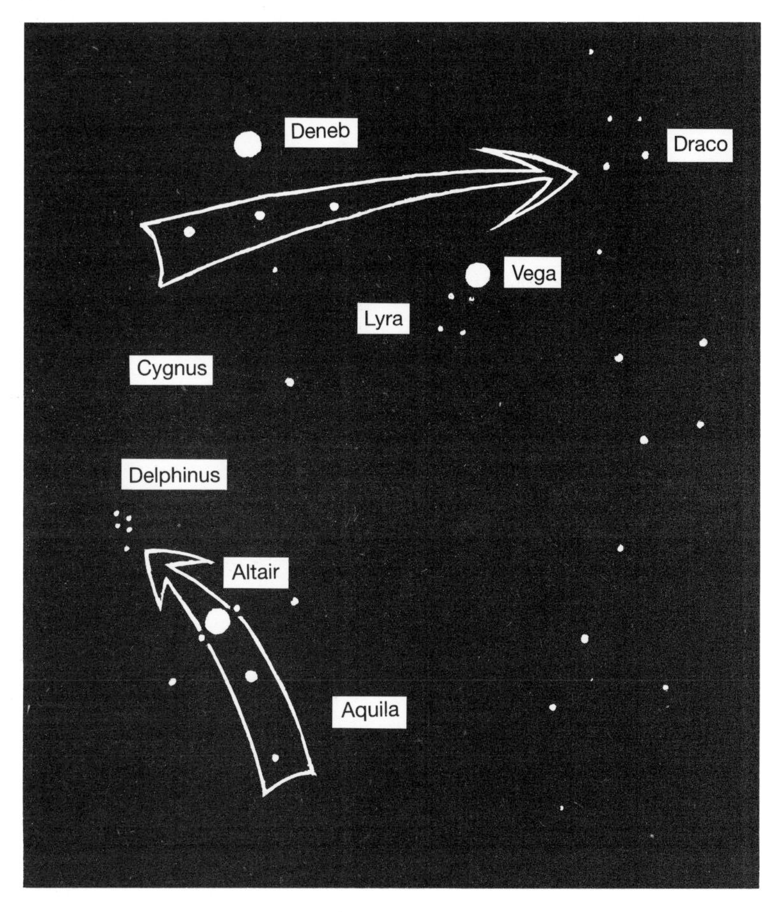

Figure 1.3. The 'Summer Triangle'. Use the star alignments shown here as arrows that point to other asterisms, as described in the text.

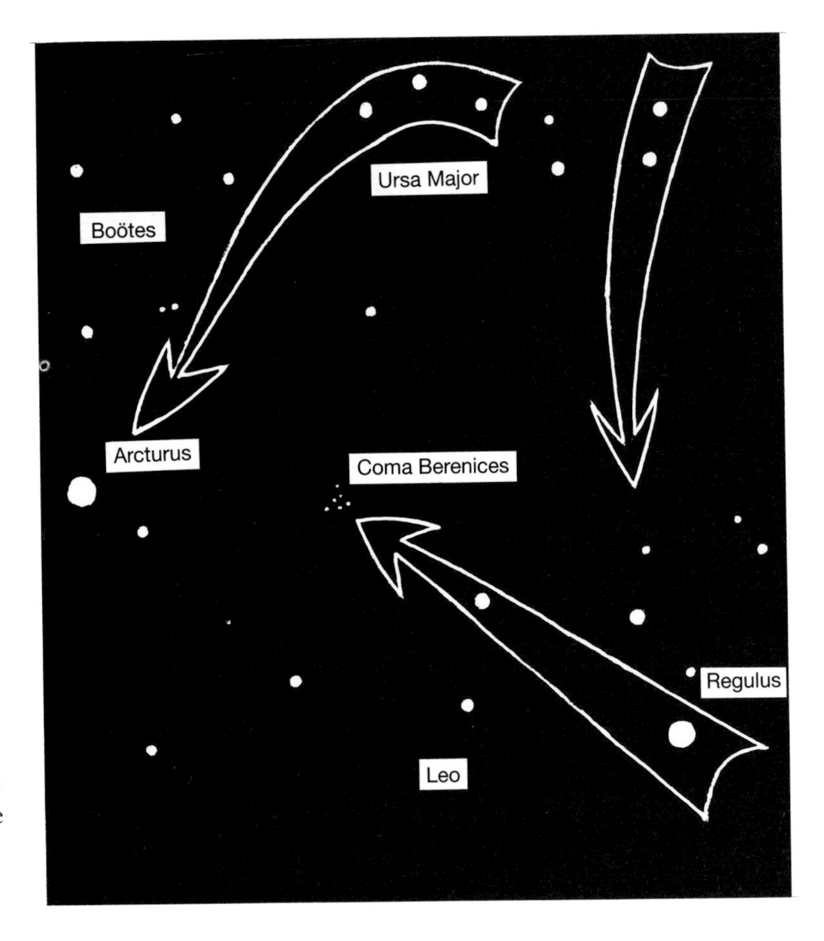

Figure 1.4. Guide stars in the Big Dipper lead your eye to bright *Arcturus* in Boötes, and to the head of Leo, the Lion. Leo (head extreme right, hindquarters lower centre of chart) shows the way to the Coma star cluster.

If it is a summer evening, try this: facing southward, look up at the large segment of sky that hangs overhead from southern horizon to the zenith directly above. The evening's three brightest stars will form a huge triangle (amateur astronomers refer to it as 'The Summer Triangle') that extends over a large portion of this segment of sky. The Summer Triangle appears on the chart in Figure 1.3. Find and recognize this sprawling configuration of the three bright summer stars.

Each of these summer stars is the highlight of a major constellation, which you should now be able to discern fairly easily.

The northern pair are *Vega* (upper right corner of the triangle), to which the compact parallelogram of little *Lyra* is attached, and *Deneb* (upper left) from which the 'cross' of Cygnus extends downward into the central space of the Summer Triangle. The southern apex of the triangle, *Altair*, lies at the heart of a smaller cross, the constellation Aquila.

Once you have recognized the shapes of these constellations, you are ready to use a powerful finding-aid that all amateur astronomers use to navigate among the stars in the night sky.

Many of the principal constellations contain lines or chains of stars that form pointers, directing the eye toward other groupings

Figure 1.5. Orion's belt as a signpost to the Hyades Cluster. Use Orion as a guide to other star groups nearby in the winter sky.

that you may wish to locate. For example, let us say that you have become familiar with Cygnus in the summer sky. You can find the dimmer and more subtle constellation Draco by following a signpost in Cygnus. An imaginary line drawn northwestward through the three bright stars that cross the constellation's axis will point directly at the small quadrangle that delineates the head of the serpent Draco. (You will find that one of the four 'head' stars of Draco is a neat binocular double.)

Try following another signpost. Begin this time with the Summer Triangle's lower apex-star *Altair* and its T-shaped constellation Aquila; an imaginary line passing northeastward along Aquila's central spine, through bright *Altair*, will point at the very neat little summer constellation Delphinus, the Dolphin. A part of the Dolphin's charm lies in the fact that it is one of the few constellations compact enough to be viewed in its entirety within a single binocular field. The larger constellations should, of course, be sought out and observed with the unaided eye.

Some further examples are illustrated in the accompanying charts.

Figure 1.4 shows the Big Dipper in part of a spring evening's sky, with the widely used lineup of the 'bowl' stars as a guidepost to

Leo, and also the dipper-handle-to-*Arcturus* pointer. Similarly, if you were searching for the Coma Berenices star cluster, you would find that a line mentally projected from *Regulus* in Leo the Lion through the uppermost star of the Lion's hindquarters would point conveniently at the faint Coma grouping.

Figure 1.5 centres on the best known winter constellation, Orion. The indicated directional arrow guides your eyes along the 'belt' stars towards the Hyades Cluster. This is only one of several guides that will suggest themselves when you study a map of Orion and the surrounding sky region.

As you begin your first explorations of the night sky with the naked eye, comparing the patterns on a star chart with actual constellations overhead, you will quickly develop your own favourite signposts. When a new constellation has been located on the chart, you will note patterns of bright or well-known stars whose alignment points the way toward the target of your search. Soon you, like most amateur astronomers, will have the entire sky linked together in an imaginary network of such interconnections.

Knowing the stars

A few months before the writing of this chapter, a young woman reporter was released after a long confinement in a Third World prison. As a political prisoner she had languished in her tiny cell for many years on no specific charge, and without sentence.

When at last she was free, a radio interviewer asked her how she had retained hope during those interminable years.

The woman replied that in her childhood she had been taught to recognize one star, the flashing blue jewel *Vega*, in the summer constellation Lyra. She found that in its season 'her' star could be viewed through the single tiny window of her cell. The return each year of that star with which she was so familiar served her as a thread of contact with the world outside. The appearance of friendly *Vega* framed in the small aperture kept alive her desire to step outside someday, to stand and look up at that sky.

No person should go through life without knowing at least a few individual stars by name. To the amateur astronomer, dozens of the prominent stars of each season should become intimate friends.

The intense iridescence of white *Sirius*, for instance, is a perennial delight to those who stargaze under cold winter skies. The intense sapphire *Vega* and the red giant *Antares* are welcome friends in the short nights of summer. At all seasons, *Polaris* is sought out by northern hemisphere skywatchers who wish to locate both the celestial pole and the bearing of geographic North. As the number of well-known specific stars increases, the observer gains a rapidly improving facility in finding his or her way among the constellations.

You will greatly enhance the pleasure of star identification if you learn some basic facts about each major star's physical charac-

Table 1.1. *Deep space with the naked eye*

Season	Constellation	Star	Distance (light-years)	Comments
Summer	Scorpius	*Antares*	388	Red supergiant, 700 times diameter of the Sun!
Summer	Lyra	*Vega*	25	Twice surface temperature of Sun.
Summer	Cygnus	*Deneb*	1470	Highly luminous, looks nearby, actually very distant.
Summer	Aquila	*Altair*	16	Very hot star.
Autumn	Auriga	*Capella*	36	Actually a closeknit double sun.
Autumn	Taurus	*Aldebaran*	52	Large, but cool star.
Winter	Orion	*Betelgeuse*	354	Giant red star, variable in brightness.
Winter	Orion	*Rigel*	906	50 000 times Sun's luminosity!
Winter	Canis Major	*Sirius*	9	Very hot star, very nearby.
Spring	Leo	*Regulus*	69	5 times Sun's diameter, 160 times as luminous.
Spring	Centaurus	*Rigil Kent*	4	Binary – closest visible star (similar to Sun in size and type).
All seasons	Ursa Minor	*Polaris*	823	Close binary pair.

teristics. What kind of a sun is Orion's reddish luminary *Betelgeuse*? Is the brilliant summer star *Deneb* a relatively nearby object in interstellar space?

In the earliest days of our experience as amateur observers of the heavens, we, the two authors of this book, began the practice of consulting astronomical reference works to learn the unique features of particular stars that attracted our interest. Many handbooks contain tables of data like the brief sample shown in the Table 1.1. (We especially like the *Observer's Handbook* edited by Roy Bishop of the Royal Astronomical Society of Canada.)

Information of this kind has an amazing effect on one's reaction to the stars of the night sky. For example, *Antares* when first seen in the summer constellation Scorpius is nothing more significant than a bright reddish point of light. After reading the first line of our data-chart, however, you will ever afterward recognize this object as the unique monster-star whose diameter is so enormous that, if our Sun were replaced by *Antares*, the Earth's orbit would lie deep within the great distended body of the star itself!

Nowadays, when most of us live in urban areas, an increasing majority of people have no personal knowledge of the night sky.

Light pollution and other factors of city life have reduced the immediacy of the cosmos to so great an extent that most people are unable to identify even one constellation or name a single star beyond our Sun.

When you have spent half a dozen enjoyable nights outdoors acquainting yourself with the visible Universe to the level described in this chapter, you will already have become a relative expert. In a world where the constellations overhead are *terra incognita* to the average person, your first intrepid ventures among the stars will give you the right to consider yourself a cosmic explorer.

1.3 Road maps to the Cosmos

In your investigation of the constellations you have become an explorer. To advance in this adventure you will want to begin acquiring essential equipment. What item represents a first priority? Should your initial expenditure be for astronomical binoculars or for your first telescope?

We think there is a far more basic acquisition for the beginning amateur.

Probably the most essential piece of astronomical equipment that you can own (after your eyes themselves) is a good star atlas. No would-be amateur astronomer can make any progress without at least simple charts.

The joy of star charts

A good star atlas is the sky observer's most practical tool, the key to finding positions of specific nebulae, clusters, galaxies, double stars and all other deep-sky features.

The amateur's favourite atlas usually becomes much more, however, than merely a useful tool. Aesthetically, star-charts are wonderful things; like other works of art, when fondly handled they challenge the imagination and expand the mind. *Don Quixote*'s creator Miguel de Cervantes felt that hidden magic:

Plate 3. A good star atlas is the sky observer's most useful tool.

> Journey over all the universe in a map,
> without the expense and fatigue of travelling,
> without suffering the inconveniences of heat,
> cold, hunger or thirst.

Yet with a fine atlas in hand, who can resist the inspiration to journey beyond the colourful universe on paper into the depths of the real Universe in space? You will find that your charts will compel you outdoors to the eyepiece of the telescope.

If you are a beginner, you may find it easiest to use an extremely basic star map like that which appears each month on the centre pages of magazines such as *Sky & Telescope* and *Astronomy.* These simple charts show the current season's visible constellations, with the positions of the few dozen most prominent deep-sky wonders. For the novice we also highly recommend Terence Dickinson's *Nightwatch* (Camden House, 1989), which covers the night sky in 20 maps that plot about 2000 stars, with lines connecting the stars that outline the principal constellations. A brief descriptive note accompanies each deep-sky object, right on the map itself.

The serious amateur, especially when he or she begins to explore with a telescope, will soon find it necessary to acquire a really detailed star atlas. The best of these show all stars to at least the limit of naked-eye visibility, and accurately plot a good selection of several thousand celestial objects. Advancing levels of detail and sophistication are represented by the five major star atlases described in Table 1.2.

Going deeper: using the charts

You will want to gain a ready familiarity with the manner in which the charts in your celestial atlas relate to the actual night sky. At the

Table 1.2. *Representative star atlases*

Atlas	Publisher	Comments
Dickinson, Terence *Nightwatch* (2nd edn)	Camden House 1991	Clear simple charts for the beginner, with connecting lines to aid in identification of constellation patterns. Plots all stars (approx. 2800) down to 5th magnitude, plus major deep-sky objects.
Ridpath, Ian (ed.) *Norton's 2000.0 Star Atlas*	Wiley 1989	A widely used classic: the first level of really serious mapping for the amateur. 8700 stars down to magnitude 6.5, plus hundreds of deep-sky features. The large sky-segments covered by each of its charts make it unusually handy.
Tirion, Wil *Cambridge Star Atlas 2000.0*	Cambridge University Press 1991	A beautifully drafted intermediate atlas covering 9500 stars to magnitude 6.5, plus more than 800 deep-sky selections. A special feature is the set of 12 monthly star-charts to aid in planning observing sessions.
Tirion, Wil *Sky Atlas 2000.0*	Sky Publishing 1981	Regarded nowadays by many observers as the single best star atlas. Its 26 large-scale charts plot 43 000 stars down to magnitude 8.0 (far below naked-eye visibility) plus 2500 deep-sky objects. This atlas is printed in multicolour format to represent various classes of objects. Too detailed to be easily used by the beginner.
Tirion, Wil, *et al.* *Uranometria 2000.0*	Willmann-Bell 1988	The ultimate amateur's atlas: covers the sky in 473 super-detailed charts, including 332 556 stars down to magnitude 9.5, with many thousands of deep-sky objects. Very useful for locating variable stars, tracking asteroids, etc., but the beginner will be lost in its complexities.

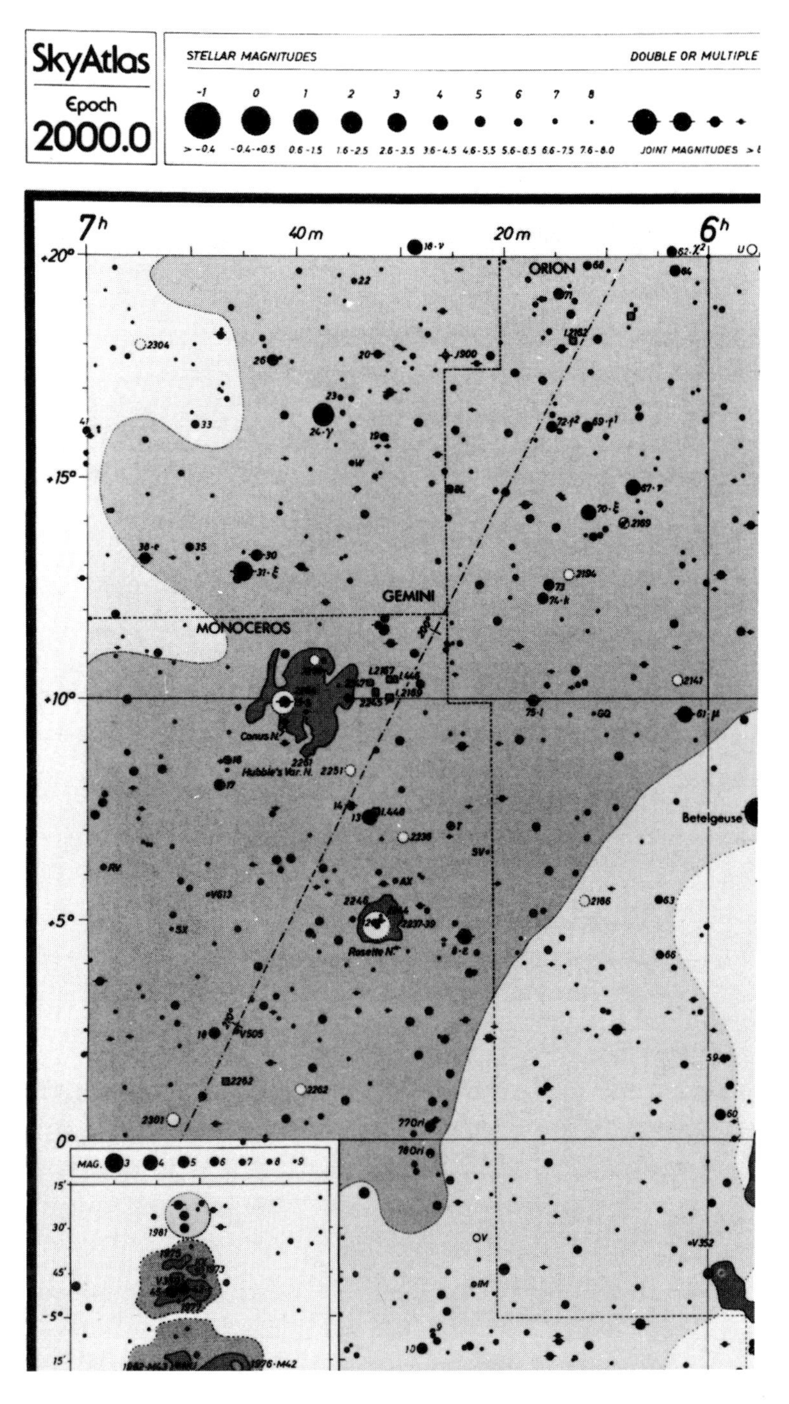

Figure 1.6. A portion of one chart page from Wil Tirion's *Sky Atlas 2000.0*. (Copyright 1981 Sky Publishing, all rights reserved. Used by permission.)

most general level, star charts are conventionally laid out with south at bottom. Thus the bottom edge of a full-sky chart represents (at least approximately) your southern horizon. The chart's right edge depicts the part of the sky that lies to the west, and the left-hand edge the east.

In a precise way, the positions of celestial objects are plotted on a grid of *right ascension* (RA) and *declination* (Dec.) co-ordinates.

These lines (see the partial chart-page, Figure 1.6) are the key to locating faint objects telescopically. The vertical lines on the chart represent right ascension, the distance in hours and minutes of time eastward from a starting point in the constellation Pisces.* The horizontal lines are declination, or distance in degrees above or below the celestial equator.

Catalogues of double stars, nebulae, galaxies, etc. (Chapters 3.5 to 3.8 on the deep-sky are an example) indicate the position of each object in terms of RA and Dec. The nebulous star cluster NGC 2244, for instance, is at a position described as 06h 30m/ 04° 54'. Using these co-ordinates, you can locate the cluster and its surrounding nebula on the chart segment illustrated here. The object's right ascension, the first pair of figures, can be found on the scale of hours and minutes along the chart's top edge. The second pair represent a position on the vertical scale of degrees and minutes on the chart's left margin. Imaginary lines drawn through the indicated points on the two scales will intersect at the location of NGC 2244.

Locate the desired object in this way on the chart, note its relationship to nearby patterns of bright stars, then find the object in the night sky by using the same bright patterns for guidance.

Most atlases nowadays follow more or less closely a standard convention of symbols to represent separate classes of objects on the charts. A short line bisecting a star symbol indicates that the star is double (or multiple), a dotted circle indicates an open star cluster, a cross within a circle is a globular cluster, a small ellipse a galaxy. Thus, at a glance you can discern the nature of each plotted object.

The brightest stars of each constellation are labelled with letters of the Greek alphabet. The designation is roughly, but not always precisely, in order of brightness, each constellation's principal star being Alpha (α), its second brightest Beta (β), the third Gamma (γ) and so forth.

The brightness or 'magnitude' of stars is represented by varied sizes of black dot. This feature of the charts and the corresponding aspect of stars in the night sky itself is a concept that is critical in the amateur astronomer's attempt to identify landmarks in the heavens.

The magnitude of the stars

Among the earliest and most useful skills that a beginner stargazer acquires is a familiarity with stellar magnitudes.

The stars' apparent magnitudes are their relative brightnesses, as they appear to us in the night sky. The few dozen most visible stars

* This zero-point is the location along the celestial equator at which the Sun crosses from the southern to the northern sky, at the Vernal Equinox.

(the highly prominent stars in the major constellations) are described as being of the *first magnitude*, or magnitude 1. *Deneb* in Cygnus and *Pollux* in Gemini are typical first-magnitude stars. A few stars, brighter than those examples, are given a negative magnitude-value; brilliant *Sirius*, for example, is mag – 1.46.

All fainter stars visible to the unaided eye are classified into five distinct magnitude steps, which the experienced observer can perceive visually. Each fainter class bears a higher number. Thus, a *second magnitude* star (typified by β in Ursa Major or α in Ophiuchus) is one step fainter than a star of magnitude 1. The next perceptible step, the *third magnitude*, is typified by the star γ in Boötes or θ in Ursa Major. To gain a feeling for the brightness steps, compare these several specific stars in a dark sky, without optical aid.

Normally the faintest naked-eye magnitude is the sixth. In practice, however, the attempt to locate sixth-magnitude stars from suburban observing sites nowadays merely brings home a sad truth: for most of us, modern light-pollution has wiped all the faintest stars from our night sky. There are many people in or close to cities who without optical aid cannot see any star dimmer than about *fourth magnitude*. Magnitude 4 examples include α in Pisces and 70 in Ophiuchus. Try these, from your location.

As you experiment with locating and viewing stars of various magnitudes you will quickly develop a feeling for the intended values of the differently sized dots in your star atlas.

Of course, the magnitude scale continues down to much fainter levels, below naked-eye visibility. The dimmest stars revealed by binoculars are ninth or tenth magnitude. A 20 cm telescope increases the observer's range to about magnitude 14.

It is useful and satisfying to acquire an accurate personal sense of at least the most prominent visual brightness levels. When considerable skill is achieved, you will perhaps become capable of estimating fractional steps in brightness. Thus, a star that is judged to be halfway between second-magnitude α *Ophiuchi* and third-magnitude γ *Boötis* will be assigned the fractional magnitude 2.5.

Perhaps Miguel de Cervantes, if he were using a modern star map as the springboard for one of his voyages of the imagination, would be entertained by niceties such as these.

1.4 Your eye is a starship

We hear people say that they would love to explore the night sky, if only they owned a suitable telescope.

Yet why wait for a telescope, when you are already equipped with the means for dramatic adventures in deep space? Your eyes are a space vehicle that can carry you much further than you might imagine into strange places in the Universe.

Probably the best way to discover the cosmos is through simple naked-eye spotting of deep-sky objects. All you need is a really dark sky; get as far as possible away from city lights. In the undisturbed darkness of a rural meadow or even in a relatively unlit suburban garden your adventure can begin.

Look up at the sky

To gain a feeling for the overall layout of the Universe, you need only gaze skyward. In fact, to do so is the most useful initiation to a lifetime of observing, because no telescope can provide so comprehensive a vision of the cosmic landscape as your eye unaided.

Late on a summer evening, after the final glow of twilight has drained from the sky, look directly overhead at the gatherings of stars. You will immediately recognize the three bright corners of the Summer Triangle, which seems to overlie a region of denser starlight than the sky to east or west. If your sky is properly dark, you will see the mottled ribbon of the Milky Way originating at a point in the north, arching through the triangle formed by *Deneb*, *Vega* and *Altair*, and straggling southward to where its densest region hangs like a pale cloud of light above the horizon.

To ancient observers this glowing ribbon that bisects the heavens must have defied explanation. Among the classical Greeks, astronomical theorists Anaxagoras and Hipparchus had begun to understand that stars were suns lying at remote distances in space. Yet even to such thinkers the narrow band of misty light dividing, branching and regrouping as it meandered across the high vault of heaven remained a puzzle.

Plate 4. In a dark sky, the unaided eye reveals the star-crowded plane of our Galaxy.

As we gaze into that crowded lane above our heads we benefit from the insight that Galileo acquired when he viewed the Milky Way through his crude seventeenth-century spyglass: the 'milk' consists of faint stars that glitter against a backdrop of myriad fainter stars, that myriad itself backlit by a dim veil that we guess to be a hundred million stars more pale and remote.

Nowadays, we understand the true nature of this clotted agglomeration of stars in our night sky. As we stand looking upward from our summer meadow we are gazing into the crowded central plane of our home galaxy, peering from our position out in the galactic boondocks inward toward nearby arms of the galaxy's enormous flattened spiral. Let your eyes drop southward along the starlit path. When you encounter the crowded density of the great starclouds in

Sagittarius you will in fact be looking directly toward our system's 30 000-light-year-distant centre.

Star-hopping into deep space

Photographs throughout this book show many of the sky's most notable 'showpiece' objects. You may be surprised to learn that you do not need a telescope to locate some of these famous star clusters and nebulae. The experienced amateur stargazer finds these subtle and intriguing mysteries of the night sky without optical aid (or with the slight aid of binoculars) by using nearby bright stars as signposts that point the way.

For example, Figure 1.7 suggests three star-hopping explorations that might be your astronomical entertainment on an evening in autumn or late summer. The constellations depicted are those that will extend from overhead toward the northeastern sector of the sky on a September night.

Each of the small insets, a close-up of an indicated region of the large chart, is a binocular view of the simple chain of stars that your eye can follow in order to locate a delicately spectacular deep-sky wonder. If the sky is truly dark and transparent all three of these remote objects can be seen with the naked eye. You will want to use binoculars, however, for the initial hop from star to star as illustrated in the little close-up views shown here.

Stepping eastward from bright *Deneb*, as illustrated in Figure 1.7, you will easily find the star cluster Messier 39 (M39), which appears to the unaided eye as a dim, misty patch, and to the binoculars as a little triangular clump of stars. M39 is a true stellar family, a physical association of thirty or forty suns lying about 800 light-years distant from us.

Yet with no optical aid at all you can plunge much deeper into the remoter neighbourhoods of our galaxy: Figure 1.7 shows the way to the famous Double Cluster in Perseus, easily spotted by eye in a moderately dark sky. This massive twin agglomeration is ten times as distant as M39. We are reminded of a classic joke about the skeptical courtroom lawyer who asked a rustic witness how far he could see. The witness, if he was familiar with this striking naked-eye star cluster, might honestly have reported that he could easily see something at least 8000 light-years away.

You can boast of much greater distance vision than that: the galaxy M31, which you will be able to find by hopping north-eastward up the bold chain of stars in Andromeda as depicted in the third inset in Figure 1.7, lies far beyond the boundaries of our own system, over 2 000 000 light-years distant. The eye is a powerful vehicle for travel into the depths of space!

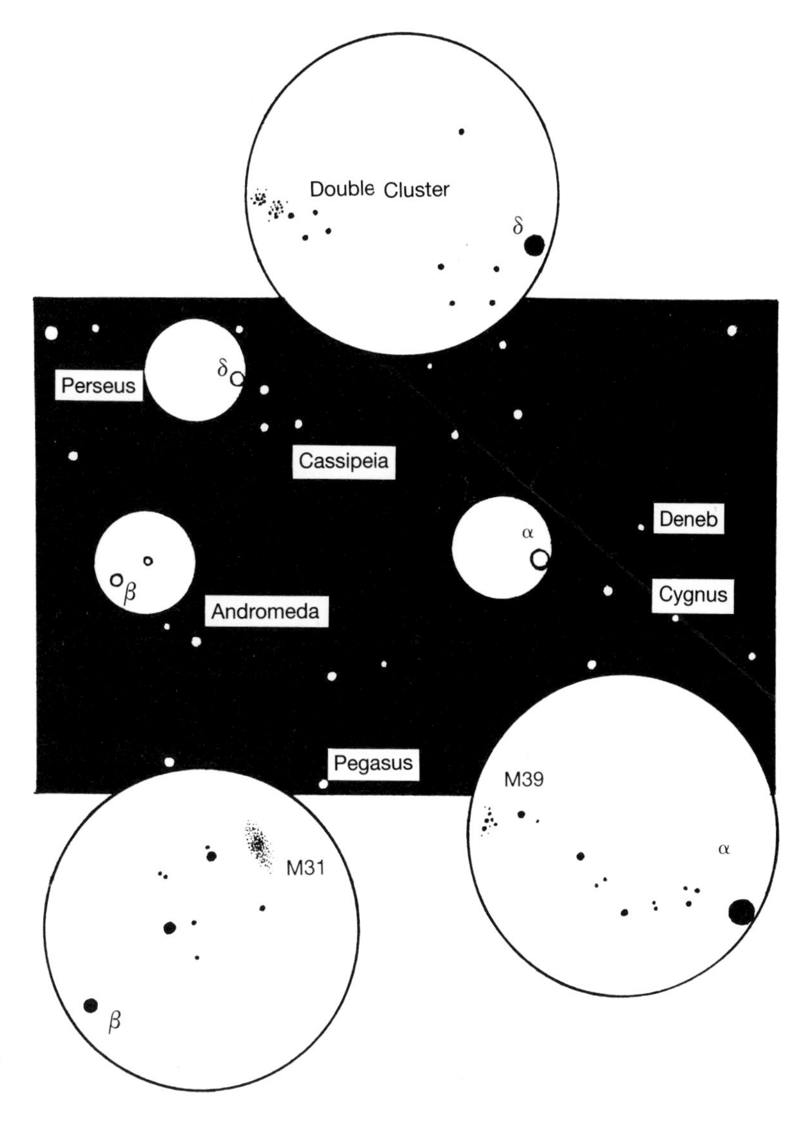

Figure 1.7. The binocular fields adjacent to Cassiopeia, Andromeda and Cygnus show star-chains that signpost the way to nearby deep-sky objects.

Exotic destinations

Once you have learned the technique of using star patterns to guide your eye in the heavens, your skill in cosmic destination finding will quickly grow. In regions of the sky where the uninitiated observer can see nothing of special interest, you will be able to locate the delicate glow of a nebula or the subtle sparkle of a distant star cluster. And you will need no telescope for these excursions into deepest space.

Locate the target objects first on the charts of your star atlas. You will develop the habit of recognizing on the map the chains of bright stars that will prove useful as stepping stones toward the precise sky-location that you seek.

Table 1.3 lists a few suggested destinations for voyages with the

Table 1.3. *Some objects for the unaided eye*

Object	Location	Description
M7	Scorpius	A large, bright star cluster, easily located by eye. Slightly spoiled for northern observers by its very low position in the southern part of the sky.
M8	Sagittarius	The Lagoon Nebula: the unaided eye glimpses it as a confused little clump of stars, smudged with haziness.
M13	Hercules	Globular cluster: visible as a fuzzy patch of light, but only in an exceptionally dark, transparent sky.
M31	Andromeda	Galaxy: surprisingly easy to spot, although more than 2 million light-years distant. Probably the most distant naked-eye object.
M33	Triangulum	Galaxy: some keen-eyed observers have seen it with unaided eye.
M34	Perseus	Open star cluster that appears as a compact puff of mist.
M42	Orion	Gaseous nebula surrounding central star of Orion's sword. Its fuzzy appearance contrasts with the sharp points of adjacent stars.
M44	Cancer	The Beehive Cluster. A hazy patch that very sharp-eyed persons may see mottled with individual stars.
M45	Taurus	The Pleiades, a brilliant knot of bluish-white stars, easily resolved without optical aid.
Mel 111	Coma Berenices	The Coma Star Cluster, larger but less conspicuous than the Pleiades.
NGC 869 and 884	Perseus	The Double Cluster: to the eye, a little cloud halfway between Perseus and Cassiopeia.
Scutum Starcloud	Scutum	A very large, dense cloud of light, shaped like a shield. Actually a bright lobe of one of our Galaxy's nearby spiral arms.
Mizar/Alcor	Ursa Major	The star ζ, in the Dipper's handle, and its very close companion – an easy double star for the unaided eye.
Epsilon Lyrae (ϵ Lyrae)	Lyra	Another double, but exceedingly challenging compared with the above. Very young observers and others with keen visual acuity see it clearly as a pair without optical aid.

unaided eye. Although binoculars may be useful for an initial attempt at some of these objects (and will of course bring you close-up for an enhanced view), all are at least dimly visible to the eye alone, if observed in a truly dark night sky far removed from the pollution of city lights.

1.5 Buying a telescope

Among enthusiasts who purchase astronomical telescopes there appear to be two distinct groups.

In the first group are those whose use of their newly acquired instruments quickly dwindles, after an initial period of keen interest, to almost nothing. There must be thousands of costly telescopes in the world whose only function is to gather dust in attics or closets. The second group of buyers are those whose use of the telescope gradually escalates to the level of a lifelong addiction. These are the serious amateurs whose nightly observations over a period of many years carry them across the void of space into mysterious regions of our Galaxy, and beyond.

Perhaps the chief cause of the former case is that people buy a telescope too soon; that is, before taking the first steps to acquaint themselves with the heavens, as described in previous chapters. Without basic knowledge of the constellations and also the use of charts or a star atlas, the beginning observer quickly runs out of interesting targets for observation. The Moon and a bright planet or two may be fascinating, but they represent only a tantalizing hint of the potential wonder that awaits discovery in the depths of space that can be explored with even a modest telescope.

When your interest has reached at least the level suggested by the preceding chapter, the time has come for purchase of a good telescope. For you, the potential of the instrument will become almost unlimited. This is the stage at which the Universe really opens its doors and allows the amateur astronomer to see with his or her own eyes the hundreds of glittering clusters, interstellar gas and dust clouds and remote galaxies pictured throughout this book.

We stress the phrase 'a good telescope,' because buying a cheap and second-rate instrument (whatever its size) may prove to be a disappointment – a mistake that all of us tend to make during our first cautious attempts to acquire astronomical equipment.

Nevertheless, we would like to express a mildly heretical viewpoint: even the cheap and rickety little department-store refractor, if that is all you happen to possess, is not to be despised. Our own first efforts, half a lifetime ago, began with such instruments. Although the mountings were frustratingly shaky and the optics barely passable, they provided nights of discovery that still live in memory, a quarter-century later. In those earliest ventures it was our imagination and our rapidly growing observational skills that provided us with those unforgettable thrills. On those memorable evenings long ago we discovered the truth that Thomas Webb had expressed a century earlier: there is no telescope so small and poor that it will not reveal something never seen with the unaided eye.

But purchase the best telescope that you can afford. A good telescope is not necessarily large, fancy or expensive; a small instrument that is sturdily mounted and optically excellent may prove more satisfactory than an impressive giant that is shaky and astigmatic. Nevertheless, since *lightgathering power is one of the most important requirements* for most kinds of astronomical observation, all experienced telescope-users agree that it is better to have an instrument of larger diameter rather than smaller, for optimum views of faint objects.

With the above consideration in mind, undoubtedly the best value for money is to be obtained by building a telescope for oneself. How an amateur can do this is outlined in a later chapter. The present chapter deals with the purchase of commercially manufactured equipment.

Common telescope types

Most commercially available telescopes are of one or another of the optical designs described below.

Which is best for you? The answer is that for general observation a good instrument of any class – refractor, Newtonian reflector or a mixed lens-mirror design – will serve you well. Success depends far more on the observer's skills and on favourable sky conditions than on details of the instrument itself.

Yet as you advance to 'serious' amateur astronomy, you may discover that your interests have become specialized, with a particular focus on one kind of observational project. You may develop a special passion for photography of deep-sky objects, or you may become fascinated by the subtle features of planetary surfaces. Your interest may lead you toward variable star observation, sunspot watching or the timing of occultations. A few astronomers eventually discover that birdwatching is what they really enjoy.

The following guide shows the general appearance and the optical configuration of each common type of telescope and includes comments on the areas of specialization for which each is best suited.

The astronomical refractor

Optical design
Main ('objective') lens brings light rays together at a focal plane, where the resulting image is magnified by a small eyepiece lens. The objective is a two-element or three-element lens designed to bring all wavelengths of light to approximately the same focus – essential in a true astronomical instrument. (Single-element objectives are virtually useless.)

Advantages of the type
Superb resolution of planetary detail, close binary stars, etc. This is true especially in very long-focus versions of the refractor and in the complex three-element types. Rugged optical setup; will stand much handling, transport, etc., without misalignment.

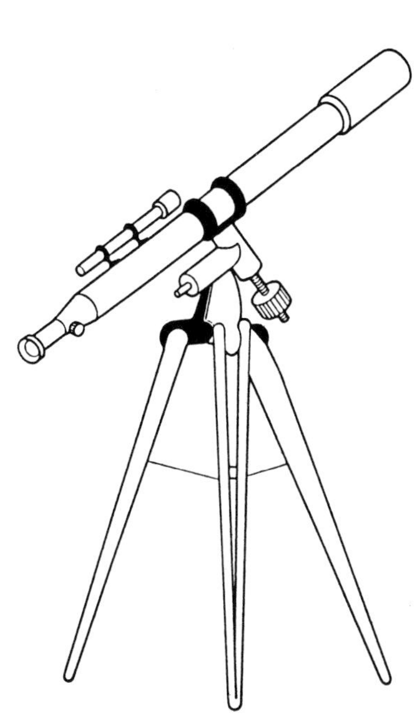

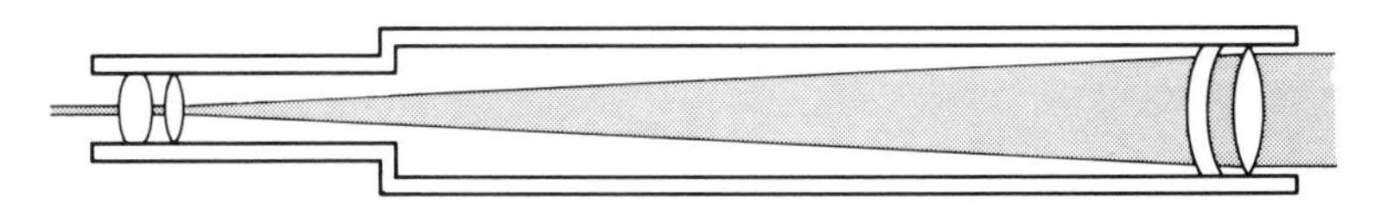

Figure 1.8. The astronomical refractor.

Disadvantages
High cost, except in small sizes. Cost-limited aperture will result in less light-grasp for deep-sky objects that can be afforded in other types. Imperfect focus of colours at ends of the spectrum may reduce usefulness as photographic instrument. (Not a problem if the telescope is to be used visually.)

The Newtonian reflector

Optical design
A curved, front-surface mirror at the bottom of the optical tube collects and focuses incoming light rays. The light beam is reflected back up to a diagonal mirror that directs it out through the tube-side to the eyepiece. *Note*: Newtonian reflectors come in configurations that vary from extremely short focus (focal length only 4 or 5 times mirror diameter) to fairly long focus (focal length 9 or 10 times mirror diameter – referred to as f9 or f10). The latter are best for high-resolution work on planets and double stars. The former are used mainly for deep-sky observation.

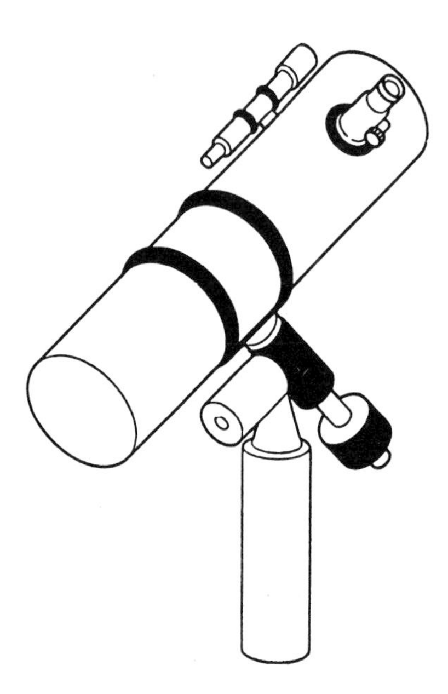

Advantages of the type
Inexpensive optics; this means that a large aperture can be afforded for optimum viewing of dim galaxies, nebulae, etc. Good colour correction. Unlike an objective lens, the Newtonian mirror does not scatter and separate the various light wavelengths. This feature makes the simple reflector useful as a good celestial camera.

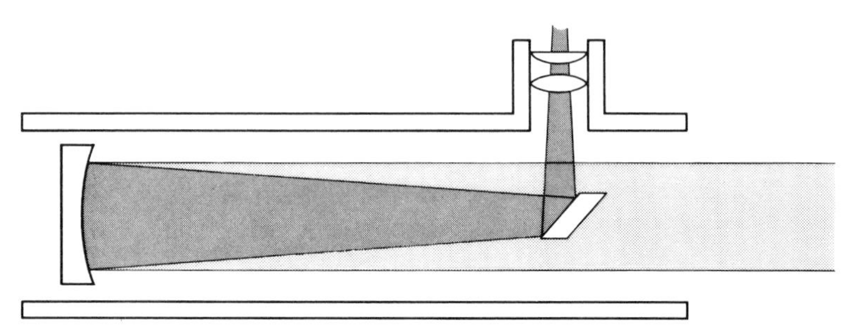

Figure 1.9. The Newtonian reflector.

Disadvantages
The obstruction of incoming light beam by secondary mirror can result in reduced fineness of image relative to the refractor. This problem becomes noticeable in Newtonian reflectors of very short focus. Optical setup is somewhat more subject to misalignment than the refractor's. Needs occasional maintenance and adjustment.

The catadioptric reflector

Optical design
Incoming light passes through a corrector lens (the precise form varies between Schmidt–Cassegrain and Maksutov types). The beam is focused by a main mirror and an amplifying secondary mirror. The 'folded' light-path gives long focal length in a very short tube.

Advantages of the type
Extreme compactness and portability. Smallest versions fit into a tiny handcase. Very large apertures are still portable. Well-made catadioptrics can yield superbly corrected images; fine choice for

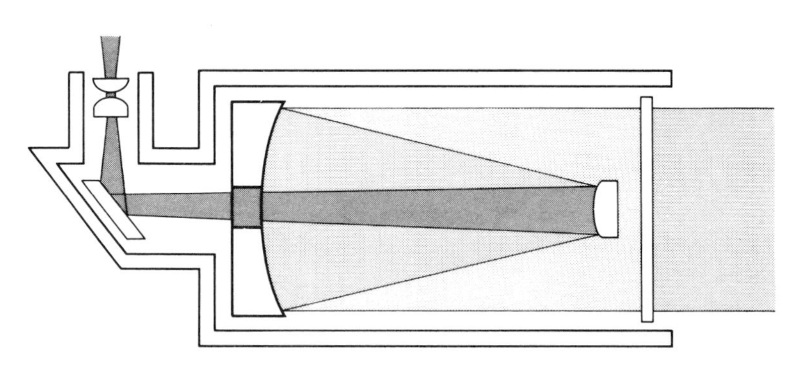

Figure 1.10. The catadioptric reflector.

both planetary and deep-sky work. Combination of optical qualities and mechanical configuration make the catadioptric unusually handy for celestial photography. Compactness and upright image make these instruments the best choice if terrestrial as well as astronomical use is planned.

Disadvantages
Costly, relative to the typical Newtonian reflector. For some observational projects that require critical resolution (for example, close binary stars), the catadioptric's large secondary-mirror obscuration renders stellar images that may be inferior to those of a fine refractor.

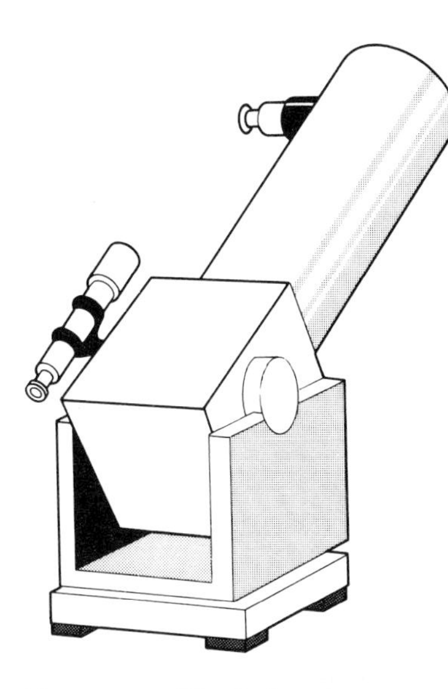

Figure 1.11. The Dobsonian reflector.

The Dobsonian reflector

Actually a conventional Newtonian type, but in a simple tube and mounting usually built of plywood, and depending for its mechanical movements on Teflon-lined bearing plates. For detailed design information see Chapter 6.3.

Advantages of the type
The Dobsonian style of telescope is commercially available nowadays at a very low price for surprisingly large apertures. Some of these instruments represent the *best buys* available if deep-sky observing is the purpose.

Disadvantages
The mounting does not permit astrophotography. Also, Dobsonians, having as a rule very short-focus Newtonian optics, perform poorly where fine resolution is a critical requirement.

Figure 1.12. A specialized instrument: the rich field reflector. This hand-held telescope combines very low magnification with a superwide field for dramatic views of clusters, star-clouds, etc.

The new super-corrected refractors

In recent years a minor revolution has occurred in the growing popularity of super-refractors. Using superbly corrected objective lenses that incorporate fluorite elements, or more commonly a three-element design called an 'apochromat', these telescopes can combine a relatively short focal length with undoubtedly the best planetary resolution available in any type, of comparable size. These complex refractors have gained popularity in spite of their one major disadvantage, their extremely daunting price tags. Top-quality apochromats cost typically at least twice the price of a good two-element refractor.

Considerations about performance

The novice who is about to try out his or her telescope for the first time is usually excited by its manufacturer's claims about the instrument's high magnifying power. Even quite a small instrument will probably have eyepieces that yield magnifications in the range of 100x to 200x (the conventional symbols for enlargement of the image by 100 and 200 diameters). Experienced observers know the truth about magnification. It is that, for many kinds of objects described in this book, *the most important eyepieces in a telescope's repertoire are those which yield the lowest magnification.* Most deep-sky objects are large but exceedingly faint. It is only when the telescope is set at a relatively low magnification that it yields a large enough field of view and a bright enough image to show galaxies, clusters and diffuse nebulae to good advantage.

Box 1.1 Basic facts about magnification

The focal length of your telescope's main lens or mirror will be found inscribed on its tube or on the focusing mechanism. Each separate eyepiece will also have its own inscribed focal length. The magnification of the telescope depends on the eyepiece that is being used. Eyepieces of small focal length yield high power; those of large focal length yield lower power. The exact magnification is discovered by means of the following arithmetic:

$$\text{magnification} = \frac{\text{main focal length (mm)}}{\text{eyepiece focal length (mm)}}$$

Thus, on a telescope of 1000 mm focal length, a 10 mm eyepiece gives 100x and a 40 mm eyepiece gives 25x.

Very low magnifications are ideal for deep-sky objects. There is, however, a practical lower limit. As magnification becomes smaller, the 'exit-pupil' (size of light beam that passes into your eye) becomes larger. When this beam exceeds a diameter of about 7 mm, it is too large to be received by the eye without some loss of light. To calculate exit-pupil diameter, divide the diameter of the telescope's principal lens or mirror by the magnification yielded by a given eyepiece:

$$\text{exit pupil (mm)} = \frac{\text{telescope aperture (mm)}}{\text{magnification}}$$

Thus, if your object lens has a diameter of 100 mm, and you are using a 25x eyepiece, the exit pupil will be 4 mm. On the same telescope, you could efficiently use a magnification as low as 14x, which would yield a 7 mm exit pupil.

Besides magnification, the other principal specification is aperture. Where telescopes are concerned, the term 'aperture' is used to describe the diameter of the main lens or mirror. Since lightgathering power is proportional to aperture (twice the lens diameter means *four* times the light grasp), all experienced telescope-users agree that it is better to have an instrument of large diameter than small, for optimum views of faint objects.

Some experts, when asked *how far* one can see with a given telescope, reject this as a nonsense question. But it is not a nonsense question: the ultimate range of a lens or mirror in deep space is dependent on its aperture. A beginner's small refractor or reflector in the 6–8 cm aperture class is capable of showing the brightest galaxies in the Virgo Cluster, about 40 million light-years distant. An aperture of 15 or 20 cm will reveal the fainter images of systems like NGC 7331 or NGC 6907, in the distance range of about 50 to 100 million light-years. With a 50 cm aperture, you may be able to glimpse galaxies of the Hercules Cluster, over half a billion light-years away.

Useful though a large aperture is, however, it is not essential. I have enjoyed good views of all the objects in Charles Messier's list, as well as hundreds of NGC objects of all classes, with a favourite old refractor of 12.5 cm aperture. In many observing sessions nowadays my choice of telescope is a diminutive 9 cm Maksutov reflector (a mixed lens–mirror type). On nights of sufficient sky transparency it is capable of revealing most of the visual features of the Orion Nebula, NGC 205 (the fainter of the Andromeda Galaxy's two close companions) and even a part of the elusive Veil Nebula in Cygnus. Although Jack Newton uses a relatively large (60 cm) reflector for photographs such as those that appear in this book, he also observes visually with a 20 cm Schmidt–Cassegrain telescope, and with an 8 cm refractor when portability is needed.

Box 1.2 Aperture and limiting magnitude

There is a considerable difference between the *theoretical minimum* magnitude for a telescope of given aperture size, and the magnitude of the dimmest objects that most people will see easily with that aperture. The table gives approximate values for both. (Diffuse objects such as nebulae and large galaxies are typically more difficult to observe than their nominal magnitudes suggest.)

Aperture (cm)	Theoretical minumum magnitude	Faintest magnitude easily observed
5	10	6
10	12	8
20	14	10
40	16	13

Box 1.3 Aperture and resolution

Resolution of fine detail increases with aperture. An empirical formula (Dawes' Equation) roughly indicates the minimum double star separation (SEP) that a given size of telescope should resolve:

$$\text{SEP (arc-seconds)} = \frac{12}{\text{Aperture (cm)}}$$

Aperture (cm)	Closest double (arc-seconds)
6	2
12	1
20	0.6
30	0.4
50	0.24

Selecting eyepieces

At first glance, the telescope's eyepiece may seem to be its smallest and least significant component. Yet every astronomical observer quickly discovers the critical role that a suitable (or unsuitable) eyepiece plays in the instrument's performance.

The principal function of the eyepiece is to magnify the tiny image that is formed at the telescope's focal plane, so that you can study it easily. As Box 1.1 has illustrated, magnification is determined by the focal lengths of the main mirror or objective lens and the eyepiece. Because some celestial objects (for example, nebulae, open clusters) require low magnification, whereas others (planets, binary stars) must be observed at high power, the telescope must have a selection of eyepieces in a range of focal lengths.

Eyepieces vary not only in focal length, but also in design. The simplest types give acceptable, and perhaps even excellent, performance when used with refractors and long-focus Newtonians. Other classes of telescopes, however, yield best results with more complex eyepieces. Often a new instrument whose image quality or colour-correction seems poor is not intrinsically bad, but only inadequately served by the manufacturer's choice of eyepieces.

Experimentation with various eyepiece designs is costly, but may be worth the expense in terms of dramatically enhanced performance. In fact, some amateurs become eyepiece fetishists, who regard the telescope itself as merely a device on which to experiment with new eyepiece designs.

The market is flooded nowadays with innovative commercial eyepieces, some of which aim to improve image quality at high magnification, while others are formulated to give the widest possible field of view at low to medium power. Box 1.4 describes the several types that are standard and most commonly used. The cost of an eyepiece is usually proportional to the complexity of its design.

Box 1.4 Common eyepieces

Design	*Characteristics*
HUYGENS 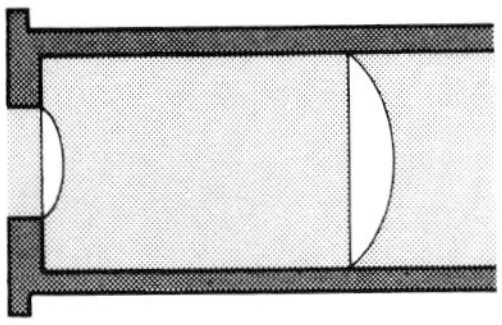	One of the simplest and least-costly eyepieces on the market. Performs very well on long-focus refractors, but shows increasing image distortion as the telescope's focal ratio becomes shorter. Recommended for projection of the solar image, because of its lack of heat-vulnerable cemented elements.
KELLNER 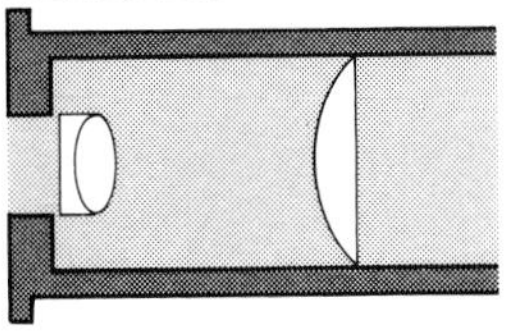	Field lens is single as in Huygens type, but the eye lens is a cemented doublet. Yields a wider field and better-corrected image (especially off-centre) than the simple Huygens. On very short-focus reflecting telescopes this design still gives considerable edge-of-field distortion. A good buy, nevertheless.
ORTHOSCOPIC 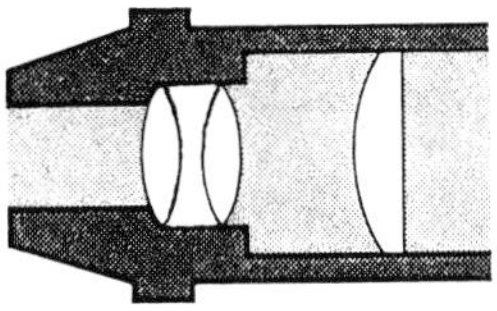	There are many variations. Usually has field lens plus cemented multiple-element eye lens. Best of the standard eyepieces for short-focus telescopes, yielding good corrections and fine image. Most orthoscopics are characterized by better eye relief than the simpler types. (Eye relief is the distance back from the lens that you can place your eye, and still view the whole telescopic field. Large eye relief results in low-strain observing.)
PLÖSSL 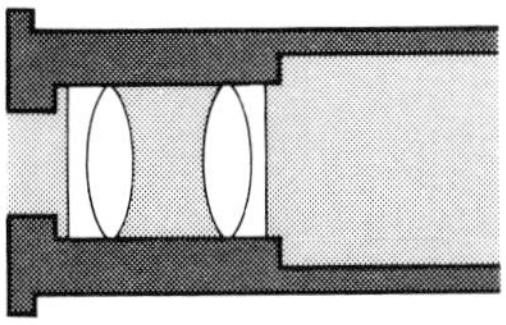	A superbly corrected type, highly recommended for 'fast' Newtonians (those with short focal length). The best made examples are quite costly. Inexpensive versions sometimes display an annoying degree of internal reflection. Plössl eyepieces give a pleasingly wide field and generous eye relief.
ERFLE 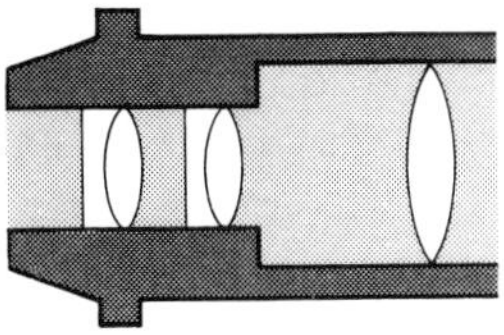	Designed to give a superwide field, this standard type is usually much less expensive than some of the newer deep-sky oculars currently on the market. Its disadvantage is considerable image distortion near the edge of field.
BARLOW 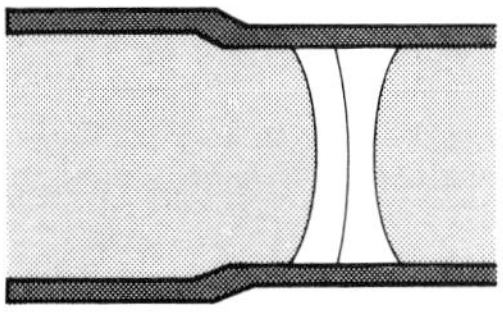	Not really an eyepiece, but an amplifying lens used in combination with normal oculars. The Barlow increases the effective focal length of the optical system, causing each eyepiece to yield higher magnification (usually by a factor of 2, 2.5 or 3). The Barlow is valuable for planetary and binary star observation, and can be employed to adapt a short-focus 'rich-field' telescope for more conventional use. Must be of high quality; an indifferent Barlow may ruin definition!

At the time of writing prices quoted by major suppliers ranged from about $20 for Huygens or Ramsden types to $70 for very good quality orthoscopics.

Specialized designs to yield superwide field and extra-fine correction can cost more, in fact, than the entire price of some telescopes!

The telescope's mounting

We can't resist a pun that is inescapably apt: an astronomical telescope stands or falls on its mounting.

A sadly large percentage of all the telescopes purchased every year by beginning amateurs are doomed to be discarded after a few months of rather frustrating use. Often the new instrument that fails to give satisfaction does so in spite of quite good optics. The

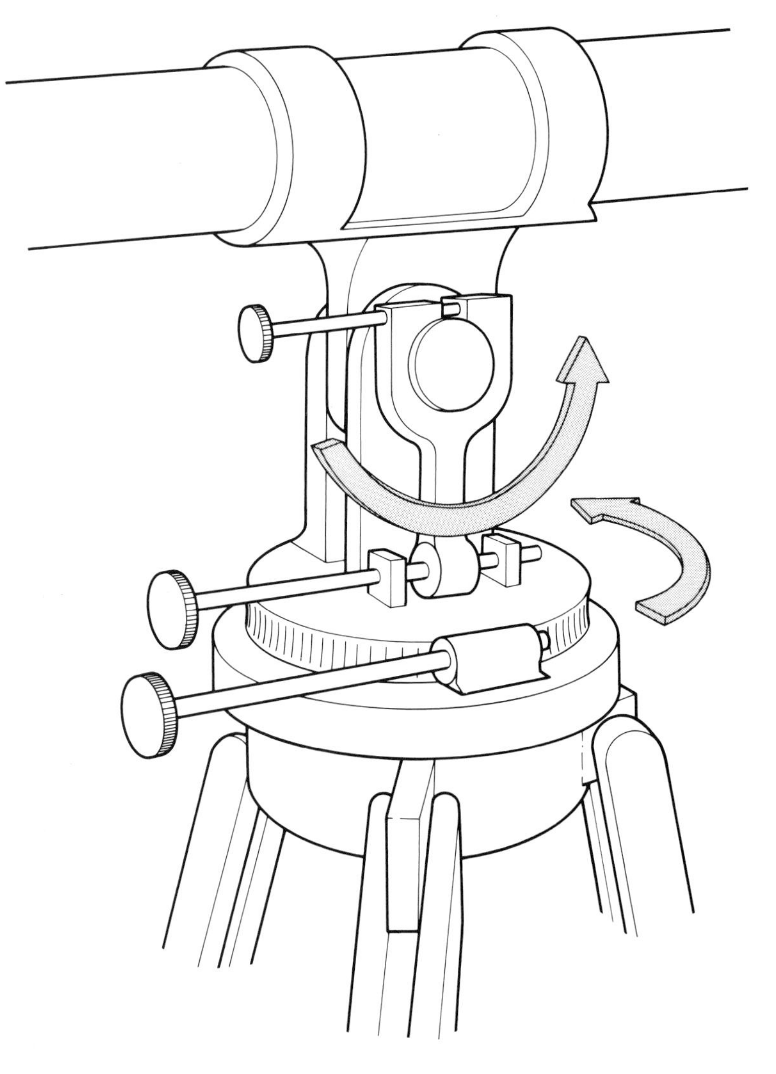

Figure 1.13. An altazimuth mounting permits only vertical and horizontal movements.

point is that the finest optics in the world are useless for astronomical work if the telescope is inadequately mounted.

Good mountings: altazimuth

The simplest usable mountings – *altazimuth* types – have two separate axes of rotation, arranged as shown in Figure 1.13. The vertical axis – the *azimuth* axis – permits movement of the telescope to left or right, sweeping along a line that is parallel to the horizon. The other axis – the *altitude* axis – permits movement up or down, at right angles to the horizon.

The altazimuth mechanism is not really an astronomical mounting, in that its movements are not co-ordinated with the actual diurnal movements of celestial bodies across the sky. Nevertheless, it can be a serviceable device for visual observing, if it is sturdily constructed and equipped with precise controls for micrometer-like slow motion.

Both the mounting and the tripod or pier on which it stands must be rigid enough to support the telescope without vibration or unwanted movement. This is a stringent criterion; remember that an imperceptibly small tremor of the telescope's optical barrel will become a major earthquake when magnified 150 or 200 times.

Similarly, the controlling mechanism must be of high quality. The least trace of backlash or roughness of motion, when highly magnified, can make precise aiming of the telescope impossible. The best altazimuth mountings are controlled by high-quality worm

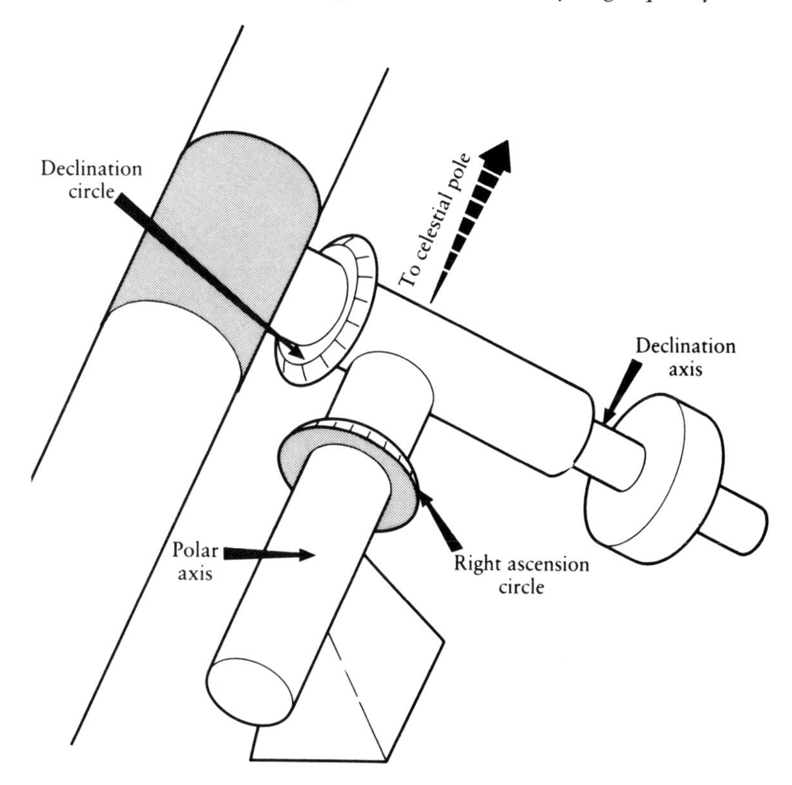

Figure 1.14. Orientation of the equatorial mounting and the positions of the two setting circles.

gears, or by a good tangent-arm arrangement (a prudently oversized metal control-arm driven by a finely threaded screw). The poorest mountings may consist of nothing more than free-moving axes with securing clamps to fix them at selected positions. A crude device of this sort is simply not good enough.

Good mountings: equatorial

The equatorial is the true astronomical mounting. It comes in a variety of forms (the 'German' or two-shaft type, the rotating fork, etc.), all of which have the following essential feature in common: there is a principal axis of rotation – the 'polar' axis – that is held at an angle parallel to the rotational axis of the Earth itself.

The equatorial configuration offers two advantages: (a) automatic following of the path of a celestial body across the sky, simply by rotating the telescope about its principal axis, and (b) location of objects through the use of scaled 'setting circles', because the equatorial telescope's two movements are in the planes of the right ascension and declination lines on astronomical charts.

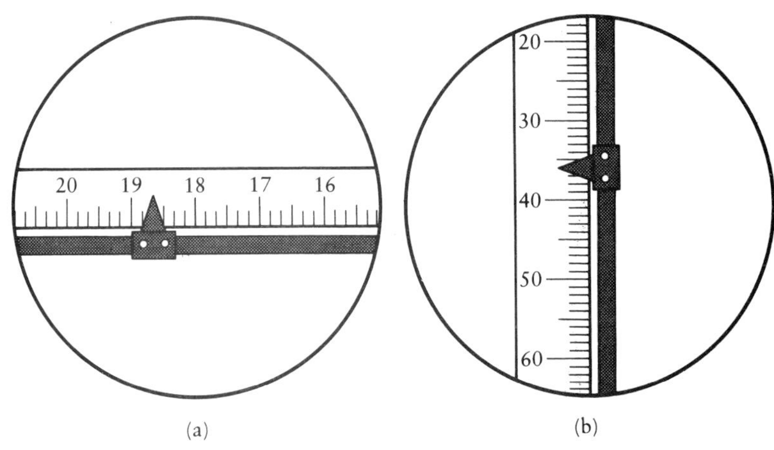

Figure 1.15. (a) A part of the right ascension circle. Pointer indicates the RA of *Vega. (b)* A part of the declination circle with pointer indicating Dec. of M13.

It is essential to have an equatorial mounting if long-exposure astronomical photographs are to be attempted.

Everything that has been said about the required sturdiness of the altazimuth mounting applies also to the equatorial. The best equatorials are quite massive, vibration-free and fitted with smooth-running controls. Mountings of this kind may also be equipped with motor drives, to move the telescope continuously about its polar axis at a rate that exactly compensates for the Earth's rotation, thus keeping a celestial object stationary in the field for many minutes or hours.

Beware of constructing or buying equatorial mountings that are spindly, small-shafted or configured in such a way as to place the telescope's centre of gravity far outboard of the mounting's pierhead or tripod-top. In a telescope that is intended for serious astronomical work, 'overkill' should be the keyword; make the mounting ridiculously massive and rigid, rather than undersized. The present

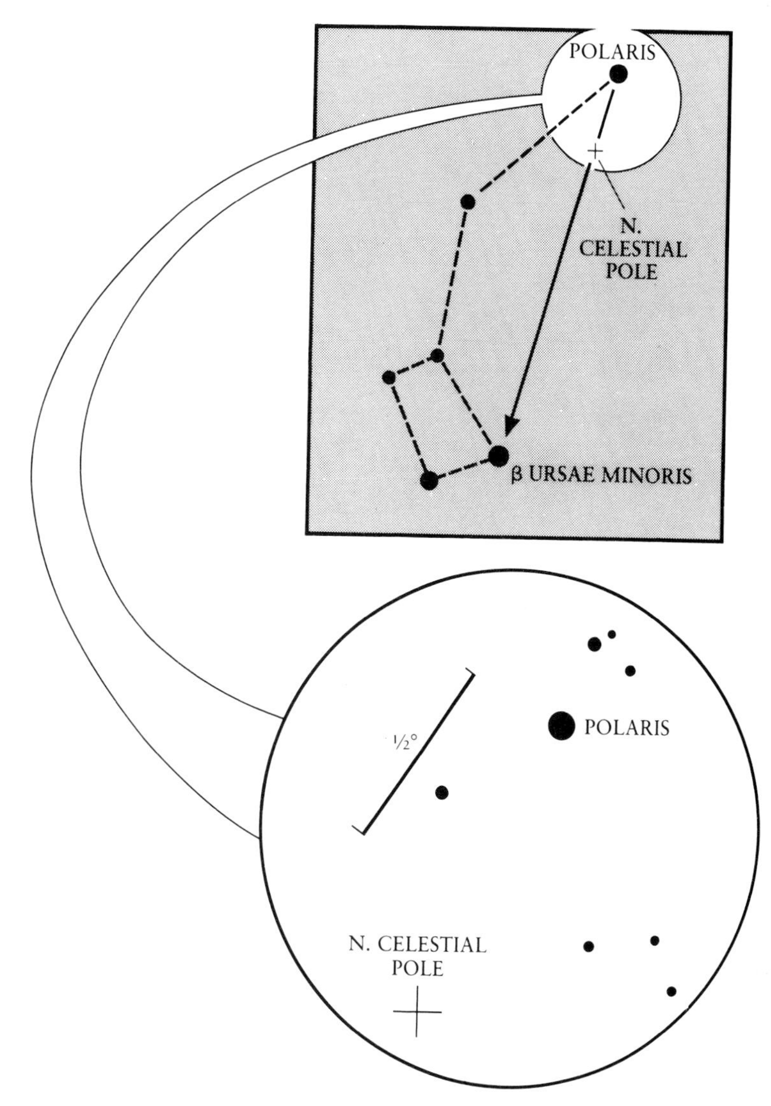

Figure 1.16. Finder view of the north celestial pole.

writer's own 12 cm refractor, with 5 cm stainless steel axis-shafts turning in huge roller bearings, all mounted atop a 50 kg cast-iron pier, can be tripped over or fallen against without disturbing an observation that is in progress!

If an equatorial telescope is to be used to advantage, it must be set up correctly, with its polar axis aligned in such a way that it points directly at the celestial pole (see Figure 1.16). For casual observation a rough polar alignment is adequate. In the northern hemisphere the celestial pole lies so closely adjacent to the star *Polaris* that it will usually suffice merely to adjust the vertical angle and the direction of the telescope's polar axis so that it points to this star. Set up in this way, the instrument can be guided along the track of any celestial object with only occasional adjustments in declination to keep the object centred in the field.

When you want to attempt guided photographs, you will need a much more precise alignment of the mounting. A detailed procedure for perfect equatorial setup will be found in the section on astrophotography.

1.6 Develop your observing skills

During the 1985–86 approach of Comet Halley, people in the northern hemisphere had for the most part to view the comet with binoculars or telescopes. Its brighter apparition (permitting naked-eye observation from dark sites) was a spectacle reserved for observers at more southern latitudes. In the first months of the great Halley show, amateur astronomers found themselves in demand to give members of the public their first glimpse of the approaching comet's subtle glow.

When we turned our telescopes upon the still-distant object and invited people to enjoy a look through the eyepiece, many of us encountered a situation that is not uncommon at public star parties. To our own eyes, Comet Halley was an easily detected globe of

Plate 5. At the earliest stages of its visibility in 1985–86, Comet Halley was not an easy object to observe.

misty light, clearly resolved into details such as a bright starlike core, a nebulous coma and a hint of the emerging tail. Our inexperienced friends, however, often exclaimed: 'I can't see anything at all; there's nothing there!'

On such occasions an unmistakable fact is evident: astronomical observation is a learned skill – a series of techniques that one develops through practice.

Learning to see

For a good view of the heavens, the sky must be truly dark. People are sometimes enticed out for their first serious look at the night sky by the glorious sight of a full Moon. This is ironic, because on a brightly moonlit night there is little to be seen except the glaring Moon itself. To see the delicate mist of a star cluster or the subtle smudge of a nebula, choose an observing time in the moonless half of the month.

Experienced stargazers are aware of two important techniques that allow the eye to penetrate most effectively into deep space. One is the adaptation of the pupil to darkness: in a brightly lit room indoors, the iris (the part of the eye that works like a diaphragm in a camera) is closed to nearly its minimum opening – a matter of perhaps two or three millimetres. Outdoors in the night-time darkness, the iris does not expand immediately. Only after an adaptation period of at least ten minutes* does the pupil enlarge to its full six or seven millimetre aperture, and only then does the eye become an effective astronomical instrument. As the eye gradually adapts, increasingly faint details within the telescopic field of view begin to emerge, like the image on a photographic print in the developing tray.

Figure 1.17. Aristotle (384–322 BC) first noted the principle of 'averted vision' – the enhanced sensitivity of the edge of the eye's retina to very faint stars, etc.

Since the light of even a small battery torch will instantly destroy the eye's adaptation, most amateur astronomers use a dim red light for the reading of starcharts, etc. The eye reacts least to light of that wavelength.

The second visual technique, known since ancient times, was first described in a treatise by Aristotle in the third century BC. The great Athenian philosopher noticed that a faint object like the 'nebulous patch' in Andromeda is invisible if one stares directly at it; yet it can be glimpsed from the corner of the eye. The reason for this is that the central part of the retina in the eye is relatively insensitive to low light levels, while the peripheral part of the retina is quite highly sensitive to low levels of incident light. Every amateur astronomer quickly discovers the technique of *averted vision.* To see the elusive glow of a deep-sky object or the dim tracery of detail on a planet's surface, look slightly to one side of the object's position in the telescopic field.

* Some experts claim that dark adaptation requires up to one hour of observing time in uninterrupted darkness.

What magnification?

The correct eyepiece for a specific observation will depend on the nature of the astronomical object and also on 'seeing' conditions (the steadiness and transparency of the atmosphere). Experimenting constantly with a range of eyepieces, the experienced observer develops a feeling for what is likely to succeed best in each situation.

For planets, planetary nebulae and close double stars the suitable magnification may be quite high. Yet it will always be found that a moderate power that reveals detail crisply and cleanly is more effective than a superhigh power that causes the image to blur and break down. Sometimes a magnification of 40× or more per centimetre of telescope aperture can be pressed into service. More commonly 20× per centimetre yields optimum performance.

For dim, extended objects such as galaxies, star clusters and diffuse nebulae, much lower magnification may be most serviceable. Certainly in the case of large galactic clusters, the lowest power and widest field are mandatory; at higher magnification these sprawling clusters are unrecognizable.

Plate 6. Large objects such as this cluster (the Pleiades) require low magnification and wide field.

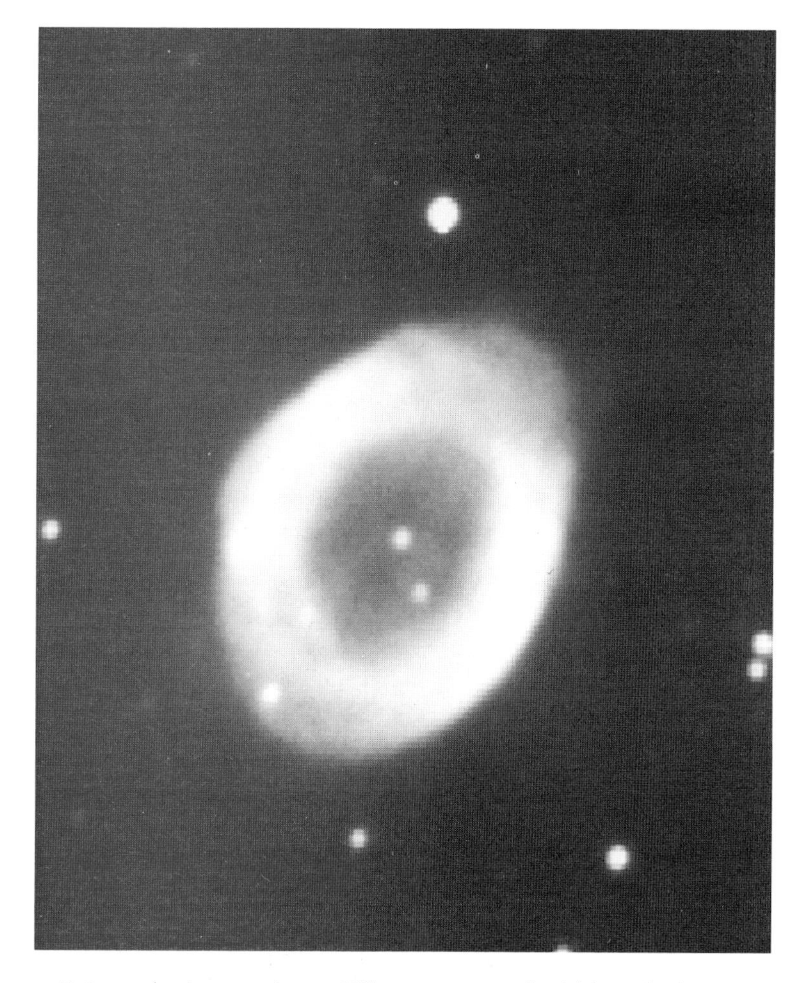

Plate 7. Planets, planetary nebulae and other small, bright targets need high power.

Faint galaxies require a different approach. Although they may be most easily found by sweeping with a very low-power eyepiece, they often appear brightest when studied at high magnification. A power of about 2× per centimetre of aperture is useful for sweeping, but 5× or 6× per centimetre will probably show a better contrast between the galaxy's pale surface and the surrounding sky.

There are some objects that require study with a range of different magnifications. An example is Messier 42, the Great Nebula in Orion. Very low power is needed to show the full extent of this huge nebulosity, but somewhat higher magnification will reveal the complex, mottled details of the nebular core. Still higher power may serve (conditions permitting) to resolve all of the components of the nebula's central cluster, the Trapezium.

A note on filters

With the overwhelming spread of light pollution in recent decades, many visual observers have felt a growing sense of despair. Near any city of considerable size nowadays, streetlighting, shopping centres and the headlights of ten thousand automobiles have caused deep-sky objects to vanish forever from the night sky.

During the late 1970s a new item of equipment became available that gave the pastime of nebula-watching at least a partial reprieve. Several major telescope manufacturers now market a variety of these light-pollution filters. Designed to fit into the cylindrical barrel of your telescope's eyepieces, these coated optical flats transmit certain wavelengths of light, but partially block other selected wavelengths. The aim is to mask the light of mercury vapour lamps and similar light pollutants, while fully accepting the subtle glimmer of a pale nebula.

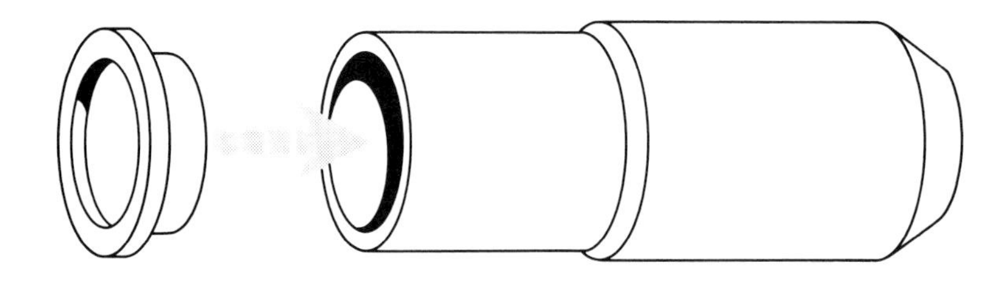

Figure 1.18. The Ultra-High Contrast filter.

Among the companies that supply effective light-pollution filters are Celestron International, University Optics, Meade and Lumicon. We have found the Lumicon *Deep-Sky*, *Oxygen-III* and *Ultra-High Contrast* filters extremely effective, particularly for emission nebulae. The first is intended for photographic use, and the third for visual observations in conditions of severe light pollution.

An intriguing experiment can be tried with these filters. It is possible, by simply holding the filter up to one's eye (without a telescope), to pick out certain objects in the night sky that elude detection in any other way. The vast, sprawling form of the North America Nebula in Cygnus, for example, is rarely seen visually through a telescope. It can, however, be spotted by eye with the aid of a filter.

Concentrated observation

When beginners or those with only a casual interest are invited to look through a telescope at some faint nebula or remote galaxy, they are usually satisfied after a ten-second glance. 'There's not much to see, really', they may comment.

By contrast, the serious amateur develops a habit of spending much time over each object, giving the eye (and the brain) a chance to adjust to the hidden subtleties of the telescopic image. Remember: the light of a galaxy has travelled for some millions of years to reach your eye tonight; surely it is worth more than a few seconds' observing time!

One useful technique for concentrating the mind on what the eye is seeing is the 'bits' approach. This is the practice of mentally identifying each 'bit' of information that is present in the image under study.

At first glance a galaxy will appear only as a featureless, ghostly presence. The fact *that it is there* is one bit of information. Closer examination may reveal a second bit: the shape is not circular, but elongated in a specific direction within the field. With prolonged attention and the use of averted vision more bits of data emerge: a contrast of brightness between the galaxy's nucleus and disc, a vague glow that comes from outlying spiral arms, perhaps a delicate hint of obscuring dustlanes among the brighter material of the galactic surface.

Another highly effective method of training the eye to detect all the details the telescope is capable of delivering is to sketch what you see. Crude, rapid pencil-drawings are all that is really necessary. The act of drawing forces the eye to see much that would be missed at a casual glance.

The practice of visual observing is one of astronomy's most rewarding experiences. Whether you use a telescope, or binoculars, or only your eyes, a direct view of an open cluster's sparkling stardust is more alive and immediate than any photograph. The details of a remote galaxy portrayed on film go far beyond what the eye will see; the eye, however, conveys the excitement of a real physical contact with something that lies millions of light-years away in space and time.

1.7 Sketch what you observe

For centuries before the invention of photography, astronomers (both amateur and professional) recorded what they observed by making simple drawings.

Galileo's discovery of the lunar craters, for instance, is on record in the form of his precise and elegant little watercolour sketches of how the Moon appeared through his telescope in the autumn of 1609. The spiral structure of a typical external galaxy (Messier 51) was first presented to the scientific world in the form of a celebrated telescopic sketch by the Earl of Rosse, in 1883. Painstakingly careful drawings of the Great Nebula in Orion by Harvard astronomer G.P. Bond in the nineteenth century are still, even in our era of sophisticated astrophotography, among the most accurate depictions of the object's visual appearance.

In the twentieth century, professional astronomy still makes use of drawings made by expert visual observers. During the 1973–74 apparition of Comet Kohoutek, Skylab astronaut Edward Gibson prepared detailed pencil sketches of the comet's nucleus, coma and tail-structure. The drawings were transmitted by television-link to earthbound astronomers for analysis.

There are several reasons why the amateur astronomer should adopt the longstanding practice of telescopic sketching.

Perhaps the most important reason is that this activity forces the observer really to observe. Try it; you will find that the effort to portray a telescopic field accurately on paper will result in your seeing the smallest and faintest aspects of an object, in a way that you have never noticed them before. Another reason for the usefulness of the amateur's sketchpad is that it can turn out to be his or her only permanent record of observations that have been made.

Some objects, in fact, cannot be observed as well photographically as in a visual observation that is recorded by means of a sketch. Planetary details that flash into view during only a transient second or two of superlative 'seeing' may never be captured on film; such details are routinely drawn by experienced pencil-draftsmen.

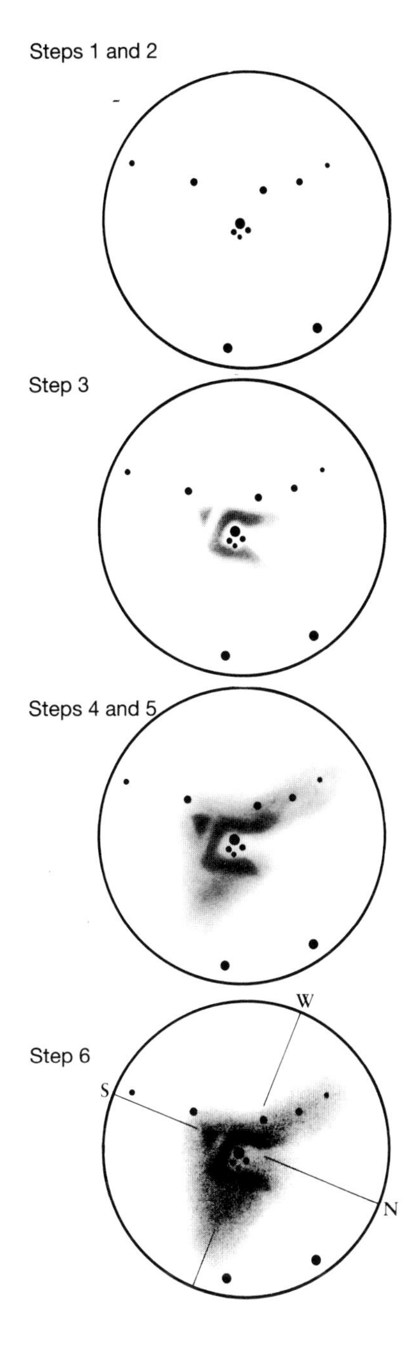

Figure 1.19. Telescopic sketching: Steps 1 and 2 – set principal stars into position and adjust their relative sizes. Step 3 – pencil in the brightest areas of nebulosity. Steps 4 and 5 – shade in fainter outlying nebulosity; smudge with finger to blend the pencil markings. Step 6 – indicate field orientation.

But I can't draw!

You may not be capable of painting a *Mona Lisa.* Nevertheless, anyone can make a simple pencil-representation of the pattern of stars, nebulosity, etc. that constitutes a deep-sky object's telescopic appearance, or the tracery of lines and smudges that may be glimpsed on a planet's surface.

Figure 1.19 shows a very basic sketching routine that involves six steps. While observing and drawing, use a dim red-filtered hand-torch to illuminate your sketchpad; white light will seriously degrade your telescopic night vision.

Step 1 On a card or notebook page that has been inscribed with a circle to represent the telescopic field (three or four inches (8–10 cm) diameter is handy), place a dot to represent the point of each star that is visible. This is the step that will establish a precise framework for the drawing. Position the stars as accurately within the field as possible. It is helpful to see the positions of two or three principal stars in terms of fractions of the whole field width. For example, a bright star's position may be seen to be halfway from the centre to the left-hand field edge. Another may lie one-quarter of the distance from that first star to the field centre. And so forth. Fainter stars may be positioned in terms of their spatial relationship to the brighter ones.

Step 2 Adjust the sizes of the stellar 'dots' to represent their relative apparent brightnesses. The faintest stars in the field will be left as minimal points, those that are marginally brighter will become slightly larger pencil marks, etc. The brightest field stars will be enlarged to become substantial circular spots.

Step 3 Using the edge of a soft pencil-point rub onto the drawing a dark, shaded mark at the position of the deep-sky object's brightest segment (if the object is a nebula or galaxy). If the object is a star cluster, dot in the cluster's principal star mass at the appropriate position within the stellar framework that you have prepared. That framework serves as your guide to the correct location and shape of the object that you are sketching.

Step 4 If fainter nebulous detail can be perceived beyond the object's bright core, rub this lightly in with a light application of the pencil lead. In some complex objects there may be even a third level of hazy structure, which you will represent with a still lighter shading with the pencil.

Step 5 Using a very light, cautious grazing motion, gently rub the shaded markings that represent the deep-sky object. This simple technique blends the grainy pencil markings into a soft, continuous gradation of tones that quite elegantly reproduces the subtle glow of a nebula or galaxy. (Note that this process of 'smudging' is useful also for the portrayal of faint dark planetary markings, atmospheric belts, etc.)

Plate 8. Compare this 20 s CCD image of M42 with the sketch in Figure 1.20 of the same object.

Step 6 Discover, and indicate on your drawing, the east–west and north–south orientation of the field. This orientation can be established either by watching the direction of movement as stars drift across the telescopic field, or by moving the telescope about its polar axis and noting the resultant shift of the stars.

Needless to say, the above sequence of steps does not constitute the only acceptable approach to drawing at the telescope. It is a very personal matter; there are as many different sketching techniques as there are amateur observers. Some amateurs use chalk (or white pencil) on black paper, as a more realistic way of depicting the bright gloss of celestial objects against the dark backdrop of the night sky. Others use a delicate pen-stipple technique. A few ambitious observers even attempt paintings in watercolour or oils. The degree of technical elaboration and also the level of scientific precision attempted will depend on each observer's own needs.

Yet here is a word of advice: you should probably opt for the simplest and quickest technique that you can develop. If you adopt a style that demands a high technical standard, you may find yourself completing very few drawings. Accepting a more basic approach (perhaps rudimentary soft-pencil sketches on small file cards), you will discover that notebooks full of drawings will be as easy to produce as verbal descriptions.

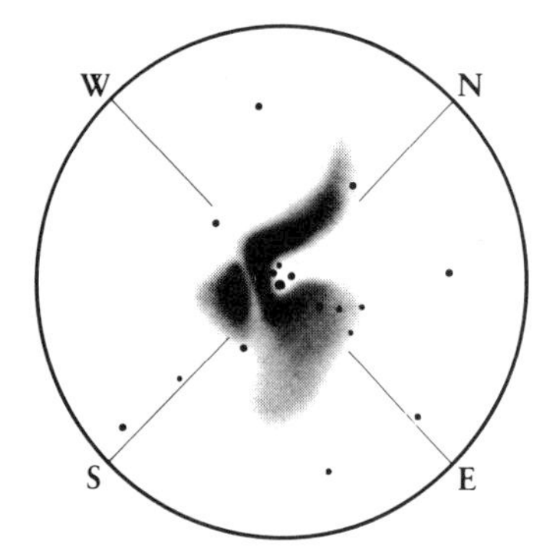

Figure 1.20. M42: The Great Nebula in Orion, sketched during observation with a Celestron-90 telescope at 80×.

Moon and planets

Solar system objects are obvious targets for drawings. Telescopic sketches can provide much more than aesthetic satisfaction; they are valuable records of information as well.

Begin with a small blank disc, circular for Mars at opposition, a slightly flattened ellipse for Jupiter or Saturn. Use a very soft pencil to record the subtle hints of dark surface detail on Mars or the

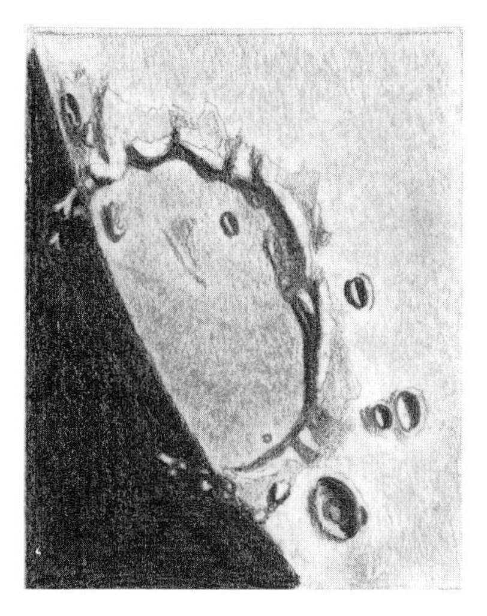

Figure 1.21. This typical page from a lunar sketchbook illustrates a simple pencil technique that will serve to capture interesting details of the Moon.

complex tracery of the atmospheric features of Jupiter and Saturn. Planetary drawings done in series (on several subsequent nights) will provide a sort of animated view of the planets' rotations and other changes.

Perhaps the most rewarding project of all – but also the most complex – is the sketching of the lunar surface. Even a small telescope reveals such intricate details of the Moon's landscape that you will be wise to concentrate, in your drawings, on the features of one very small region at a time, rather than the full expanse presented in the telescopic field of view. You may want to focus at first on the accurate portrayal of a single large crater or a specific range of mountains, or whatever. Use a soft pencil to reproduce the areas of dark shadow, an eraser to pick out white highlights, a sharp, hard pencil to delineate fine points of detail.

Your scrapbook of drawings

There are a few written annotations that should be added to every sketch done at the telescope. What facts you record will depend on what you feel you may need to remember about the observation at some future date. The usual notes include the object's identity, the date and perhaps the time of its observation, and the aperture and magnification used. I find it useful also to indicate the approximate diameter of the overall field, as sketched. When consulting the drawing later, it is quite significant to know whether the array of stars that you are looking at covers a little circle only a quarter-degree in width, a full degree, an unusual three-degree rich-field view, or whatever. For this purpose it is essential to have some

Plate 9. The first step in lunar exploration is to familiarize oneself with the Moon's most prominent features.

notion of the true field-diameter yielded by each of your various eyepieces. This can easily be discovered simply by timing the drift of a star (close to the celestial equator) across the centre of the field, from its eastern edge to its western edge.

A little arithmetic gives the field in minutes of arc:

$$\text{Field (arc-minutes)} = 15 \times \text{Time (minutes)}$$

A drawing done at the telescope is an excellent form of observational record-keeping. Years later, to your amazement, you may recall the sound of wind in the trees and the fragrance of the night air that you unconsciously experienced during the minutes when you were concentrating on your simple sketch. Although this freakish degree of recollection is not an astronomically significant phenomenon, it is certainly a delightful bonus attached to the activity of eyepiece sketching!

PART 2 Solar System voyages

2.1 Our neighbourhood as a system

Long ago the present writer as a child paid a visit to the Dominion Astrophysical Observatory on its mountaintop site near Victoria, British Columbia. I remember my overwhelming first impression as if it had occurred only a week ago: the silver-painted dome rising among dark fir trees, the cold cavernous interior and the towering framework of the 72-inch (183-cm) telescope directed upward toward a ribbon of starry sky glimpsed between shutters overhead. In the huge silent dome that opened onto a silent Universe I discovered the mystique of a great observatory.

Yet that evening's greatest revelation came by way of a simple graphic display on a wall in the observatory's foyer. It was a comparative view of our solar system's principal members, very similar to the drawing presented here in Figure 2.1. I stood for half an hour in front of the little backlighted panel of glass on which it was painted, stunned by my first realization of our Earth's minuscule size relative to its parent body the Sun, and by my discovery that our local system included planets as gigantic as Saturn or Jupiter, and as tiny as little Ceres.

That evening's visit spawned a frenzy of telescope building and nightly planetary observation. Very soon I knew the experience of viewing the shimmering, banded disc of giant Jupiter through lenses that I had assembled in a cardboard tube. Saturn, no longer just a sketch on the observatory wall, revealed its unbelievable brownish-yellow globe and pale rings through my crude optics. Even our little neighbour Mars occasionally showed a tantalizing hint of *something* (what, I could never quite discern) on its rusty surface.

Our system of planets began to seem familiar. But after a few decades of observation they were destined to reveal a whole new level of strangeness when the Voyagers gave us our first truly close-up views of worlds that we thought we knew well.

Figure 2.1. Sun and planets to scale. Mercury, Venus, Earth, Mars and Ceres are miniatures beside the gas giants Jupiter, Saturn, Uranus and Neptune. Pluto is the tiny dark body beyond the edge of frame.

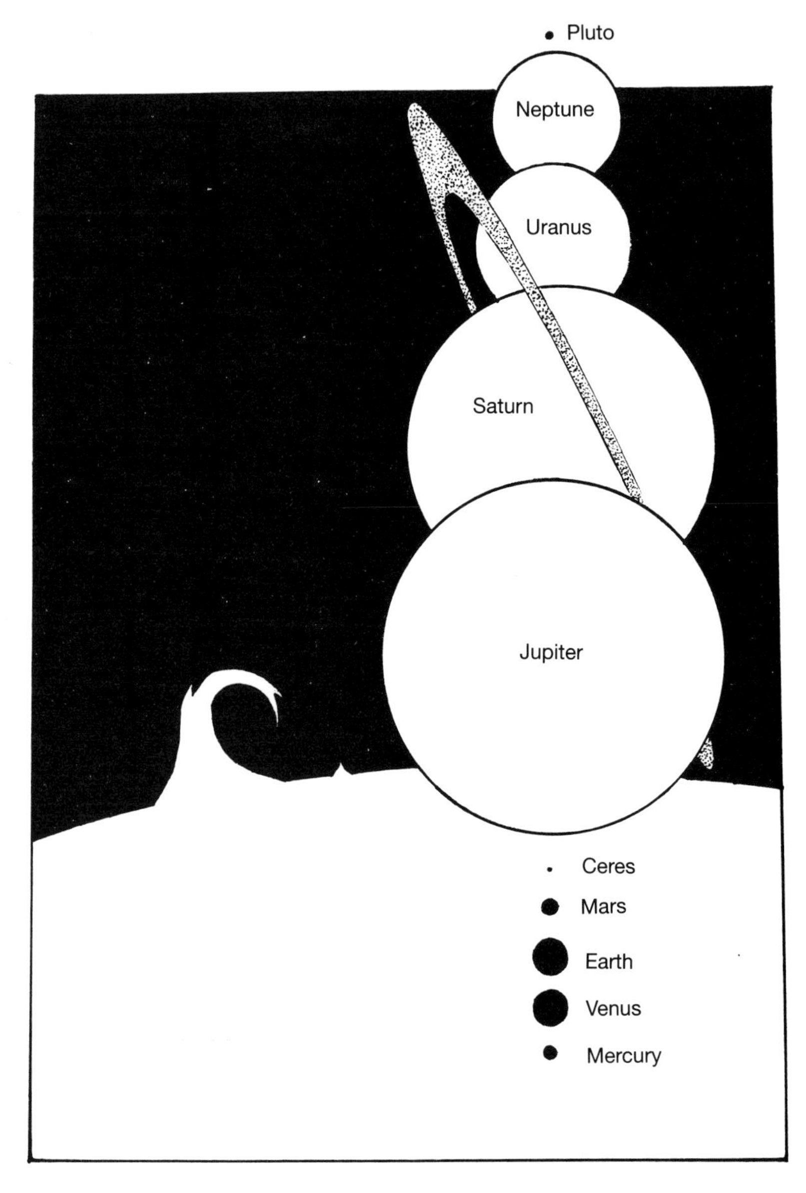

A voyager's overview

When you observe the individual planets your experience will be greatly enhanced by a sense of the overall layout and scale of the solar system.

Perhaps nothing has so graphically developed our feeling for the 'geography' of our system as the drama of the Voyager spacecraft. During more than 15 years the media have chronicled the adventures of NASA's two planetary explorers. These years have been a most effective public travelogue, taking all of us from the familiarity of our Earth–Moon backyard to the region of the giants, Jupiter and

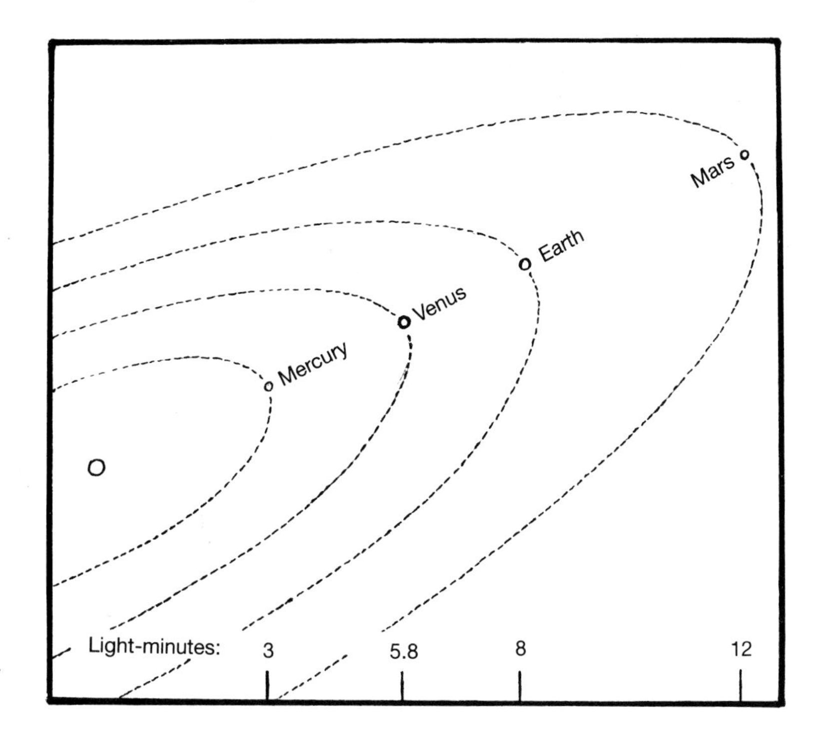

Figure 2.2. Distances within the inner solar system, in terms of light-minutes from the Sun.

Saturn, and then into the less-well-known realms of the distant outer planets.

After the Voyager 2 launch in 1977 we needed to be patient for only a brief two years before the craft arrived at Jupiter, to send back awesome colour views of that turbulent planet and its family of rough-skinned moons. Four years after departure, Voyager 2

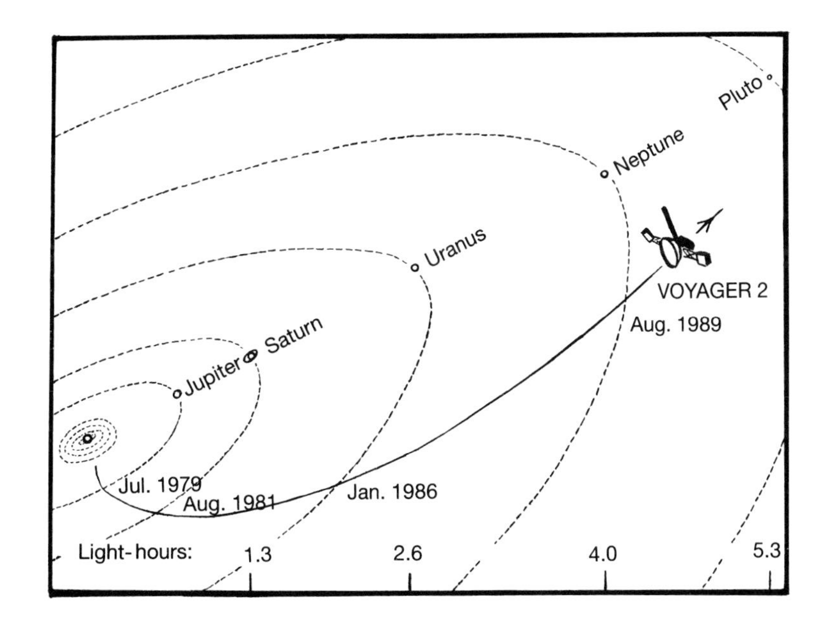

Figure 2.3. The outer solar system. Distances in light-hours from the Sun.

skimmed over Saturn's rings. Five more years were to elapse before the spacecraft reached the eerie blue world Uranus, and the little robotic explorer had been under way for twelve years when it glided through the dark cold outer emptiness where Neptune orbits.

A better sense of the scale of this adventure may perhaps be conveyed by describing the interplanetary distances in another way: by imagining a ride of the crest of a light-wave.

Light crosses the 'little' distances within the inner solar system in just a few minutes. From the Sun, light reaches Mercury in a mere three minutes, the Earth in eight minutes, Mars in twelve. Riding our light crest beyond the orbit of Mars we move out into the solar system's truly wide open spaces. The light which required brief minutes for the Sun–Earth journey rushes onward for more than an hour and a quarter to reach Saturn. By the time it encounters Neptune it has travelled four hours (also, of course, the length of time for Voyager's radio signals to return from Neptune). Icy little Pluto, when at its furthest from the Sun, receives light that has been more than five hours on its way from the Sun.

Your own planetary explorations

The number of objects within our solar system that can be usefully studied with a small- to medium-sized amateur instrument is surprisingly large. An objective lens as small as 7 cm will show all the major planets except one (Pluto), at least eight of the planetary

Table 2.1. *The solar system: 25 worlds on our doorstep*

Name of body	Diameter (km)	Mean distance from Sun (km)	Observing notes
Sun	1 392 000		Must be observed by projection, or through adequate solar filter. Much detail visible to small apertures, and considerable day-to-day change.
Mercury	4 878	58 000 000	Always viewed near horizon at either dawn or dusk. This small, rocky planet shows moonlike phases and, in large instruments, some hint of surface detail.
Venus	12 104	108 000 000	Dramatic phases visible to smallest telescopes, but an opaque atmosphere permits no view of surface features.
Earth	12 756	150 000 000	Don't forget that this planet is worth close study, too!
Moon	3 476		With tens of thousands of easily resolved, named features, our own satellite repays many years of telescopic exploration.

Table 2.1. *The solar system: 25 worlds on our doorstep* – continued

Name of body	Diameter (km)	Mean distance from Sun (km)	Observing notes
Mars	6 787	228 000 000	Not easy to observe, but at close approaches it is perhaps the most intriguing planet. Ice caps and albedo features can be detected with very small instruments; finer surface details with large aperture.
Ceres Pallas Vesta Juno	1 000 608 555 247	All approx. 370 000 000	At close approaches, these large bright asteroids are easily tracked with a small telescope as they move rapidly across background starfields.
Jupiter	142 800	778 000 000	Easiest of all planets to observe with modest instruments. Resolved into a disc of considerable size at medium power. Shows obvious atmospheric belts and zones. Even binoculars reveal the planet's four 'Galilean' moons.
Io Europa Ganymede Callisto	3 630 3 140 5 260 4 800		All visible with good binoculars or small telescope. The rapid orbital movements of these Jovian satellites cause an endless parade of phenomena such as eclipses, shadow-transits, etc., all easily observed by the beginning amateur.
Saturn	120 000	1 427 000 000	Saturn's celebrated ring system is visible at moderate power in simple instruments. Better telescopes reveal 'Cassini's Division' and faint atmospheric belts.
Titan Rhea Tethys Dione	5 550 1 530 1 060 1 120		Saturn's large moon Titan is a bright, obvious feature in small telescopes. Of the nine largest satellites in this planet's system, only the three additional bodies listed here may be detected without considerably larger instruments.
Uranus	50 800	2 869 600 000	Visible with binoculars. A moderate telescope can resolve its pale greenish-blue disc.
Neptune	48 600	4 496 600 000	Can be tracked with large-aperture binoculars. A good telescope barely resolves its tiny disc.
Pluto	3 000	5 899 900 000	Many amateur astronomers never glimpse this small, remote body. It can be located (as a pinprick of light) with a 20 or 25 cm telescope.
P/Encke	1–2(?)	Highly variable	A periodic comet that is usually spotted by amateur observers during its close Earth-approaches at 3.3-year intervals.
P/Giacobini-Zinner	2(?)	Highly variable	Also a frequent cometary visitor, returning at 6.5-year intervals. (Fine apparition in 1985.)

satellites, hundreds of minor planets or asteroids and a good percentage of all the comets that are newly discovered each year. It is little wonder that many amateurs specialize for a lifetime exclusively on solar system astronomy.

A brief 'menu' for solar system observers is shown in Table 2.1. The 25 bodies listed and described in this outline are good observational fare for the beginning astronomer. They may suffice to provide a full programme of study over a lifetime. Advanced amateurs will eventually track fainter asteroids and comets, and will perhaps engage in specialized projects involving photography, photometry and the precise timing of occultations. Detailed guides to Sun, Moon and individual planets will be found in subsequent chapters.

2.2 Exploring the Moon

Imagine a soft, warm summer evening about an hour after sunset. The air is fragrant and the sky has deepened to the velvety purple hue of full twilight. Like a bright mother-of-pearl pendant the first-quarter Moon swims in the gathering darkness overhead. On such a night you will automatically look toward this nearby world in space. Who has ever been able to resist doing so, since human beings first looked skyward?

Historically, the Earth's natural satellite has been the most heavily monitored of all celestial bodies by amateurs. This is not surprising. In our solar system, where some of the remoter planets are several light-hours distant, the Moon is so close by that its light takes little more than one second to reach the Earth.

Today, after manned lunar missions have left footprints (and other manmade litter) on the surface of our nearest neighbour, some of the traditional mystique has perhaps been lessened. Yet to most of us, the detailed topography of the Moon remains a vast and complex area for personal exploration and discovery. The degree of pleasure in lunar study with a telescope will be found to be proportional to the level of knowledge and the care of observation that the individual observer is willing to attain. Nevertheless, all of us begin as casual sightseers on the Moon; the best beginning is a broad overview.

Eventually, after a general familiarity with the lunar geography has been gained, systematic projects of detailed study will kindle a fuller degree of wonder and satisfaction.

First steps on the Moon

Binoculars show the broad lunar geography – the smooth plains called *maria* or 'seas', the largest of the craters and the mountainous highlands. A modest 5 cm telescope at 50× reveals almost a surfeit of details, including individual peaks, valleys, cliffs, craterlets and rills (surface cracks or faults).

Plate 10. Lunar detail photographed through a typical three-inch (76 mm) refractor.

Figure 2.4. First-quarter Moon (zenith prism orientation). *The Maria*: I *Mare Crisium* (The Sea of Crises), II *Mare Foecunditatis* (The Sea of Fertility), III *Mare Serenitatis* (The Sea of Serenity), IV *Mare Tranquillitatis* (The Sea of Tranquillity), V *Mare Nectaris* (The Sea of Nectar), VI *Mare Vaporum* (The Sea of Vapours). *Lunar mountains*: A. The Caucasus Mountains, B. The Apennine Range (a dramatic V-shaped highland on the boundary between the two phases), C. The Pyrenees Mountains. *Some outstanding craters*: 1. *Aristoteles*, 2. *Eudoxus* (this and the previous form a dramatic close pair), 3. *Endymion*, 4. *Plinius*, 5. *Manilius*, 6. *Macrobius*, 7. *Theophilus* (contains a dramatic central peak), 8. *Albategnius*, 9. *Maurolycus* (high power will reveal small craters on its floor), 10. Alpine Valley.

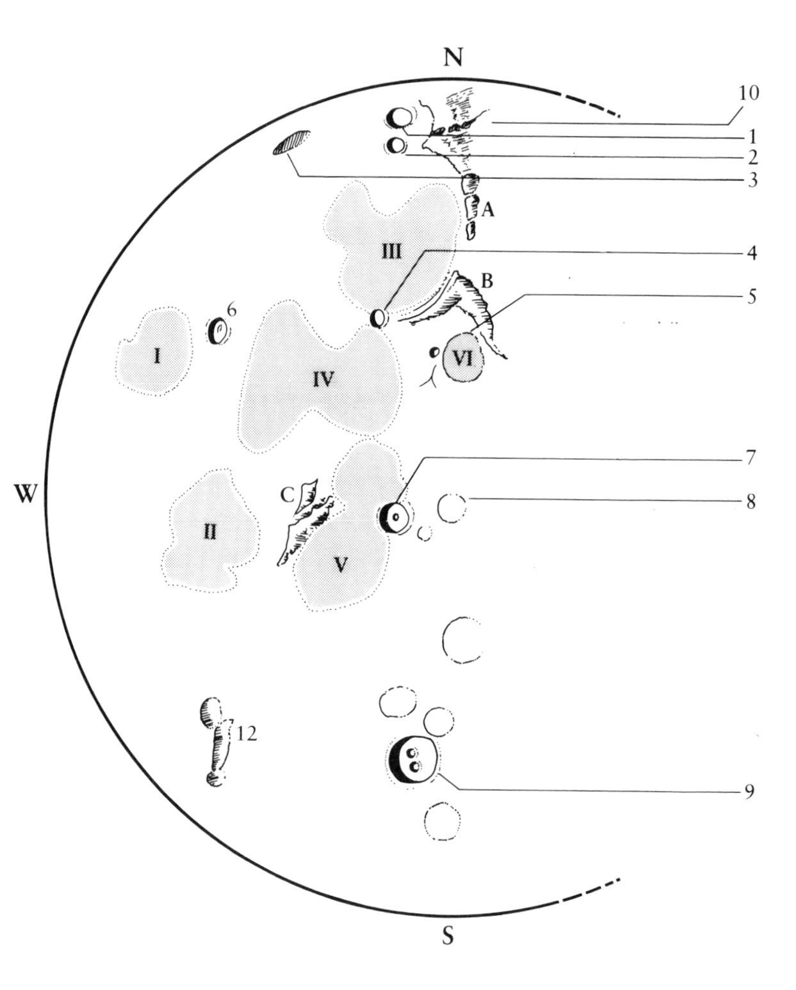

The key to enjoyable observation of the Moon is a knowledge of its named topographic features. One begins this process of familiarization by getting to know the broadest general regions, the huge dark plains or *maria*. The simple framework of their Moon-girdling pattern will serve as a guide to the locations and identities of smaller features such as mountain ranges, individual craters, and the many intriguing oddities of topography that you will discover.

Begin your explorations with a very basic lunar map, similar to those of Figures 2.4 and 2.5, which simplify the lunar geography to its most obvious features. During your first attempts to find your way about the Moon, a more detailed atlas will probably cause confusion. Later, when you have gained a degree of intimacy with the best-known features, a good lunar atlas will be useful for the identification of finer details.

What are the special techniques that enhance lunar observation?

The first and most critical is the choice of a suitable phase for best resolution of surface details. A full Moon (or the Moon at a phase approaching or just past full) appears surprisingly featureless, even when viewed through the largest telescopes. This is because the lunar surface, sunlit from directly overhead, is uniformly bright,

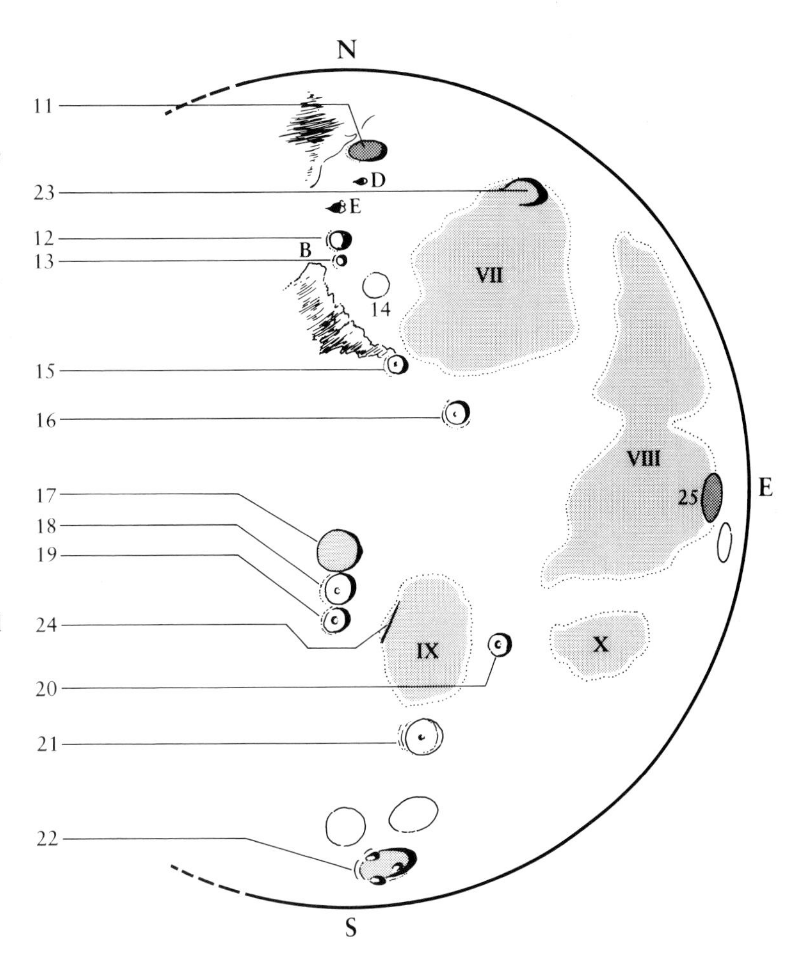

Figure 2.5. Last-quarter Moon. *The Maria*: VII *Mare Imbrium* (The Sea of Rains), VIII *Oceanus Procellarum* (The Ocean of Storms), IX *Mare Nubium* (The Sea of Clouds), X *Mare Humorum* (The Sea of Moisture). *Lunar mountains*: B. The Apennine Range, D. & E. *Pico* and *Piton* (striking solitary peaks; observe with high power), F. Spitzbergen Mountains (a small, but dramatic range). *Some outstanding craters*: 11. *Plato* (with smooth, dark floor), 12. *Aristillus*, 13. *Autolycus*, 14. *Archimedes*, 15. *Eratosthenes*, 16. *Copernicus* (perhaps the most perfect, and well known of the lunar craters), 17, 18, 19. *Ptolemaeus, Alphonsus, Arzachel* (a prominent chain of three large craters), 20. *Bullialdus*, 21. *Pitatus* (small isolated peak near its centre), 22. *Clavius* (huge crater with walls and floor marked by numerous smaller craters). *Miscellaneous smaller features*: 23. *Sinus Iridium* (a vast ringed plain, open on its southern side), 24. *Straight Wall* (a striking cliff-face, 65 miles/100 km in length and 800 feet/245 m high), 25. *Grimaldi* (conspicuous dark ringed plain).

without highlights or shadows. By contrast, if you train your glass on the Earth's satellite at quarter, or smaller, phase you will be awed by the dramatic bowls of the deep craters and by the towering thrust of high mountains whose presence is revealed by the oblique lighting, and by the resultant shadows cast by every feature large or small. Every experienced moonwatcher finds it exciting to spy out tantalizing hints of peaks and ranges that lie just beyond the *terminator*, in darkness, and then to watch over a period of hours as those features emerge into full sunlight, to reveal gradually their full shapes.

When the Moon's first quarter phase has passed, users of large telescopes especially will find it useful to employ a lunar filter in their eyepiece. (Unlike the Sun, our satellite offers no threat to eyepiece-mounted filters.) With a dense green or blue filter in place, the glare of even the full Moon can be reduced sufficiently to permit observation of some detail.*

* Is the unfiltered image of a full Moon hazardous to the eye? Although most experts say it is not, the present writer is not entirely certain. On occasion, after a brief view of the full Moon through a 30 cm telescope, I have felt discomfort and a degree of visual impairment for the following hour!

Probably no celestial body lends itself so easily to the beginner's first attempts at astrophotography. This is because, requiring an exposure of only about one-tenth of a second on fast film, the Moon can be 'snapped' through a fixed (that is, undriven) telescope. A camera body attached directly to the telescope's focuser, without eyepiece or camera lens, will take a neat little Moon-image of about half-inch (1 cm) diameter on 35 mm film.

Moonwalking with your telescope

The landscapes of the Moon are so accessible that even at relatively low magnification (a moderate 50× to 75×) you can become very nearly an astronaut walking on the lunar plains and hillsides. As you gain increasing knowledge of specific topographic features, your telescopic excursions on the Moon will become highly focused; you will visit and revisit regions that will have become intimately known favourites on our satellite.

Imagine for instance a night of steady 'seeing' (i.e. low atmospheric turbulence) when the Moon is at first quarter. You have alighted at about the midpoint of the mountainous terminator, the shadowed and highlighted border that separates lunar night from day. Travelling slowly northward you skirt the bright rampart of the Apennine Range (the V-shaped highland labelled 'B' on Figure 2.4, our sketch map of the first-quarter Moon). To the east these mountains rise into a glare of sunshine; to the west spreads the dark expansive plain of the *Mare Serenitatis*. A little further north you find yourself crossing a level gap before meeting the high, pointed headland of the Caucasus Mountains, along whose craggy spine you advance until you reach the abrupt rim of a deep chasm, the Alpine Valley. While hesitating on the lip of this canyon, you may decide to increase the telescope's magnification. Then, if seeing conditions are favourable, you may glimpse details far down in the shadowy depths of that intriguing gorge.

Such is the experience of moonwalking with your telescope. To enhance that experience you will want to be able to recognize the principal geological phenomena that characterize the landscapes of your adventure. The following is a description of the kinds of terrain through which you will travel on the lunar surface.

Features of the Moon's surface

There are several general classes of topographic features on the Moon. Basic to any detailed telescopic exploration of the lunar surface is an ability to recognize the kinds of features at which one is looking. These principal features may be described as follows.

Plate 11. The Moon is most advantageously observed at quarter (or smaller) phase, when the terminator reveals shadowed and highlighted details.

Maria ('Seas')
The singular form of this term is *mare*, pronounced 'máh-ray'. These are the very large, dark regions that can clearly be seen from the Earth with the unaided eye. They are relatively flat lava plains on which few significant craters, mountains or other features appear to the telescope. The maria are enormous regions; even the small *Mare Crisium* is over 560 km in breadth, and the more typically sized *Mare Tranquillitatis* is about twice that size.

Craters
These are small, circular depressions with sharply raised perimeters,

ranging from about 5 km to 60 km in diameter. Smaller features of this kind are called craterlets. Analysis of data gathered by the Apollo missions has suggested the likelihood that most craters are a result of meteorite impact. The smallest of such features visible in a moderate-aperture telescope at highest magnification will be about a kilometre in diameter.

Walled or ringed plains
These are craterlike formations on a very large scale – up to 300 km in diameter. Although some examples (for example, Plato, near the northwestern limb) appear smooth-floored, most of these huge plains are pocked with smaller craters and cleft by rilles or valleys.

Mountains and mountain ranges
Individual peaks stand out as bright points that cast dramatic shadows on the adjacent terrain. Long chains or ranges of mountains are a prominent feature on the Moon, the 9000 m high Apennine Range in the north central region being an example. The summits of such ranges are sometimes dramatically visible, sunlit above the hidden terrain on the dark side of the terminator.

Valleys, clefts and rilles
Large, medium and small chasms in the lunar surface. The greatest of these features (for example, the famous 'Alpine Valley' in the Moon's north central region) are huge valleys several hundred kilometres in length and ten or more kilometres in width. The most elusive of the rilles are mere cracks or faults that challenge large-aperture optics.

Other features
Include '*domes*' (mysterious small surface-bulges only a few kilometres in diameter) and '*rays*' (bright streaks of material ejected from craters and lying upon the surrounding terrain).

Programmes of observation

Projects of lunar research by which an amateur can contribute scientifically useful data are very limited nowadays, and yet not wholly nonexistent. As some of the suggested areas of interest below will indicate, a serious programme of monitoring the Moon may occasionally still uncover phenomena that will be worth recording and reporting to professional lunar scientists.

Nevertheless, perhaps the amateur astronomer's main goal should be systematic studies of the Moon that aim primarily at his or her own satisfaction.

A fundamental programme for the serious lunar specialists is

Plate 12. The Moon's surface, taken with a 30 cm f9 reflector on Tri-X film.

high-precision sketching of lunar features. No other procedure gives so thorough a familiarity with the Moon's detailed topography. Most amateurs restrict themselves to techniques involving just a soft pencil (or a combination of soft and hard pencils), and the best lunar sketchbooks are those that adhere to a single scale, such as perhaps about ten kilometres represented by one centimetre. While drawing at the telescope one must, of course, protect the eye's dark adaptation by using a red filtered light to illuminate the drawing pad.

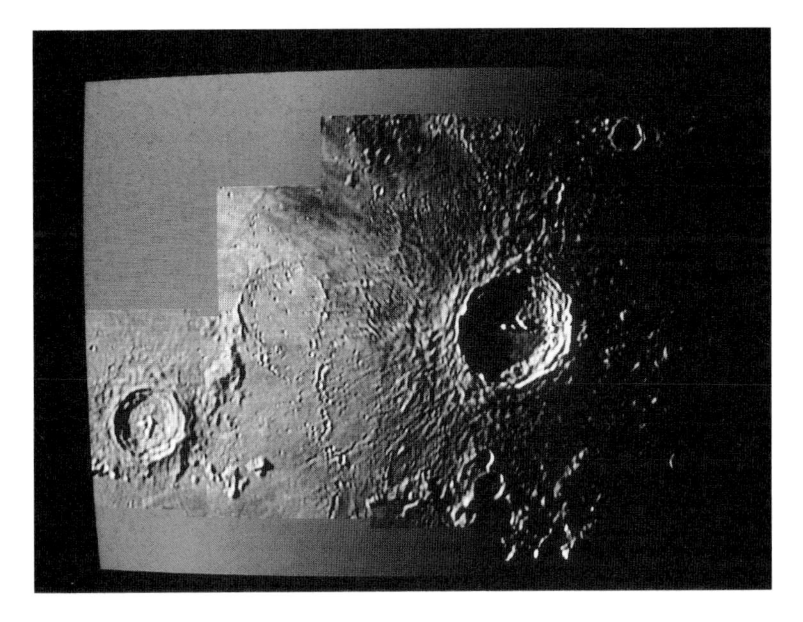

Plate 13. Lunar crater Copernicus – a CCD image by Richard Berry. This fine image approximates the best visual observing experience.

It is best to begin by focusing attention on relatively small, simple features (a single crater, a lunar valley, etc.) rather than to attempt overall mapping of large regions. An instructive practice is the sketching of the same feature at very frequent intervals, to reveal the varying aspects of the formation that emerge as angles of illumination change. A different procedure, especially for complex formations, is the gradual development of a single drawing of one feature by observing it repeatedly at times that provide similar conditions of lighting. Each new observation offers a chance for new details, or for corroboration of doubtful phenomena.

An unusually thorough and useful aid to this sort of thing is Fred Price's very practical *Moon Observer's Handbook* (Cambridge University Press, 1988). Price describes and illustrates detailed techniques for recording what you observe on the Moon in standardized drawings.

Here is a different kind of lunar project: experienced observers find themselves challenged by the study of *domes*, which are among the smallest and most subtle details on the Moon. Domes appear to be low bulges caused perhaps by pressure of subsurface gases. Because of their low and gently contoured profiles, these formations tend to escape detection except when they happen to be observed very close to the terminator, where the illumination angle is highly oblique. Many lunar specialists feel that there is still a possibility that earthbound telescopes may locate small, hitherto unknown domes. The experienced amateur who detects such a feature that cannot be identified in the most detailed lunar atlases may wish to communicate with the Association of Lunar and Planetary Observers, a society that collates reports of this kind.

Another project that has been popular with some amateurs is monitoring of the Moon for possible *lunar transitory phenomena*. Among the occurrences that such observers hope to witness and record are localized flashes of coloured light and mysterious obscurations of lunar features. This is an activity that may require an element of faith as well as patience, for many lunar experts have called into question the very existence of transitory phenomena on the inert lunar surface. Yet the detection by Apollo 14 astronauts of gaseous emissions from certain locations on the Moon have added credence to amateur reports of such phenomena.

A further incentive for systematic monitoring of the Moon for transient events is an encouraging set of statistical indications. Visual sightings of these events have tended not to be randomly scattered over the lunar surface, but highly focused on a small number of specific locations. An unusually large number of events reported by amateur and professional observers have been associated with a single topographical feature, the very bright crater Aristarchus. These phenomena have been reported in high concentrations in another locality near Aristarchus: the dark canyon known as Schröter's Valley. The culminating evidence for activity in these places may spur you to a programme of your own observational activity.

Figure 2.6. In a 'graze' the occulted star may disappear several times behind mountains on the lunar limb.

Finally, for a change of pace: yet another lunar programme is more likely to yield an easy and gratifying success. It is the watching and recording of lunar occultations – especially the fascinating *grazing occultations* that occur many times a year.

A full occultation is the total disappearance of a background star behind the Moon's eastern limb as our satellite's orbital movement carries it directly between the terrestrial observer and the star. A grazing occultation is a marginal case – a virtual miss. The intriguing aspect of such an event is that, as the star moves past the Moon's northern or southern limb, it may briefly disappear and reappear several times behind elevated features such as mountains, crater rims and lunar highlands. Predictions of times at which grazes may be visible from specific locations on the Earth can be found in the popular astronomical journals and in annual observers' handbooks.

Lunar photography is another obvious amateur pursuit; it will be described in the section of this book that deals with astrophotography.

2.3 Our Sun: the nearest star

No star, when viewed or photographed through even a very large telescope, appears as anything more than an undetailed point of light. All the stars are so immensely distant that no optical power can resolve them as discs, or show features of their surfaces.

Yet the exciting thing about the above statement is that it is untrue! One star is so nearby as to permit intimate close-up views of its secrets, even through the smallest and simplest telescopes. That star is, of course, our own Sun.

A warning to the curious

If you are a novice at solar observation, and if your curiosity regarding our planetary system's central star has been aroused, please read the following precautionary comments before setting up your telescope. Of all the astronomical activities described in this book, the exploration of the Sun is the only one that includes a strong element of actual physical danger.

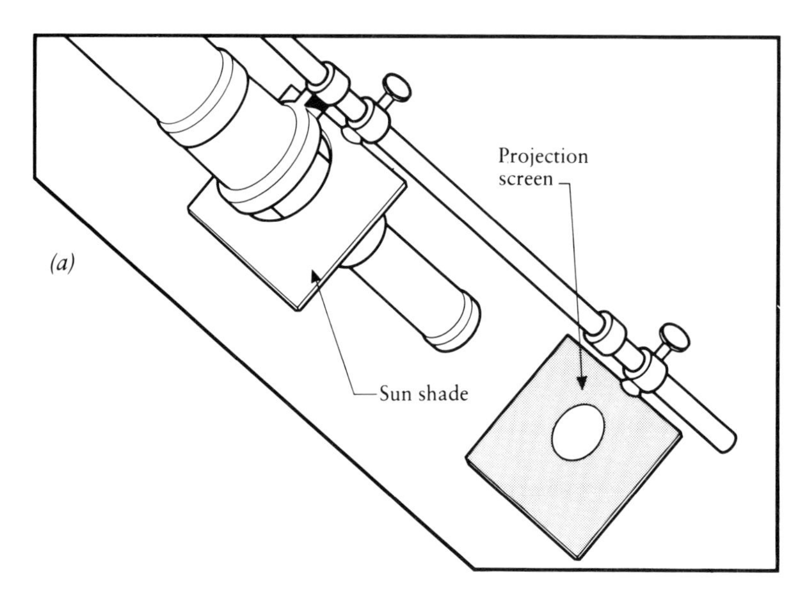

(b)

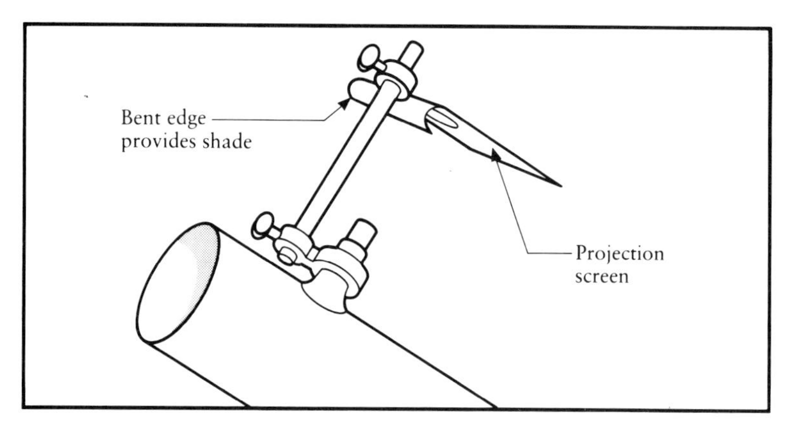

Figure 2.7. *(a)* Refractor set up for projection of the Sun. *(b)* Solar projection arrangement for a Newtonian.

The standard precaution is this: *never look directly at the Sun with any optical instrument* (including the eye) unless a suitably designed and carefully manufactured solar filter is used.

Once, in a careless moment, the present writer removed the objective-lens filter from a small solar telescope, and capped the eyepiece without first turning the instrument away from the Sun. Within two or three seconds the focused rays had melted a smoking hole through the little dustcap. What if it had been my eye, at that eyepiece!

If your telescope is a refractor or a simple Newtonian reflector, probably the easiest way to view the Sun safely is to project its image through the eyepiece onto a white card or a screen attached to the telescope, as illustrated in Figure 2.7. Remember, however, that the Sun must be located and its image directed onto the screen without your looking through either the telescope or its finder. Cap the finderscope, to protect its delicate crosshairs from destruction.

The shadow of the telescope's tube projected onto the ground will aid you in orienting it toward the Sun.

The optics of catadioptric telescopes become dangerously overheated if they are used for solar projection. A better approach is a good professionally designed sun-filter over the full aperture. Safe versions of the 'objective filter', made either of heavily aluminized flexible Mylar or of metal-coated glass, are available from most telescope suppliers. Although the sort of filter that is made of coated Mylar plastic film can be recommended for its low cost, these inexpensive versions usually give a rather unpleasing bluish colour to the Sun's image. More costly glass solar filters are often coated to produce a more natural golden Sun. The difference is a matter of aesthetics rather than of practical importance.

Beware: do not confuse the objective filter described above with the small glass filters that some manufacturers intend for use in the eyepiece. Eyepiece filters for sun observation are never fully safe. The intense 'laser' of a focused solar beam can crack them without warning.

Using either of the safe observing techniques one may view not only the Sun's most dramatic feature – sunspots – but also a number of more subtle details. Even a small telescope is capable of revealing the *faculae* (bright strands of superluminous gas), especially near the edge of the solar disc, where there is an apparent darkening of the surface. Also, when atmospheric conditions are not too turbulent, the general granulation of the solar surface is sometimes quite apparent.

Observing sunspots

Our Sun is an active and perhaps even rather variable star. In spite of its closeness to us in space, it is a star whose activities and variabilities are still imperfectly understood. Some physicists wonder, in fact, whether our 'dependable' Sun is capable of occasional, very extreme, declines and outbursts.

Of the indicators of solar activity, sunspots are the most easily monitored by amateur observers. These turbulent, short-term features of the visible surface (the *photosphere*) are carried across the solar disc by the Sun's rotation, which has a period of 26 days at the equator, but 36 days in the polar regions. When viewed close to the Sun's edge or limb, large spots reveal their characteristic dished or concave profile. The visual impression of a low-centred 'saucer' (called the *Wilson effect*) is easily noticed with a small telescope (see Figure 2.8*b*).

The smallest visible sunspots appear merely as structureless black dots. Larger spots show more detail: a very dark central region, the *umbra*, will be sharply differentiated from a much lighter surrounding structure called the *penumbra*. In very large spots (which often appear in pairs or groups) the umbra may be divided by

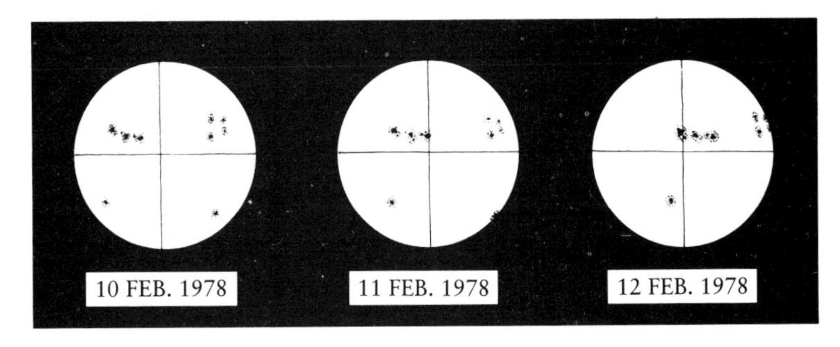

(a)

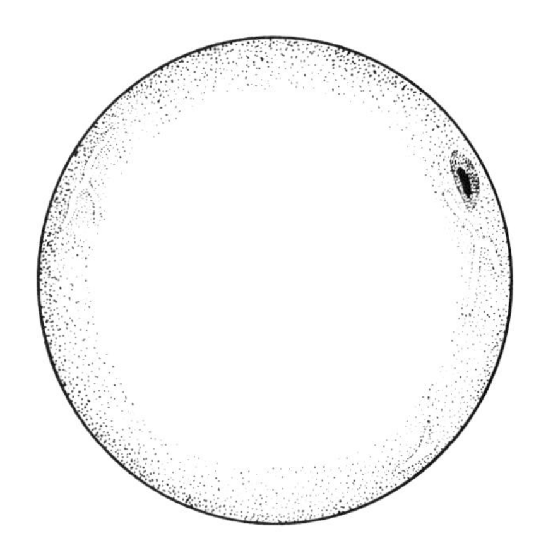

(b)

Figure 2.8. *(a)* The projected solar image on three consecutive days, showing the Sun's east-to-west rotation. *(b)* Some features of the projected Sun: the Wilson effect, limb darkening, faculae.

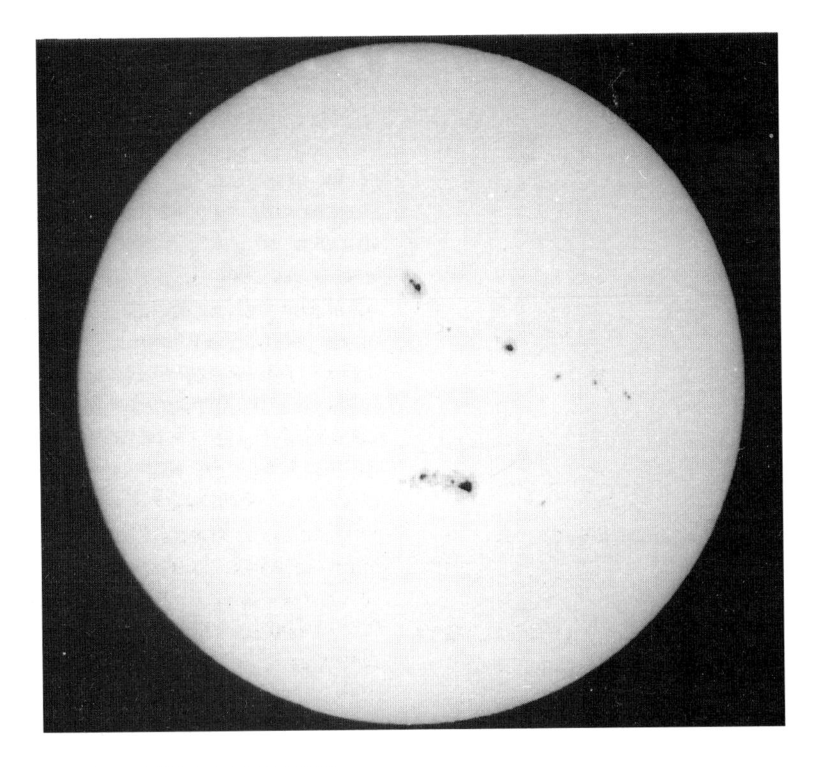

Plate 14. A very small telescope equipped with solar filter or projection screen shows the sunspotted photosphere.

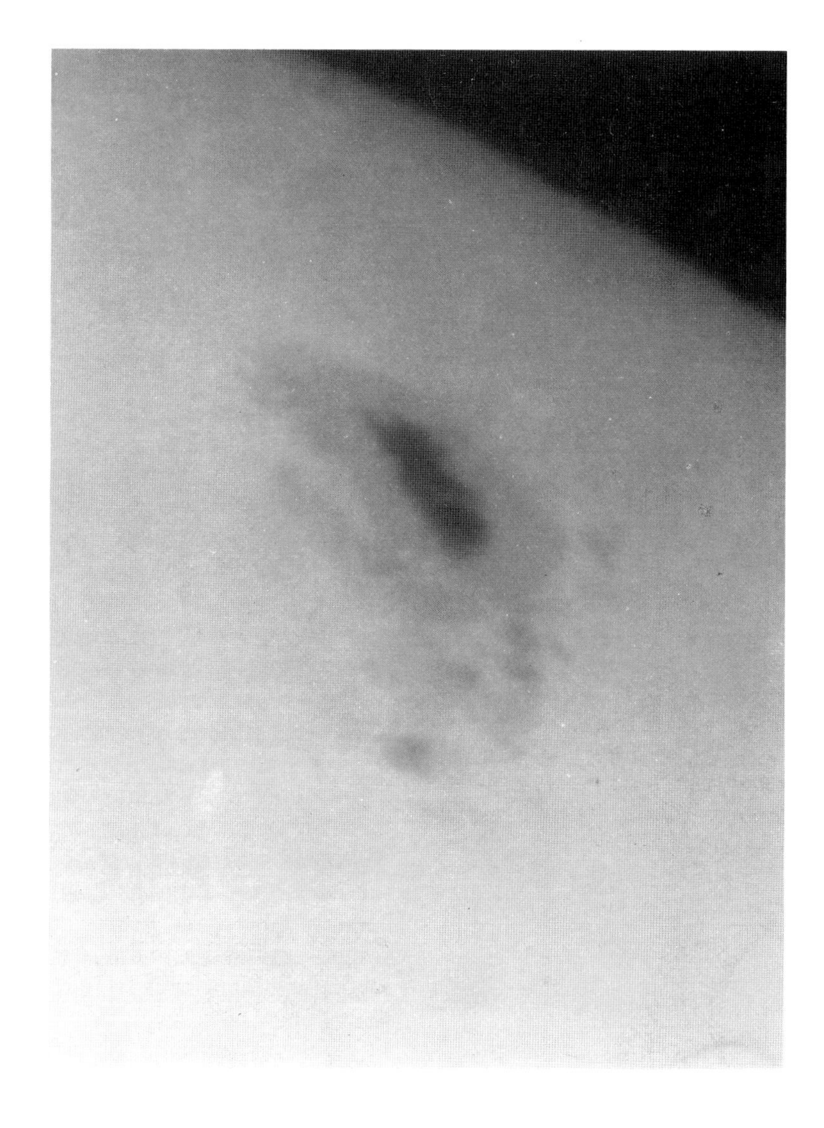

Plate 15. Sunspot at high magnification shows dark umbra and surrounding penumbral detail.

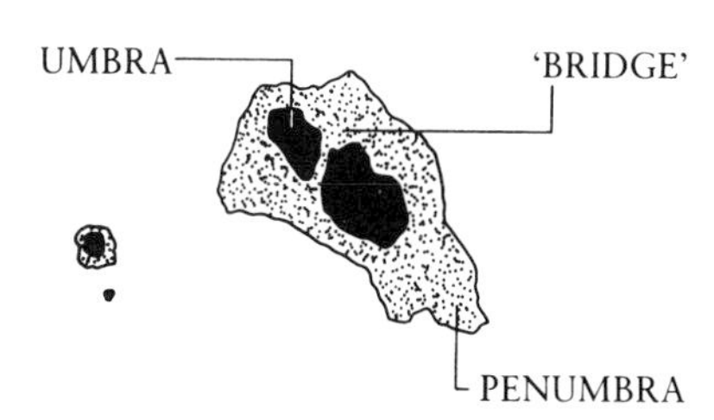

Figure 2.9. Typical sunspots (large and small).

bridges of light material. Observation of such a feature over several days may show the fascinating spectacle of the umbra dividing and changing into a multiple spot.

The number of sunspots at any given period is indicative of the measure of solar activity. There is a well-known periodicity or *solar cycle* that yields large numbers of sunspots in some years and small numbers (or none at all) in other years. The approximate time between one maximum and the next is usually about eleven years. For instance: spot activity was at a peak during 1980 and peaked again in 1991. Daily counts of sunspots and recording and graphing of sunspot-activity levels is a major project for serious amateurs. In Great Britain the British Astronomical Association and in North America the American Association of Variable Star Observers collate and make use of systematic reports received from regular sunspot observers. Because observation and reporting must be of uniform quality, the associations provide interested amateurs with instructions for standard submission procedure. Contact these official bodies for detailed information.

Solar prominences

Perhaps the most awesome impression of the Sun's surface violence is given by a view of the *prominences* that arch far into space beyond the solar limb.

Often associated with the turbulence of large sunspot groups, these vast, flamelike ejections of material rise to altitudes as great as 300 000 km to 400 000 km above the surface. Thus, the height of a large solar prominence may be equivalent to 30 Earth-diameters!

So rapidly does this giant-scale movement of gases occur that prominences may appear, disappear or alter significantly within the course of only a few days. The fully developed arch of a typical large prominence (familiar to most people through dramatic photography) is a form created by a combination of gravitational and magnetic forces, as eruptive material flows out into space and then falls back sunward.

This magnificent spectacle is not often witnessed by amateur observers; the relatively cool and dim material of the prominences is inevitably overwhelmed by the Sun's intense photosphere. Yet there are simple procedures that allow a fascinating glimpse of the enormous moving structures and their gradual alterations. One very satisfying, although quite expensive, approach is the use of a commercially made hydrogen-alpha (H_α) prominence filter. This system employs a special eyepiece unit to enhance the light of the prominences while suppressing the bright wavelengths of the photosphere, and combines this unit with an objective lens 'prefilter' to reduce overall light intensity to a safe level. Solar prominence filter systems are regularly advertised by a variety of suppliers in the pages of astronomical magazines. The cost of these items is as much, in fact, as the cost of a small telescope.

There is a considerably less-expensive alternative method for prominence observations. A standard telescope eyepiece of relatively long focus is fitted with three additional parts: (a) a hydrogen-alpha (H_α) filter threaded into the eyepiece barrel, (b) a dense red filter also threaded into the eyepiece, nearer to the telescope's objective lens and (c) at the focal plane of the eyepiece is placed a small metal *occulting disc* whose size is just slightly more than sufficient to mask the Sun's image in the telescopic field.

A commercially available H_α filter is the Day-Star H_α filter. Visually, this filter is fantastic. The prominences can be seen with the same clarity and detail as a feather at arm's length.

Roger W. Tuthill, New York, sells a 0.3 nm (3 ångström) solar prominence filter that requires mounting in a 30 mm outside-diameter eyepiece with an occulting disc to block out the disc of the Sun. The occulting disc can be machined out of brass shim stock or a tin

Plate 16. Solar prominence photographed with H_a 'Day-Star' filter.

can lid. The Sun's diameter can be easily measured by projecting the prime focus image on a ruler or on paper with ruled graduations. When you have determined the Sun's diameter in your telescope, cut the occulting disc to that size. Then cut the disc in half, trim it, and mount it into the eyepiece at its focus point, the same place as a cross hair would go. The size of the occulting disc can be tested and adjusted by trying the solar eyepiece on the full Moon, whose apparent diameter is the same as that of the Sun.

The solar prominence filter should be used at f30. My 75 mm refractor with a 3× Barlow lens provides breathtaking views of solar prominences. The Sun's disc **must** be kept **behind the occulting disc at all times or eye damage will result**.

To observe the prominences with the eyepiece described above, one must carefully guide the telescope so that the Sun's photosphere

Plate 17. Solar prominence spectroscope.

is wholly covered by the occulting disc at all times. Projecting beyond the limb of the Sun (and beyond the perimeter of the occulting disc) the solar prominences are dramatically visible as pale pink or magenta columns and arches.

BUT READ THIS WARNING: guidance of the telescope must be done with care and skill. The filtering in this eyepiece system is not sufficient to prevent eye damage if even a narrow arc of the Sun's bright photosphere is allowed to emerge from behind the occulting disc for more than a second or so. A small refractor is the safest telescope to use for this kind of prominence study, and the guiding technique should be practised first on the full Moon, before the Sun itself is observed. The serious solar enthusiast is advised to purchase a full solar prominence filtering system of the kind described earlier.

Solar prominence spectroscopes

Many books state that solar prominences are for the eyes of professional astronomers only. This statement is not true, but one *must* use common sense and good judgment.

I have built two spectroscopes and viewed hundreds of solar prominences under a variety of atmospheric conditions. The solar prominence spectroscope is a device that spreads out the sunlight into a basic spectrum. In that spectrum are all the colours of the rainbow, which is Nature's spectroscope. In the red end of the spectrum is the hydrogen-alpha (H_α) spectral line*; this line will show us solar prominences.

* The 'H-alpha line' is a narrow band at one specific wavelength (656.3 nm) in the red part of the visible spectrum in which the solar prominences produce their most energetic emission of light.

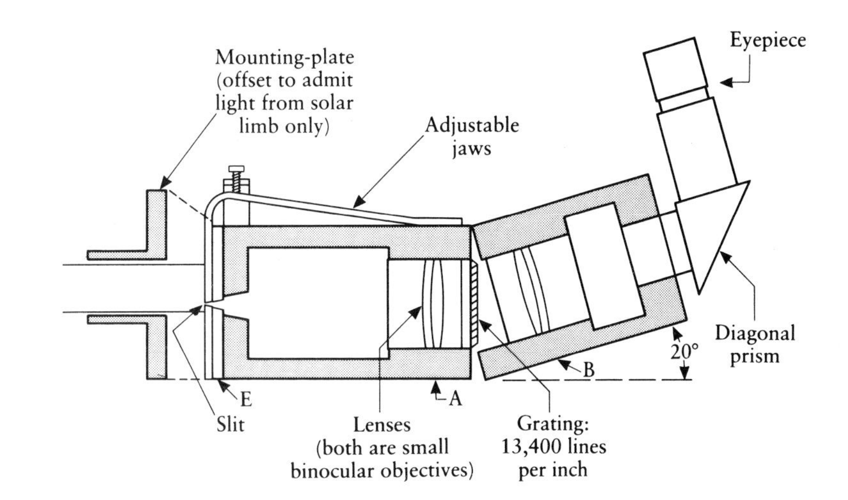

Figure 2.10. Solar spectroscope.

My first spectroscope was constructed of wood, the second of aluminium. The parts required are two achromatic objective lenses, 100–150 mm focal length and 15–40 mm diameter, a diffraction grating, slit jaws, and an ocular star diagonal.

I used the objective lenses from a pair of 6 × 30 binoculars, and a diffraction grating obtained from the Edmund Scientific catalogue. The grating had approximately 5275 lines per centimetre and was mounted in a 35 mm slide blank. My eyepiece was a 40 mm telescope ocular and star diagonal. A microscope adapter was used to mount the spectroscope on the telescope. The body could be any material that will stop light, and I chose to use wood and aluminium.

The slit is the most important part of the spectroscope. Care must be taken to ensure that the jaws meet squarely. A strip of aluminium 25 × 180 mm was used in my spectroscope, about 40 mm of which was bent over the front of the spectroscope, to form the upper jaw of the slit. The bottom jaw is a 40 mm strip of the same material. The end of each strip was carefully filed so that the outer surface formed the knife edge, as shown in Figure 2.10. The final sharpening of the slit was done with an emery stone. When the slit is sharp, it should be dulled slightly by running the knife edge at a 90-degree angle very lightly on the stone a few times. This will allow the two edges to meet squarely, and completely block out the light when the slit is closed.

The next step is to drill a hole, 2–3 mm in diameter, in the centre of the aluminium plate, marked 'E' on Figure 2.10. This plate should be about 3 mm thick, and large enough to cover the front of the spectroscope. The back of the hole should be countersunk or tapered so that the front will be sharp as a knife edge. The plate can be screwed to the front block after a larger hole is drilled in this block; the hole should be large enough to allow clear passage of the light from the plate. The bottom slit jaw should be mounted permanently to the aluminium plate, so that the knife edge covers exactly half the hole, as shown in the diagram. The top jaw of the slit should be bent over the front of the spectroscope so that it will meet squarely with the bottom jaw, as illustrated. The slit is made adjustable by means of a band of aluminium bent into a U with a thread bolt to adjust the opening of the slit.

The mounting plate on the front of the spectroscope, where the adapter goes, should be offset by 1/240th of the focal length of the telescope, which is half the diameter of the Sun at the focus. This enables the instrument to be rotated, so that the entire limb of the Sun can be examined with little readjustment of RA or Dec., keeping the slit tangential to the limb.

By unscrewing the front section of a pair of binoculars from the housing and leaving the lenses mounted, ready-mounted lenses are available. Ninety millimetre plywood can then be used to remount them in their new format. Lens A must be mounted so that its focus falls at the front surface of the slit jaws. The grating can be glued to wooden blocks behind lens A as shown in Figure 2.10. This will complete the front of the spectroscope.

The back half of the spectroscope is designed to swing, so that it may be adjusted visually on the H_α line after completion. It should be preset at an angle of approximately 160 degrees. Lens B is mounted in similar fashion to lens A except that it is reversed. The distance between lens A and lens B is not critical as long as there is room to mount the grating. The star diagonal prism is mounted in a block of 18 mm plywood, and should not be permanently fixed until final adjustments of the focus are made. The sides of the spectroscope can be any size, depending on the size of the lenses – the smaller the better because they determine the width and height of your spectroscope.

Now you can mount the spectroscope on the telescope, which should be a 50–100 mm refractor, equatorially mounted, preferably with a clock drive. With the instrument mounted, the Sun is focused on the centre of the slit. With the slit jaws open slightly, the red end of the spectrum should now be visible in the eyepiece. If it is not, the back half of the instrument can be moved up or down until the red area of the spectrum is centred. There may be longitudinal dust lines through the length of the spectrum, but these will disappear when the slit is opened more fully. There should also be dark absorption lines across the spectrum. If not, the block with the star diagonal can be moved back and forth until the lines appear and are in focus. With the lines sharp, the block may be permanently mounted. Further small adjustments can be made later with the eyepiece, by sliding it up or down in the star diagonal.

The spectroscope is now ready. A clear day is necessary, with a bright blue sky and no haze. The slit is placed, slightly open, tangential to the Sun's limb. As the slit touches the Sun's limb (the edge of the Sun's disc), a flicker of light will be seen shooting to a point down the length of the spectrum. This condition is only fleeting, and it takes a steady hand to keep the slit on the Sun's limb. When the point of light touches the H_α line, if there are any prominences at that point on the limb, the H_α line will change from a dark absorption line to a bright emission line. This will seem to burst along the line from the centre to each end. Opening the slit a little more reveals a ghostly prominence. The amount you open the slit is determined by the size of the prominence. The home-made solar prominence spectroscope affords only fair resolution at one spot on the solar limb. The delicate filaments of gas can be seen jetting away from the Sun's limb as they trace out the strong magnetic fields that direct them.

It may require several days for you to become an expert at holding the instrument on the Sun's limb, but with a little perseverance your effort will be rewarded. You will discover that the drama of the Sun's arching, cascading prominences are not accessible only to professionals at the great solar observatories; they are a spectacle for the backyard amateur as well.

The Sun in eclipse

The artificial occulting disc described in the previous section is one way of masking the solar photosphere. A far more dramatic method is provided by the exceedingly strange coincidence of the Sun's and the Moon's almost precisely similar apparent diameters, as observed from the Earth. Because the Moon's apparent size is less at *apogee* (greatest distance from the Earth) than that of the Sun, eclipses that occur when the Moon is in that orbital position are not total, but *annular*, leaving a ring of the solar photosphere uncovered.

Observed or photographed telescopically at total eclipse, the Sun reveals its prominences and many other phenomena of interest. Among these are the *chromosphere*, a thin reddish atmospheric layer just on the Sun's limb, and *Baily's Beads*, tiny dots of brilliant photospheric radiation that may be glimpsed through deep clefts or valleys on the Moon's limb. The dominant impression connected with a solar eclipse is that of the *corona*, the streaming, wispy glow of the Sun's rarefied outer atmosphere. Although this superheated envelope is a part of the Sun not ordinarily seen, it reveals its true extent during a total eclipse. If the transparency of our atmosphere co-operates during the observation, pale streamers of the corona may be glimpsed up to several solar diameters beyond the Sun's limb.

There is, of course, an added dimension to the experience of a total solar eclipse: it is the psychological aspect. Perhaps no other celestial phenomenon has had so great a power, throughout history, to inspire wonder, awe and dread. Even in our scientific age, the eeriness of an eclipse is something that almost every observer feels. When darkness and a sudden chill turn a sunlit morning into night, and birds fall unnaturally silent in the trees, a vestige of the primeval dismay can still be sensed.

Because the few minutes of totality pass very quickly, experienced eclipse watchers make a practice of drawing up an advance plan of observations. This may include naked-eye monitoring of the partial eclipse phases (through suitable protection such as a solar filter or welding glass), photography through the telescope during the first minute of totality, and visual telescopic observation of corona, prominences, etc., in the final seconds.

It has been suggested that a first-time eclipse observer may become so deeply involved in photography and other detailed projects that the immediate experience of the eclipse might be lost. Perhaps the first-time observer should leave his camera at home.

Solar eclipse photography, incidentally, is described in detail in the celestial photography section of this book.

Undoubtedly the most frustrating thing about solar eclipses is that they are infrequent and relatively localized phenomena. For some amateur astronomers eclipse-chasing becomes an adventurous addiction. While Philip Teece has observed only a partial eclipse, at home on Vancouver Island, Jack Newton is a true devotee, having travelled to places as widely separated as Manitoba, Siberia and Indonesia to record the eclipsed Sun on film. For people living in most places on Earth, the remainder of the twentieth century offers the chance of solar eclipse experiences only to those willing to travel. Table 2.2 will inform you about when and where the next ten such opportunities will occur.

Table 2.2. *Solar eclipses*

Date	Location	Type of eclipse	Approx. duration (min.)
3 November 1994	South America	Total	4
29 April 1995	South America	Annular	$6\frac{1}{2}$
24 October 1995	Asia, Borneo	Total	2
9 March 1997	Mongolia, Southern Siberia	Total	3
26 February 1998	Caribbean	Total	4
22 August 1998	Indonesia, Malaysia	Annular	3
16 February 1999	Australia	Annular	$1\frac{1}{2}$
11 August 1999	Southeast Britain, Central Europe	Total	$2\frac{1}{2}$
21 June 2001	South America, Africa	Total	3
14 December 2001	Pacific Ocean, Central America	Annular	4

2.4 Chasing a solar eclipse

Probably no astronomical activity offers the amateur so unique an opportunity for sheer fun as an eclipse of the Sun.

Total eclipses usually take place many miles away from populated areas, or at least far from the area where you happen to live. Thus, in order to enjoy an eclipse you will find yourself and other observers banding together to form an eclipse expedition. The delightful safari flavour of such an undertaking combined with the gamble of unpredictable weather will give this undertaking an element of excitement not found in most other astronomical pursuits.

What can you expect your eclipse-chasing expedition to be like? To convey some feeling for this very special observational adventure, we present notes on three past solar eclipses. The first two describe Jack Newton's experiences in Siberia and Indonesia – indications of how far afield the addicted eclipse hunter will venture to enjoy the brief spectacle of totality. The third anecdote tells of the frustration and joys of an eclipse for which observers travelled to Mexico.

If the mood of these adventures stirs your interest, consult the schedule of eclipses in the preceding chapter. Perhaps you will be on location to participate in the next event.

Siberia, 31 July 1981

Our expedition took us first to Leningrad and the Pulkova Observatory. It was the presence of the late Dr Bart J. Bok that opened doors and provided the royal treatment afforded us by our astronomer hosts.

We were met by the director, Kirill Tavastsherna, and the former director, Vladimir Krat. We were treated to an informal lecture and a tour of the complex. The observatory consists of several buildings. The principal telescopes comprise a one metre heliostat, a large computer-controlled meridian circle*, and a 660 mm (26 in.) refrac-

* A heliostat is a mirror mechanically driven to reflect the Sun's image to a fixed point, in spite of the Earth's rotation. A meridian circle is an instrument designed to measure the altitude of a star with great precision.

tor. The observatory had been completely destroyed during the Second World War and has since been restored.

The next stop was Yerevan, at the foot of biblical Mount Ararat, in Armenia. There we visited the Byurakan Observatory and were warmly received by the director, Victor A. Ambartsumian and his associate director, L. V. Mirzoyan. The primary observatory houses a 2.5 m Cassegrain. A one metre Schmidt with full-aperture grating is used to discover flare stars. The grounds were large and sprawling, with living quarters for the astronomers.

The flight to Bratsk from Moscow was the same distance as a flight from Los Angeles to New York. In Bratsk, Dr Bart Bok gave the group a talk on the Milky Way.

The next morning, we were awakened at 03.45 and bussed to our waiting boats. It was a one-hour boat ride to our camp on the central path of totality, about 500 km northwest of Lake Baikal. The site had been well prepared by the USSR eclipse organization, and accommodated groups from Japan, Canada, the USA and several eastern-block countries, including East Germany.

The sky started out clear, the temperature 24°C (75°F), but at the onset of totality the temperature drop caused very thin cloud to form, the effect of which was almost negligible. Totality lasted for 112 s, but shadow bands (parallel ripples of darkness that sometimes briefly appear on the ground immediately before totality) were not seen. However, a few members of our group had stayed in Bratsk and saw strong shadow bands and had a perfectly clear sky.

The Siberian eclipse proved to be one of the most interesting and enjoyable expeditions.

Plate 18. Total eclipse, Java, 11 June 1983. A one-second exposure through Celestron-90 on Ektachrome 400.

Plate 19. Eclipse expedition, Java: the observing site.

Java, Indonesia, 11 June 1983

This expedition took me to Hong Kong, Singapore, and Indonesia, in the Southern Hemisphere. We arrived in Jakarta on 9 June and flew on Garuda Airlines to Jogjakarta that same day. On arriving in Jogjakarta, we boarded rented buses to take us to Salatiga, about 80 km inland.

Our eclipse site was located on a military base on the outskirts of Salatiga. This was a flat region surrounded by active volcanoes. The evening prior to the eclipse, the sky cleared and we were able to enjoy the southern sky.

On the morning of the eclipse, the sky was crystal clear, and strong shadow bands were seen prior to totality. The total eclipse lasted nearly 5 minutes and two coronal horns were seen.

Mexico, 11 July 1991

John Hicks of the Royal Astronomical Society of Canada's Toronto Centre has provided us with the following impressions of the 1991 total eclipse, for which he travelled to Baja California, on Mexico's west coast:

The hotel telephone beside my bed rings loudly. Startled, I lean over to look at my watch. It is 3:00 local time in Puerto Vallarta, Mexico.

For a minute I question my intelligence, coming all this way to see a solar eclipse. But I realize this is it – July 11 1991, the day of the Big One, and the culmination of three years of planning. Still in a state of somnambulism I struggle to the door to see if the others in the expedition are up and about down the halls. A wave of hot humid air hits me as soon as I open the door,

Plate 20. Solar Eclipse, Mexico, 11 July 1991.

and I begin to dread the unimaginable that must certainly await us out in the Baja desert.

Soon I am aboard our chartered La Tur jet bound for San Jose Del Cabo with 175 completely ecstatic eclipse chasers, along with tons of optical equipment, all fine-tuned and ready to capture a spectacle deep in the mountains near Santiago, Baja California ...

After what seems like a journey across a continent we swing around a mountain and there is Santiago ... By now eclipse fever has taken over completely and we all feel like a deep-sea fisherman with a 200-pound tarpon on the line.

I ask myself, what must the locals in this little town be thinking? Perhaps we are here for a religious ceremony, a sacrifice or a revelation. I also question my associates about the armed militiamen I see in a troop-carrier with flak apparel and AK-47s. In an ominous way they park under a mesquite tree – a perfect spot for an ambush. My mind rushes away with wild visions as someone yells, 'This is it, folks. Set up whatever equipment you are carrying on the field in such a way that you are 10 metres apart from one another'.

Meanwhile a half-ton truck rolls up in a wave of dust. Out jumps a Mexican in white, rolling 45-gallon drums across the field. What is this, I wonder; but another man in white begins to pour bags of ice cubes into the upended drums. Then, behold: cases of beer are split and bottles carefully placed in the ice. I recognize the 'Corona' label on the old-fashioned brown bottles!

Staggering with the combined effects of the heat and a welcome bottle of the Mexican 'Corona', I unpack my equipment – first the scope, then the tripod, camera, air-bulb and solar filter. I am all hands, and nearly drop the nickle-plated optical glass filter that will screen out the searing heat of the sun. 'Clumsy fool!' I mutter to myself, realizing that the loss of the delicate coating on that filter would mean the loss of my also delicate eyes.

The shiny blue telescope barrel is so hot that I cannot touch it, and I'm worrying that the mirror may become misaligned in this heat. I wonder too if I can last three hours to film the whole event without heat-stroke. But all is ready; I check everything over and over again.

And then someone yells, 'First contact!'

We are underway. It has really begun, and the whole field is seized with concentration. Sitting comfortably on a Home Hardware kneeling pad, I watch the moon quietly slide into the image of the sun. It takes a perfectly round bite out of the top of the sun, and begins to gobble up the sun spot groups on its surface. One by one they disappear. I easily shoot half the 36-exposure film capacity and notice with every minute the air is getting cooler.

It is also getting grey, that is, the *air* is getting grey, not just the colour grey but a feeling of 'grey'. I tear myself away from the spectacle in the eyepiece to look around me. It's getting dark, and the mountains to the west are purple now and darkening down their long flanks. I think: it is really going to happen – right here in this unique spot on Earth – the total eclipse of the century. Suddenly I miss my wife, and wish she could be here to share all of this with me. I curse the fact that I didn't include her. My emotions drift into feelings of love, warmth and some sadness not being able to share this with her.

Suddenly someone screams: 'Second contact approaching! Get ready for filters off'.

I can't wait, and against years of knowledge of the sun's capacity to burn one's retina out, and what should be my better judgment, I grab the nickle-plated glass filter from in front of the objective. A blinding shaft of light pours through the viewer, turning the delicate cross hairs in the lens to silver gossamer threads. Carefully I place my right eye near the corner of the lens and see the spectacle of my life. The tiniest sliver of slivers of the last of the sun is pouring out pure nuclear energy, and the sky about it is swirling in ghostly silver light as light cirrus clouds play under it. The scene is heavenly; if ever one could get a tiny glimpse of what we think or imagine is heaven, this must surely be it.

For a split second we are all aesthetes, not just astronomy buffs. It is clear we have all come here to revel in the mystery of the universe, and in particular the beauty of it. Then it happens – in the splittest of split seconds, the sun goes black.

My partners alongside me are raving now and bellowing all kinds of mindless words, trying to describe the undescribable. I begin to laugh, I don't know why, but perhaps it is the total release of all my controlled emotions. The whole field is alive with the clicking of shutters.

Someone screams out at the top of his voice: 'The prominences, the prominences, look at the prominences!' I pounce into the eyepiece again and witness the most bizarre of the bizarre – the lizard-tongues of soft pink leap out around the now dark limb of a black sun. They are the softest pink in the palette of pinks, and probably not humanly reproducible.

After seven minutes, without warning a blast of concentrated light bursts from the solar limb as the moon places one of its valleys on the exact edge of the sun – the Diamond Ring effect. I see it in the eyepiece, and wonder if I still have any semblance of a *fovia* (the eye's sensitive focal point), as I have looked into 'forbidden' light. Quickly I reach for the winder on my camera. It rests at 36 – I'm out of film!

Yet I have seen the most precious spectacle of the eclipse, and without a filter. Walter, next to me, smiles and says: 'We saw it, and that's all anyone can hope for. Film could never record what we just saw'.

2.5 Elusive Mercury, shrouded Venus

To ancient watchers of the skies the behaviour of two planets appeared rather mysterious. Mercury and Venus were enigmatic: unlike Mars, Jupiter and the rest, this rapidly shifting pair did not traverse the sky each night, rising in the east and drifting overhead to an eventual setting in the west. Instead, they seemed forever tied to the horizons, always disappearing in the west shortly after the setting of the Sun or hovering low in the eastern sky when the Sun was about to rise there.

The invention of the telescope brought an instant revelation about Mercury and Venus.

When Galileo first turned his small glass toward Venus in 1610, and when Johannes Hevelius studied Mercury a few decades later, they witnessed a highly significant spectacle. At modest magnification the two horizon-loving planets displayed a range of phases similar to those of our Moon. These phases, combined with the two planets' Sun-hugging positions at all seasons of the year, suggested not only that they were objects in orbit about the Sun but also that they were *inferior* planets – bodies whose paths lay inside the circuit of the Earth's orbit.

You can use even the smallest modern telescope to experience the discoveries of Galileo and Hevelius. Our composite sketch (Figure 2.11) gives an accurately scaled impression of the strikingly similar appearances of the Moon, Venus and Mercury as viewed through a modest instrument at average magnification.

Although a repetition of this discovery of the inner planets' phases is easy, detailed observation of these two worlds is unexpectedly difficult. Centuries of telescopic work have yielded nothing like the wealth of detail sketched and photographed on, by comparison, Mars or Jupiter. The best images of Mercury and Venus produced by ground-based instruments have provided only vague and shadowy hints of topographic or atmospheric features. Space probes that explored the inner solar system during the 1970s transmitted views of these planets that finally revealed Mercury's cratered landscape and Venus' cloud-wrapped mountains and plateaus. Nevertheless, to the earthbound amateur these worlds remain a difficult

Figure 2.11. Telescopic appearance of the Moon, Venus and Mercury, their apparent sizes indicated to scale.

observational challenge; even the finest terrestrial optics will normally reveal little beyond the rapidly changing phases that intrigued Galileo and Hevelius.

Yet, diligent observers occasionally report (and sketch or photograph) traces of something more suggestive. It is this fact that may spur you to undertake regular and methodical studies of the two inner planets during every interval in which they appear above your horizon.

Observing Mercury

If you have never glimpsed the innermost planet through a telescope, you should not feel uniquely deprived. You may in fact be amazed to learn that many experienced amateurs have not viewed Mercury, in some cases even with the unaided eye.

The rocky little world, orbiting rapidly at only about one-third the Earth's distance from the Sun, makes fleeting appearances in the twilight, never far above the horizon. Unlike the bright major planets that will be seen at a casual glance high in the midnight darkness, Mercury must be carefully stalked by an observer who knows where and when to find its pale glimmer in the low haze of dawn or dusk.

Although there are six brief appearances of this kind each year, Mercury is well seen only once or twice in that span of time. Some especially favourable *elongations* (periods of Mercury's separation from the Sun that will prove most advantageous to the observer) are listed in Table 2.3.

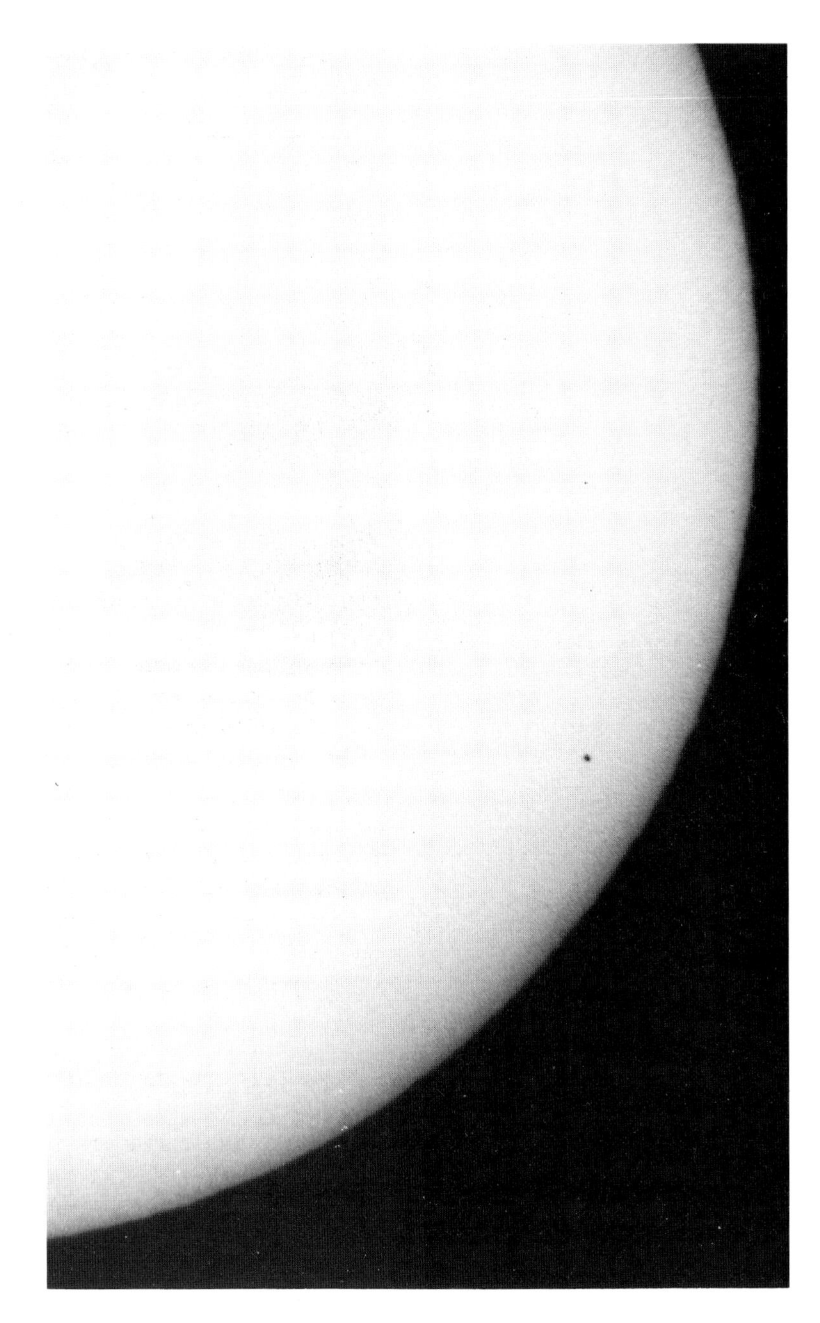

Plate 21. Mercury in transit across the Sun's disc.

Because of its very rapid orbital motion Mercury's phases change quickly; this fact is the little planet's chief attraction. Its tiny disc requires a relatively high magnification of 150× to 200×. From full phase to thin crescent phase it varies from only about 5 to about 13 arc-seconds in apparent size. Compared with the 40-arc-second disc of Jupiter or the 64-second diameter of Venus, Mercury presents a very tiny image in the telescopic field. The smallness of the visible disc combined with the turbulence of the air-mass through which

Table 2.3. *Watch for these apparitions of Mercury*

	In West, evening	In East, predawn
1994	(a) Late January	(a) Mid-July
	(b) Mid-May	(b) Early November
1995	Early May	Mid-October
1996	Mid-April	Early October
1997	Late March	Mid-September
1998	Early March	Late August
1999	Late February	Mid-August
2000	(a) Early February	(a) Late July
	(b) Early June	(b) Mid-November

Mercury is typically observed, near the horizon, will severely limit your chance of detecting features on the planet's surface.

Amateur observers do, however, record surface details. The drawings that they have produced over a long period tend to show similarities that suggest the reality of the light and dark regions that are being seen and sketched. Nevertheless, it is exceedingly difficult to relate these features to specific plains and cratered badlands photographed by the Mariner 10 spacecraft.

For your own programme of Mercury-watching you will want a good planetary telescope (perhaps ideally a fine refractor), although this instrument need not be larger than 10 or 15 cm aperture. Persevere until the generally featureless yellow planet yields glimpses of grey markings. Frequent sketches of these markings should be compared for evidence of consistent repetitions. In these repeated configurations you will have the satisfaction of suspecting that you are observing true aspects of the Mercurian topography.

The challenge of Venus

The other inner planet is in one sense much easier to observe, lying at times high above the horizon in the full darkness of predawn or post-sunset. Table 2.4 lists the most favourable elongations of Venus during coming years. Its phases, which vary from full to crescent like Mercury's, are satisfactorily seen with a very small telescope at a magnification of only about 40×. At its largest (in the thin crescent phase) Venus has an apparent diameter of over 60 arc-seconds.

In another sense, however, Venus is actually more challenging than the smaller and more distant Mercury. Although radar soundings by the Pioneer Venus spacecraft revealed an unusual surface characterized by gigantic plateaus and mountains higher than Everest, you will detect no trace of these features with even the finest telescope that you will ever construct or purchase. The dense greenhouse-atmosphere of Venus – largely carbon dioxide – presents a perpetually opaque and highly reflective barrier to observation of anything beneath the planet's outer gas-envelope.

Plate 22. Venus, on fine-grain positive film ($\frac{1}{8}$ s with 30 cm reflector).

Nevertheless, experienced amateurs do record variations (large-scale meteorological features?) in the atmosphere itself. Telescopic sketches by many observers throughout the past century have depicted vague darkenings and streaks of grey on the normally uniform white cloud-surface of Venus. Are these variations real or illusory? Perhaps your own long-term monitoring of this intriguing planet will give you a basis on which to form your own opinion. Some observers advise that your chances of detecting something will be increased if you use a blue-tinted filter in conjunction with your eyepiece.

Less controversial is another atmospheric effect: at the thin crescent phase a glow of sunlight may be seen extending beyond the extreme ends of the crescent, occasionally appearing to continue around almost the full perimeter of the planet's darkened disc. This phenomenon, which occurs near *inferior conjunction* – that is, when Venus is in close contact with the Sun on the earthward segment of the planet's orbit – is a simple effect of the planet's dense atmosphere being 'backlit' and glowing with dispersed sunlight.

Most Venus enthusiasts undertake to repeat purposefully an observation that Napoleon once made quite by accident. It is said that while leading his great march against Russia the emperor

Table 2.4. *Watch for these apparitions of Venus*

	In West, evening	In East, predawn
1994	April to October	November/December
1995	November/December	January to May
1996	January to May	July to December
1997	June to December	
1998		March to August
1999	January to July	September to December
2000		January to March

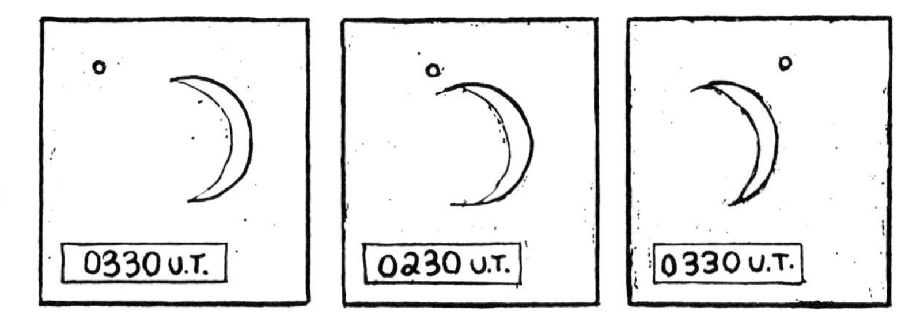

Figure 2.12. A daytime observation of Venus in close conjunction with the Moon. This kind of event provides an opportunity for easy location of the planet in daylight.

chanced to let his gaze fall upon Venus in the luminous blue of a daytime sky. The luck of seeing the planet in full daylight seemed (mistakenly, as events proved) to be a good omen.

You can locate and study our nearby planetary neighbour during the day more easily than Napoleon did. In an astronomical almanac note the day's RA and Dec. for the Sun, and the same co-ordinates for Venus. Using an equatorially mounted telescope equipped with setting circles, set the *solar-filtered* telescope on the Sun. Then offset the instrument in RA and Dec. by amounts equivalent to the differences of the Sun's and Venus' co-ordinates. Finally, remove the solar filter and locate Venus, moving the telescope slightly if necessary to bring the planet into the field of view. Your chance of success will be improved by the use of your lowest-powered, widest-field eyepiece.

Take great care while moving the telescope to avoid inadvertently re-finding the Sun after the solar filter has been removed.

An opportunity for much easier location of Venus in daylight occurs when the planet and the Moon happen to lie close together in the daytime sky. This sort of conjunction (usually predicted well in advance by the monthly astronomical magazines) permits fairly effortless naked-eye finding and observation of Venus. The sketches in Figure 2.12 record a very close conjunction of this kind that the present writer watched on an August afternoon in 1978.

2.6 Voyaging to Mars

Mars! Surely this is the most evocative name in the heavens.

What astronomer, amateur or professional, can deny the powerful mystique of the Red Planet? H.G. Wells' image of Mars as a threatening neighbour in space, C.S. Lewis' vision of a benignly populated planet, the Viking landers' photos of a stark Martian reality – these and many other sources have endowed us with a complex feeling about the Earth's nearby sister world.

Two visions of Mars

No planet in our solar system is so difficult to observe objectively. During the late nineteenth century and much of the twentieth Mars has been wrapped in a modern mythology so intriguing that we find ourselves reluctant to shake off its charming vestiges.

For many of us perhaps our earliest vision of Mars had its roots in events that occurred during the momentous winter of 1894/95. It was while Mars was close to the Earth in those winter months that Percival Lowell first observed the planet from his newly completed observatory at Flagstaff, Arizona. His book *Mars*, written in 1895 at the conclusion of this series of observations, contained revelatory descriptions of what he had discovered. He claimed to have observed a network of canals and to have watched a wave of darkening along these 'waterways' during the Martian spring that must, he argued, be vegetation thriving on a system of artificial irrigation. These

Figure 2.13. In his 'Mars Hill' observatory during the 1890s Percival Lowell began his widely publicized survey of the Red Planet.

sensational claims created a public mood about Mars that fuelled generations of amateur Mars-watching projects.

Even now, after an intervening century that has included the landings of two Viking spacecraft on Mars' desert surface, the present writer never fails to be strangely stirred by the classic photograph of Lowell at the eyepiece of his towering Alvan Clark refractor in its dome on Mars Hill.

The second vision of the Red Planet has arisen from our present knowledge of what the Vikings recorded in orbit and sensed, measured and photographed on the surface of that fascinating world.

This newer vision is superbly evoked in a book that once again bears the title that Lowell used. The 1992 novel *Mars* by Ben Bova is one of the best attempts we have seen to present an accurate impression of the frigid plains, the dry wispy atmosphere and the bizarre geology of Mars as today we know it actually to be. When the story's astronauts attempt an exploration of the Noctis Labyrinthus region we find ourselves experiencing firsthand the awesome scale of a typical Martian canyon system. With the scientists in the book

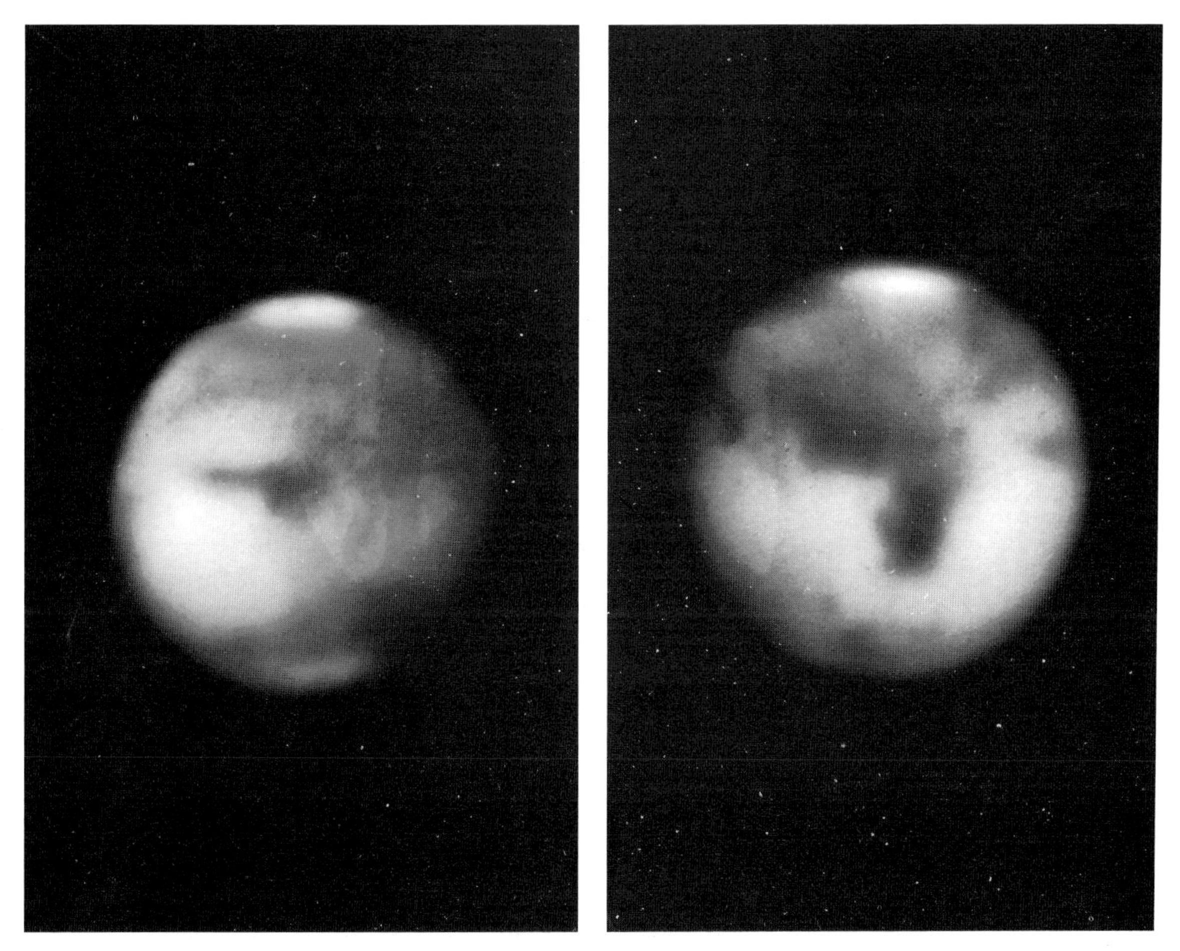

Plate 23. Mars is a challenging target for amateur telescopes. (Images by Donald C. Parker)

we ponder the conundrum of channels that seem to have been scoured by water, on a planet whose free-running water has mysteriously vanished.

It is ironic that the first *Mars*, written as scientific nonfiction, has proved to be a far less accurate vision of our neighbouring planet than the second *Mars*, whose entertaining fiction embodies much that we have learned from the Mariner spacecraft of the 1960s and the Viking landers of the 1970s.

What refinements of this vision await us, in the information that will be returned by the Mars Observer, launched in 1992?*

Your own Mars-watch

At first glance Mars can severely disappoint the observer, whether he or she is steeped in the traditional literary impressions or excited by the vivid landscapes revealed by NASA photographs. Even at the most favourable oppositions, when the little planet is seen to best possible advantage, it does not readily yield more than the vaguest hint of surface details. Mars is not an easy planet to study with the telescope.

On the other hand, few astronomical experiences are so thrilling as the occasional, momentary flash of intricate markings that emerge when transient atmospheric clarity rewards the observer with success. The most important single skill that must be developed is patience. In our experience, this willingness to view the red planet night after night, undiscouraged by long periods without results, is vastly more important than the size or quality of the telescope used.

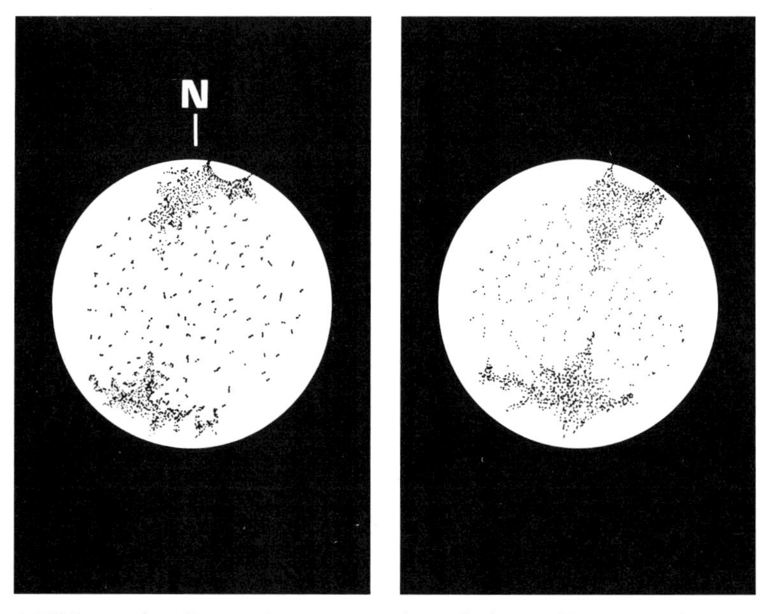

Figure 2.14. Mars observed with 90 mm Celestron on 14 and 16 March 1982. The prominent triangular feature is Syrtis Major.

* While proof reading was in progress we learned of Mars Observer's unfortunate malfunction and demise.

This is not to deny that a large instrument of good optical quality is better suited to serious Martian study than a smaller one. The accompanying sketches (Figure 2.14) show some of the characteristic detail that a very small telescope can reveal when conditions are favourable. A polar cap and large surface features are here represented as observed with an excellent little 9 cm Celestron Maksutov telescope.

Details of the Martian surface

Serious enjoyment of Mars through the telescope requires some knowledge of the kinds of observable features and also of the identities of certain individual surface-regions that will repeatedly be seen and recognized.

The *polar caps* (ice composed of water and carbon dioxide) are immediately noticeable. At close, or perihelion, oppositions the Martian south pole is tilted toward us and it is the south polar cap that we view. At distant, or aphelion, oppositions we view the north polar cap.

The dark surface markings that can be detected with amateur equipment are *albedo features*, regions on Mars where the surface material is darker than that of the general desert landscape. Although most of these features are permanent, their sizes and shapes may vary from time to time, perhaps because of the shifting of dust or sand coverings by wind.

Although the Martian atmosphere is thin (only 0.7 per cent as dense as the Earth's) and exceeding lacking in water vapour, there are occasional *clouds* that may be glimpsed as localized bright patches, most easily noticed when they lie near the planet's limb.

Figure 2.15 indicates the shapes and positions of the largest and best-known permanent features on Mars. The famous *Syrtis Major* ('Great Bog') is almost invariably the first surface detail that a novice observer identifies at sight, because of its darkness and its unmistakable wedge shape. The much smaller *Meridiani Sinus* ('Mid-longitude Gulf') is often quite obvious also. The *Erythraeum Mare* ('Red Sea') in the southern hemisphere and the *Acidalium Mare* ('Acid Sea') in mid-northern latitudes are paler regions, and yet often quite visible through apertures as small as 6 or 8 cm. Most of the other albedo features are considerably more subtle.

How do these features that you observe with your earthbound telescope relate to the topography that has been revealed to us in the spacecraft photographs? In general, the grey-green darkenings and dull yellow lighter regions that your telescope displays represent much broader features of the Martian surface than the individual plains, mountains and canyons resolved in the close-up photos.

Nevertheless, some correlation can be found between the usual telescopic map of Mars' surface and photographic maps compiled by the space probes. The bright 'gulf' of the *Hellas* region often

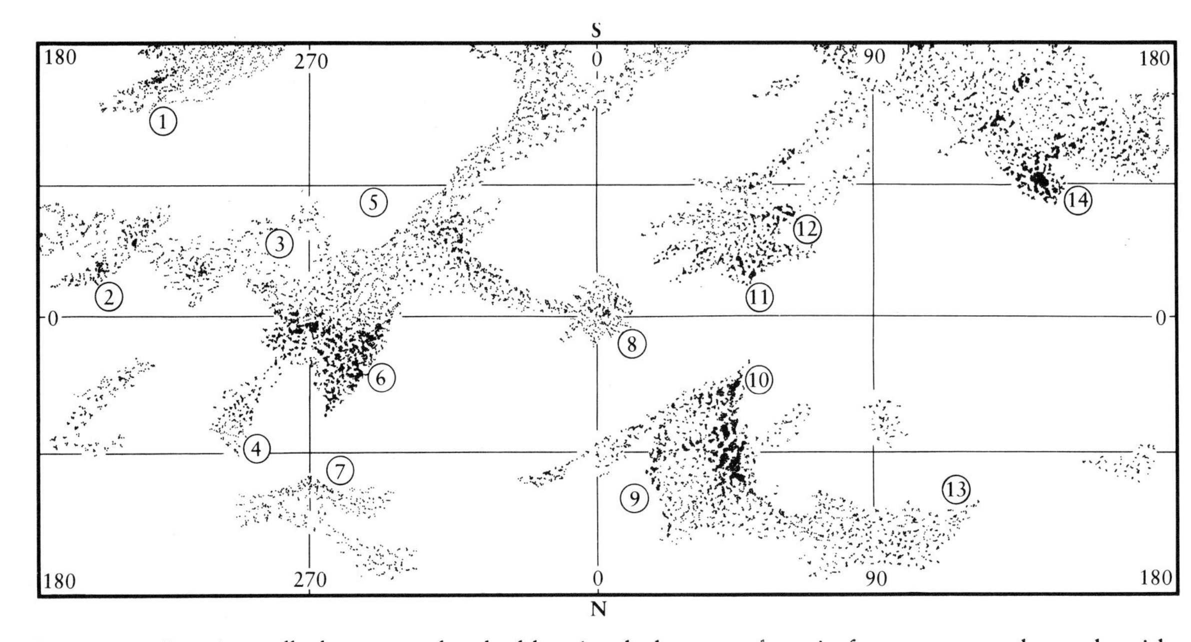

Figure 2.15. Prominent albedo features of Mars: 1. *Chronium Mare*, 2. *Tyrrhenum Mare*, 3. *Thoth*, 4. *Hellas*, 5. *Syrtis Major*, 6. *Meridiani Sinus*, 7. *Acidalium Mare*, 8. *Xanthe*, 9. *Aurorae Sinus*, 10. *Erythraeum Mare*, 11. *Solis Lacus*, 12. *Bureum Mare*, 13. *Tharsis*, 14. *Sirenum Mare*.

sketched by visual observers does, in fact, correspond exactly with the plain recorded by the Viking cameras and named *Hellas Planitia*, with its semicircular rampart of surrounding highlands. An east–west chain of dark markings that keen-eyed observers have repeatedly sketched close to the equator near the bright *Solis Lacus* seems to coincide with the shape and position of the immense canyon, *Valles Marineris*. A circular dark spot that often appears in drawings of the *Tharsis* region overlies, at least, the position of *Olympus Mons*, the most gigantic volcano known anywhere in the solar system.

A Mars sketch-book

Even a single, casual sketch of Mars can embody much interesting information. The most intriguing data, however, will be gained from series or sequences of related drawings. This, incidentally, is also true in most other astronomical connections. One may also regard these pairs or series as successive frames in an animated film, showing movement or change over a period of time. As in the case of animated movie stills, the planetary drawings will best serve their purpose if they are made as closely comparable with each other as possible. To achieve this, one should use the same telescope, the same magnification and the same filters (if any) for all sketches in a sequence.

Here are a few of the drawing projects that amateurs often undertake when observing Mars.

(a) *Long-term monitoring of the polar cap*. Sketches at relatively widely spaced intervals during the months in which Mars is approaching and passing through opposition will show a dramatic shrinking of the polar cap, as summer develops on the Martian hemisphere that is inclined toward us. If a telescope of 20 cm or greater is used, considerable details may be recorded, perhaps including narrow dark channels that divide the shrinking icecap.

(b) *Drawings to show rotation.* If the positions of surface features are sketched with reasonable precision at intervals of a few hours, the planet's rotation will easily be detected in the resulting series. Similarly, drawings made at 24-hour intervals will show a perceptible retrogression of features across the visible face of Mars, because of that planet's 37-minute-longer period of rotation compared with our own. Thus, in nightly observations that span a period of slightly more than one month one may do a sketch-map of the entire Martian surface.

(c) *Recording transient phenomena.* Occasionally an observation made on one night will include bright features not visible during previous nights' observations. Appearing like smaller, paler versions of the icecaps, these white patches are Martian clouds. A light blue filter may enhance the visibility of such features. Another phenomenon that can be recorded in a series of nightly drawings is the rapid obscuration of dark surface regions by yellow swatches of windborne dust.

(d) *Long-term study of specific features.* Drawings made over very long periods, spanning several oppositions, sometimes reveal apparent changes in the size and shape of major albedo features. (The *Meridiani Sinus* and *Syrtis Major* are regions in which unmistakable changes have been observed.) Since a small telescope is capable of providing only a very general impression of the shapes, a moderate to large instrument is needed for the study of changing details.

Oppositions of Mars

An opposition is the close approach between the Earth and Mars that occurs when the two planets lie on the same side of the Sun and on a line that intersects the Sun. For users of instruments smaller

Table 2.5. *Mars among the constellations*

	January–March	April–June	July–September	October–December
1994	(Not visible)	Pisces	Taurus	Cancer/Leo
1995	Leo	Leo	Virgo	Virgo/Libra
1996	(Not visible)	Taurus	Gemini/Cancer	Leo
1997	Virgo	Leo	Virgo	Scorpius/Sagittarius
1998	(Not visible)	(Not visible)	Leo	Virgo
1999	Virgo	Virgo	Libra/Scorpius	Sagittarius
2000	Pisces	(Not visible)	(Not visible)	Leo/Virgo

than a major observatory telescope these close approaches are the only occasions when Mars is anything more than a tiny, featureless blob of light. Although the Earth-to-Mars distances at various oppositions range from a disappointingly remote 100 million kilometres to an invitingly nearby 56 million kilometres, every opposition gives a reasonably good chance for viewing Martian surface details with a small telescope.

If you are planning telescopic studies of Mars in future years, be prepared for relatively distant (and yet quite interesting) approaches in February 1995 and March 1997. Following those rather lean years for Martian oppositions, the encounters between Mars and the Earth will become increasingly attractive at oppositions in spring 1999, summer 2001 and (most favourable of all) late summer 2003.

Of course you do not have to limit your observation of Mars to oppositions. Table 2.5 provides a guide to locating the planet during all months in which it will be readily visible. Even in its most distant orbital position Mars is an easy naked-eye object among the constellations.

2.7 The giants: Jupiter and Saturn

When we venture telescopically beyond the orbit of Mars we leave behind our intimate neighbourhood of small earthlike worlds. In the wider, colder spaces of the outer solar system we will encounter mysterious giant planets whose size and structure seem quite alien, compared with the rocky little spheres of Mercury, Venus, the Earth and Mars.

Jupiter

Jupiter is the planet most frequently observed by amateurs, in part because of the great number of months in each year during which this giant world is prominent in the night sky. It is available to the telescope for about ten months at each apparition.

Another reason for the ease with which Jupiter can be studied is dramatically illustrated in Figure 2.16a, which shows Mars and

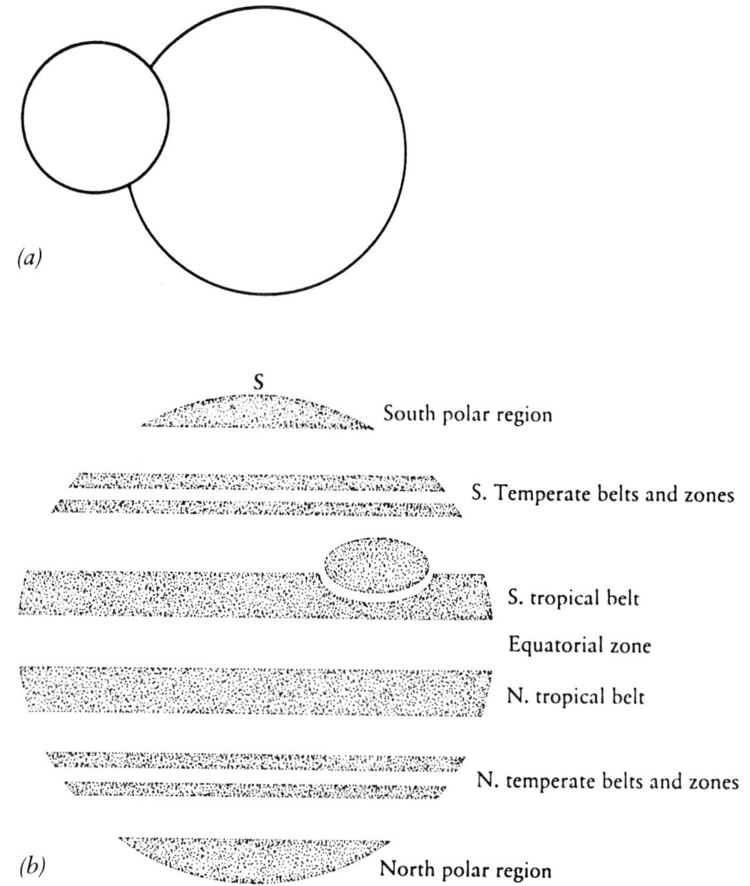

Figure 2.16. (a) Reason for Jupiter's relative ease of observation, compared with Mars; the two circles represent the planets' apparent diameters at opposition. *(b)* The principal belts (dark) and zones (light) that will be seen through the telescope.

Jupiter at their most favourable oppositions, both drawn to the same scale. Compared with Mars' 25-arc-second apparent disc, Jupiter's 48-second disc appears impressively large through the telescope. An additional attraction is the wealth of observable detail in the form of prominent, constantly changing atmospheric features.

Detailed amateur observations of the planet may include the following areas of attention.

Observing the Jovian atmosphere

The dynamics of an atmosphere as dense as Jupiter's are both fascinating and puzzling. Science fiction readers will remember the vivid impression of this unearthly environment conjured up by Arthur C. Clarke in his classic tale 'A Meeting with Medusa' (in Clarke's *The Wind from the Sun*, Harcourt Brace Jovanovich, 1972). Clarke describes a violent, soupy world in which floating-gasbag creatures ride on thermal currents, amid catastrophic electrochemical storms.

The apparent turbulence that even a modest telescope reveals is a result not only of high density and thermal imbalance, but also of

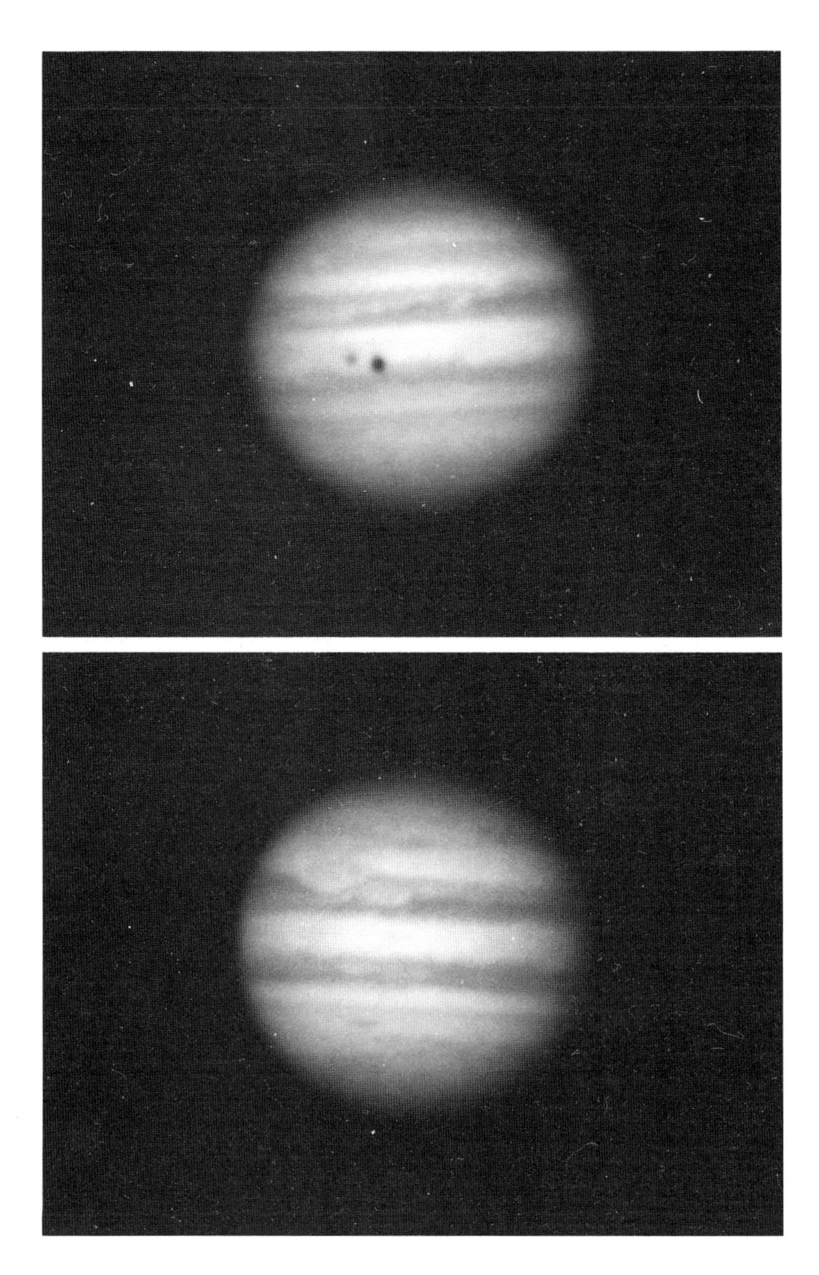

Plate 24. Jupiter, showing details typically observed visually through large amateur telescopes. Photos by Donald C. Parker.

the giant planet's unusually great rate of spin. Jupiter completes a full rotation in a little under ten hours. To the telescopic observer this means that individual features move very quickly across the planetary disc, passing from the limb to the central meridian in only two and a half hours. In fact, the high-speed rotation will cause a specific feature to shift position in the time it takes to make a carefully detailed drawing.

Although the rotation period generally is 9 hours 55 minutes, the equatorial zone rotates slightly more rapidly, in 9 hours 50 minutes. Because of this, a series of very precise sketches over a period of several revolutions will show a change of position of equatorial zone features relative to features in the adjacent belts. The shearing effect where these belts adjoin the faster-moving zone produces short-term whorls and streamers of dark turbulence that are worth monitoring with a telescope of 15 cm or larger. Small, ephemeral spots, protuberances and dark swatches that suddenly appear in the Jovian atmosphere are rarely detectable in photographs by even the largest earthbound telescopes, and yet they are quite easy to observe visually.*

Like the Earth and Mars, Jupiter has been assigned an arbitrary system of longitude lines, beginning at zero degrees longitude and progressing westward around the planet in degree-increments. Unlike the terrestrial continents or the surface regions of Mars, features in the Jovian atmosphere tend not to remain at fixed longitudes; they *advance* (drift westward) or *retrogress* (drift eastward) from the position at which they are first observed. The movement of individual features is of interest as an indicator of Jupiter's atmospheric dynamics.

It is possible for the amateur to detect, at least roughly, the positional shifts of bright or dark spots and irregularities or 'knots' in the belts. The procedure is a matter of timing the transit of a feature across the central meridian of the disc.

Although the central meridian (an imaginary north–south line that precisely divides the disc into halves) can be located by micrometer measurement, it can also be judged with a fair degree of accuracy merely by visual estimate, after some practice. As Jupiter's rotation brings an individual feature to the meridian, the observer estimates and notes the time at which the feature's leading edge lies on the meridian. The next step is to determine what longitude lies at the central meridian at that time. Astronomical yearbooks such as *The Observer's Handbook* provide tables from which these longitude positions at specific dates and times can be calculated. Since the times cited in such tables are, of course, Universal Time, it is UT that must be used when transits are observed and recorded.

The timing process repeated at subsequent reappearances of the same atmospheric feature will show, over a period of weeks or months, that the leading edge of the spot or dark knot under observation is no longer at the same longitude as at the beginning of the study. The results noted at various times can be converted into a graph that shows the longitudinal drift of features. More dramatically, the same data can be translated into a sequence of drawings of the features, with longitude-grids precisely superimposed.

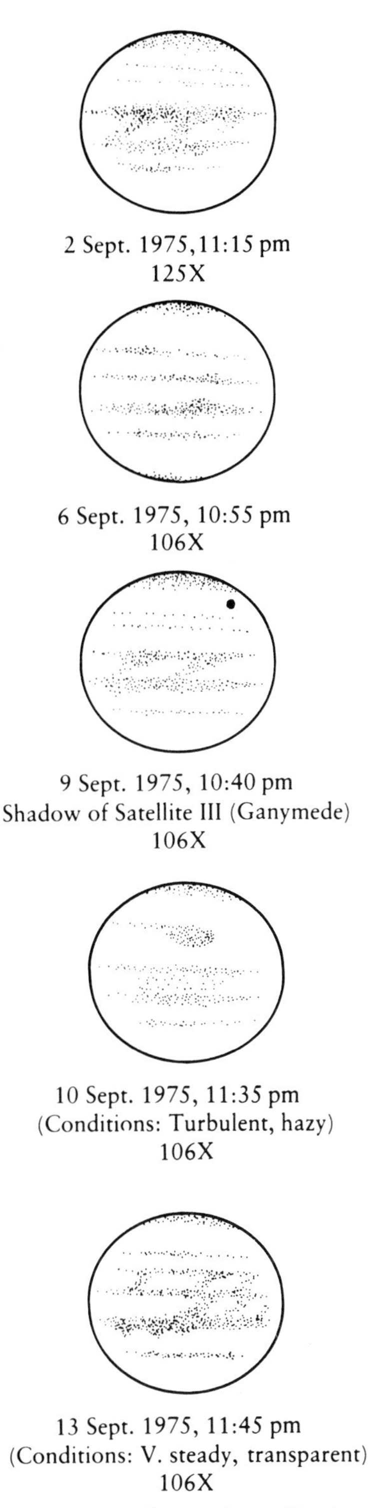

Figure 2.17. Observations of Jupiter (with five-inch (12·5 cm) f5 refractor, September 1975). Page from 1975 observing log.

* It is interesting to note that the difference in rotational speeds amounts to something like 400 km/h; this is the velocity at which the material of the equatorial zone drags itself past the resisting medium of the adjacent belts. Little wonder that the line of contact is turbulent!

Plate 25. A dramatic look at the four Galilean moons.

The Galilean satellites

When Galileo turned his earliest experimental telescope towards Jupiter at the beginning of 1610, he became the discoverer of the first-known moons orbiting a planet beyond the Earth. His notebook for January of that year contains nightly drawings of the planet with its four newly found satellites in a sequence of rapidly changing positions.

It is perhaps ironic and yet somehow appropriate that, of the sixteen Jovian moons known to astronomers today, Galileo's original four (known as the Galilean satellites) remain as the prime targets for the amateur's telescope, in fact, amateur observers' notebook-drawings of Io, Europa, Ganymede and Callisto often look strikingly similar to Galileo's seventeenth-century sketches, and embody similar data about the satellites' movements.

With the knowledge and instrumentation available nowadays to even the beginner, much more can be done in connection with these moons.

Unlike Galileo's little telescope, which showed the moons only as points of light, our modern optics can reveal something of their true form. A good refractor of only 10 or 15 cm aperture is capable of resolving the 1.5-arc-second discs of Ganymede and Callisto, if seeing conditions are really superlative, Io's one-second disc may also (barely) be resolved. It must be admitted, however, that the present writer has never observed these discs with less than a 30 cm instrument; a trace of atmospheric turbulence will defeat the attempt. Patient and persistent study of the tiny satellites' discs may occasionally result in the detection of faint hints of surface detail. A wide range of subtle shadings have been sketched by various observers over a period of many decades, especially on the planet-sized moon Ganymede. Although none of these markings can be identified with any of the ridges, clefts or craters photographed by the Voyager spacecraft, some of them at least may be actual colour variations associated with large-scale albedo features. A large telescope, excellent seeing conditions and an experienced eye are needed for the attempt to detect such details on the satellites.

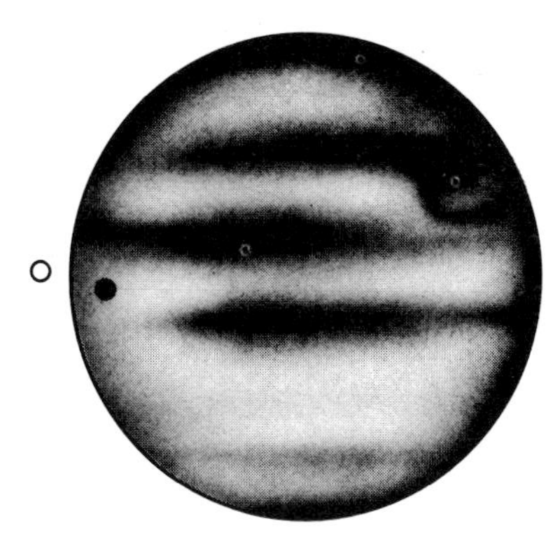

Figure 2.18. Jupiter, 12.15 am 1973. Observed with a five-inch (12·5 cm) refractor at 106×. Satellite I (Io) emerges from the limb of the planet, casting a shadow on the upper atmosphere.

Although the effort to discern surface markings will perhaps be found excessively difficult, there are other phenomena that lend themselves to easier and more dramatic projects. All of these are predictable events associated with the satellites' movements. Most can be exciting to observe with even the smallest telescopes. We now give brief descriptions of the several classes of event.

(a) *Shadow transits* occur frequently and are easily seen. They are a phenomenon upon which beginning amateurs usually stumble accidentally while observing Jupiter. What is seen is a tiny black dot proceeding slowly across the planetary disc; this is the shadow that a satellite casts upon the bright surface of the Jovian upper atmosphere (see Figure 2.18).

When Jupiter is moving toward opposition, the shadow precedes the satellite itself (that is, the shadow will begin to transit Jupiter's disc while the satellite is still approaching the planet's limb). Once opposition has passed, the situation is reversed: a satellite will begin to cross the disc first, followed by its shadow. The shadow can be seen with only a 5 or 6 cm telescope at very limited magnification.

(b) *Satellite transits* themselves are considerably more challenging, because the tiny yellowish moons tend to be overwhelmed by the bright yellow background of Jupiter's atmosphere. Nevertheless, the little disc of a satellite can often be discerned when it is traversing the relatively dark limb of Jupiter, or when it crosses in front of the brownish gases of an atmospheric belt. In this situation the Galilean satellites show their true discoid form more readily than when they are observed against the backdrop of space. We have seen Callisto's minute disc in transit across Jupiter's grey limb with a telescope of only 9 cm aperture, at quite high magnification (160×).

Because the moons have perceptible discs they move onto and away from the Jovian disc not instantly, but by a slow process of *ingress* and *egress*. A careful observer will easily see the stages of these processes. Because the inner satellites orbit at high velocity, their ingress and egress occur quickly; the whole disc of Io will slide onto the Jovian limb in only two and a half minutes, and Europa's entry occurs in approximately the same length of time. Ingress for more distant Ganymede is a ponderous seven-minute procedure, and that for Callisto, is nine minutes. The times of ingress and egress – the most fascinating and instructive segments of a satellite's transit – are predicted in tables that appear in the annual astronomical handbooks.

(c) *Eclipses* occur when the satellites move into Jupiter's shadow.

Thus, at times predicted in annual handbook tables, a moon under observation will fade and disappear, usually at some considerable distance from conjunction with the actual limb of the planet. When Jupiter is approaching opposition, its moons will fall into eclipse after emerging from behind the planet. After opposition eclipses will occur as the satellites are about to move behind the planet.

As in the case of transits, an eclipse will be a gradual rather than an instantaneous event. The light curve of a satellite as it dims toward extinction and then emerges from eclipse can be plotted with the aid of a photometer. See p. 266 for details concerning photometric work.

(d) *Occultations* are the disappearances of the moons behind the limb of Jupiter itself, rather than into the planet's shadow. Occasionally it is possible to observe a variation on this theme that is especially intriguing: it is the occultation of one Galilean satellite by another of the Galileans. In 1985, when the plane of the satellites' orbits lay precisely on our line of sight, events of this kind were frequent. The same fortuitous alignment recurs approximately every six years; at such times the major astronomical magazines usually publish details and predicted times of mutual satellite encounters that will be observable. In our chapter on photometry (Chapter 5.4) there is a description of one 1985 amateur project that applied photoelectric photometry to the study of that year's several Galilean satellite events.

Table 2.6. *Jupiter among the constellations*

	January–April	May–August	September–December
1994	Virgo	Virgo	Virgo (Not visible, Nov./Dec.)
1995	Scorpius	Scorpius	Scorpius (Not visible, Nov./Dec.)
1996	Sagittarius	Sagittarius	Sagittarius
1997	Capricornus (Not visible, Jan./Feb.)	Capricornus	Capricornus
1998	Aquarius (Not visible, Feb./Mar.)	Aquarius	Aquarius
1999	Pisces (Not visible, Mar./Apr.)	Aries	Aries
2000	Aries	Taurus (Not visible, May/June)	Taurus

Finding Jupiter

During most of the ten-month period each year in which Jupiter is visible in the sky, it is so bright an object that it is instantly and easily findable. Table 2.6, of the planet's wanderings among the constellations in coming years, will indicate where you should look.

Saturn

So captivating is the telescopic spectacle of this ringed giant that some amateur astronomers become Saturn specialists (as, admittedly, there are Mars and Jupiter specialists, also). Saturn is a planet that seems capable of revealing endlessly more subtle wonders as the observer's skills and equipment increase in sophistication.

Stunning imagery of the ring-systems provided by Voyager 2 in 1981 has greatly added to earthbased interest in the planet. In centuries of telescopic study astronomers had formed a relatively simple impression of the Saturnian rings, which appear generally as three broad concentric systems, with hints of slightly greater complexity just beyond the threshold of visibility. Nothing prepared us for the surprise of the Voyager discoveries: at close range Saturn exhibited a ring system of seven major concentric components, each resolved into scores of discrete gaps and ringlets. Among unexpected

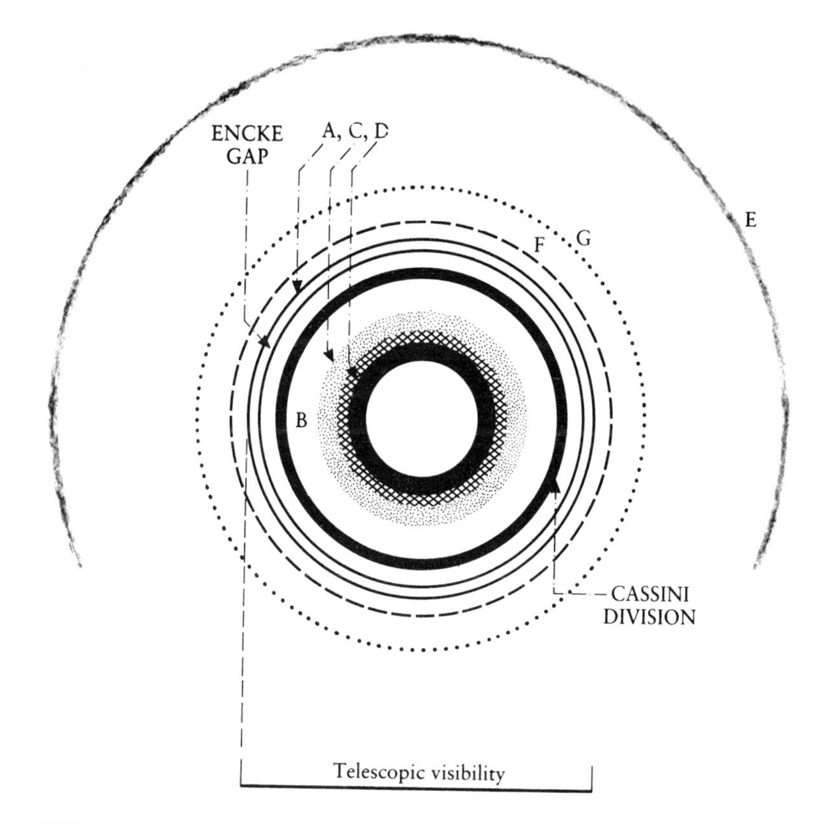

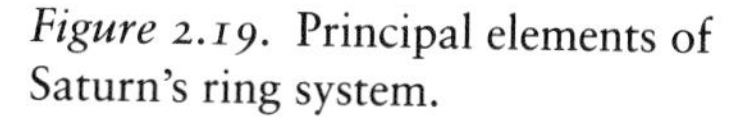
Figure 2.19. Principal elements of Saturn's ring system.

Plate 26. Saturn, photographed with rings viewed in plane (1967) and tilted to our line of sight (1971).

features were apparent 'spokes' that radiate outward across the traditional bright B-Ring, and previously undiscovered satellites embedded within the ring system. A particularly intriguing situation is seen in the planet's threadlike F-Ring, whose form may partially be due to tidal interaction with the two tiny 'guardian moons' that orbit on either side of it.

In broad outline, the elaborate architecture of rings and gaps surrounding Saturn is as illustrated in Figure 2.19.

Observing the rings

Much of the detail shown in Figure 2.19 is known only through computer-enhanced images obtained at close range by the two Voyager spacecraft. The earthbased telescopic view is more limited.

With typical moderately-sized amateur telescopes an experienced visual observer can easily distinguish the B-Ring, the narrower A-Ring and the dark *Cassini Division* that separates them. In good seeing conditions it is also possible to detect 'C' (the so-called *Crepe Ring*) and, if the telescope is excellent, *Encke's Division*, which appears as a slender discontinuity in the A-Ring. Hints of two similar discontinuities in the wider B-Ring have been sketched by some observers using very large instruments.

Plate 27. CCD image of Saturn shows ring details typically observed visually. By Donald C. Parker.

How much of this will you see with a really small telescope? The answer is: a surprising amount of detail. A good 6 cm refractor can show the B-Ring and the A-Ring clearly divided by the dark line of *Cassini's Division*, and may yield an occasional glimpse of the portion of the Crepe Ring that lies like a grey discolouration across the planet's bright disc.

The actual nature of the rings is now, of course, quite well known; they are zones of orbiting debris in the form of mixed small (dustgrain-sized) to large (meteor-sized) particles lying in a plane only about 100 m thick. There are two amateur techniques that can yield an insight into the different particle-densities of the separate visible rings. One is experimentation with coloured eyepiece filters, which tend to enhance the contrast of brightness between more- and less-reflective segments. Violet, blue and yellow filters will each affect the contrasts in a slightly different fashion.

The other project is observation of the occultation of a star by Saturn's rings. Parts of the system will extinguish the star totally, while others leave it practically undimmed. From detailed monitoring of the variation of the light from a star during such an occultation one may produce a crude graph of the light-transmitting characteristics of the separate rings. Presumably this monitoring could lend itself very well to photoelectric photometry, although the authors are unfamiliar with any such project that has been attempted by amateurs.

Successful earthbased observation of Saturn's rings is in part dependent on the orientation of the ring plane to our line of sight. This orientation changes as Saturn moves about its $29\frac{1}{2}$ year orbit.

The optimum situation, with the rings inclined at a generous 27 degrees to our line of sight, occurs at about 15-year intervals. For instance, 1987 was a year in which the presentation of Saturn's rings was at its most satisfactory, and this circumstance will recur in 2002. In the years between these best phases the angle will vary from an only slightly less dramatic slant to a situation that represents the opposite extreme: no angle at all (with the rings viewed directly edgewise).

The edge-on-view of the ring system, when it occurs, represents an interesting challenge. During the few hours when we lie precisely in the plane of the rings, they disappear altogether from view. At times near the exact edgewise orientation an experienced observer with a good telescope may be able to see the system as an exceedingly fine thread of bright material bisecting Saturn's disc; it is an eerie sight.

Saturn's moons

Saturn has at least twenty satellites. Only six of these are easily located with moderate telescopes. Those Saturnian moons whose brightness is above 12th magnitude are listed in Table 2.7 in order of their positions outward from the planet (with Iapetus outermost). In theory, all six can be observed with a telescope as small as 8 cm aperture; in practice, larger instruments are recommended.

Unlike Jupiter's moons, which are normally viewed in a relatively straight line close to our line of sight, the Saturnian moons lie in orbital planes that are considerably tilted to our view. This is because, as explained above, Saturn itself normally has its equatorial plane sharply inclined relative to our perspective. It is advantageous, therefore, to observe the satellites at times when Saturn's equatorial plane is oriented precisely edge-on. In these special years the moons participate in phenomena of the kind that make Jupiter's satellites so fascinating: occultations, eclipses and transits. Of particular interest are passages of the moons 'through' (behind or in front of) Saturn's ring system.

Table 2.7. *The moons of Saturn*

Satellite	Mag	Period of revolution (days)	Recommended telescope-size, minimum (cm)
Enceladus	11.8	1.37	20
Tethys	10.3	1.9	15
Dione	10.4	2.7	15
Rhea	9.7	4.5	8
Titan	8.4	15.9	5
Iapetus	11.0 (variable)	79.3	20

Table 2.8. *Saturn among the constellations*

	January–June	July–December
1994	Aquarius (Not visible Jan./Mar.)	Aquarius
1995	Aquarius (Not visible Feb./Mar.)	Aquarius
1996	Aquarius (Not visible Mar./Apr.)	Pisces
1997	Pisces (Not visible Mar./Apr.)	Pisces
1998	Pisces (Not visible Apr./May)	Aries
1999	Aries (Not visible Apr./May)	Aries
2000	Taurus (Not visible May/Jun.)	Taurus

Identification of a faint moon may be confirmed by observing its movement over a number of subsequent hours or nights. Its period of revolution is each satellite's unique identifying 'fingerprint'. It is helpful also to consult the finding-charts for the Saturnian moons that appear in annual handbooks and, from time to time, in the popular astronomical magazines.

Of rather special interest among the moons is Iapetus, a body that has for centuries puzzled observers because of its extreme variability. At its westernmost position relative to Saturn, Iapetus is about one magnitude brighter than when it is at the opposite side of its orbit. It was long supposed that this moon must be an elongated splinter, like some of the variable asteroids. Close-up Voyager images revealed the strange truth: Iapetus is marked by an odd blotch of dark surface material that covers a substantial portion of one hemisphere. The drop in brightness when this marking faces toward us can be noticed visually by an observer who is practised in the techniques of variable star work.

2.8 The outer reaches: planets in the dark

The final three major planets travel increasingly remote orbits, where they circle in a frigid darkness feebly lit by an exceedingly distant Sun.

These worlds (two of them gaseous giants and the third a rocky miniature) share one thing in common: unlike all the planets that lie nearer to the Sun, these three were discovered within historic, and in fact recent, times.

The circumstances of their discovery suggest a kind of graded challenge for today's amateur. Uranus, accidentally found in 1791 by William Herschel with a small reflecting telescope, was a relatively easy discovery. The much more remote Neptune eluded detection even after Adams and Leverrier had mathematically indicated its probable position, until in 1846 an astronomer named Galle used a moderate-sized telescope to locate its minute greenish disc. Tiny Pluto yielded the secret of its presence in the outer boondocks of the solar system only after a 20-year photographic survey that ended in the triumph of Clyde Tombaugh's 1930 discovery.

With your own backyard telescope you can repeat these planetary discoveries. The relative challenge of the three excursions into increasingly remote space will be proportional to the difficulties of the historical discoveries.

Uranus

When you know the location of Uranus you can find the planet in the night sky using optical aid as modest as binoculars.

If you are observing with binoculars, positive identification will depend on your detection of Uranus' movement against the backdrop of fixed stars. Because of the planet's remote, leisurely orbit the movement will be slow – typically a shift of less than two arc-minutes in twenty-four hours. Nevertheless, you will have no difficulty in seeing this shift of position, especially when Uranus is

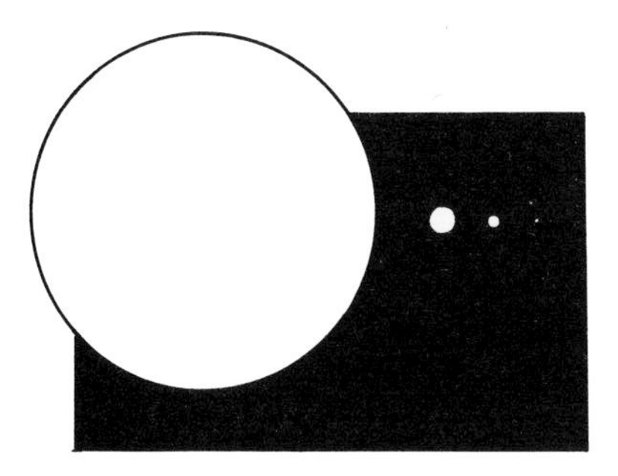

Figure 2.20. The outer planets through a small telescope: apparent sizes of Jupiter (very large disc at left), Uranus, Neptune and Pluto.

observed in very close proximity to a background star that you can use as a point of reference.

Most of our knowledge of this enigmatic blue giant was acquired during the Voyager 2 encounter in January 1986. A gaseous sphere like Jupiter or Saturn, Uranus has proved on close inspection to be much less turbulent and featureless than those planets. At a distance of only nine million kilometres, Voyager photographed a surprisingly bland, unmarked atmospheric envelope; close-up, Uranus was an hypnotically tranquil orb of unvarying blue. Eventually, computer-enhanced images revealed a faint pattern of banded clouds that indicated complex wind-systems and something like the dynamics observed on Jupiter and Saturn.

By far the most dramatic aspect of Uranus shown to us by Voyager was its regime of satellites. As the spacecraft approached the planet we were aware of only five Uranian moons; by the time of its departure we knew of a total of fifteen. The photographs that were returned to us showed the cratered little satellite Puck, the ice canyons of Ariel, the bizarre chevron-marked face of Miranda and much more geologically fascinating detail.

With your own backyard telescope, how much of this regime can you observe?

Uranus presents a tiny disc whose apparent size never exceeds about 4 arc-seconds. In spite of this, a good 8 cm or 10 cm telescope at 100x or greater will resolve the planet's bluish globe if atmospheric conditions are steady and transparent. Larger telescopes and higher magnification may prove disappointing; probably little detail beyond the mere resolution of Uranus as a planetary disc will be seen through any amateur instrument.

During past decades planet-observers' handbooks have recorded amateur (and professional) astronomers' impressions of 'markings' or darkenings occasionally seen on Uranus. Now, however, the Voyager photographs have cast serious doubt on the reality of variations prominent enough to be seen through any ground-based telescope. Yet we should perhaps remember that the Voyager encounter was a brief 'snapshot', a record of the planet's activity during only a period of days. Should your telescope appear to reveal features or variations of any kind, you will want to record what you see, and to look for the recurrence of patterns in your series of sketches. Whatever you observe or record on the marginal little disc of Uranus will always remain enigmatic, lacking the unmistakable quality of surface features on Mars or the belts and festoons in Jupiter's dynamic atmosphere.

All of Uranus' wonderful retinue of moons lies beyond easy reach of most amateur telescopes, for visual observation. However, if you are equipped with one of the very large-aperture telescopes that have become common nowadays you will find it possible to glimpse the largest and brightest of the moons. Titania (magnitude 13.9) and Oberon (mag 14.1) will be the easiest because they attain the greatest separation from the glare of the planet.

Neptune

Although it lies more than one and a half billion kilometres further from the Sun than Uranus, Neptune is (perhaps surprisingly) still a relatively easy target for binoculars. At a visual magnitude of about 8 the planet can be picked out and followed in its slow movement among the background stars.

In August 1989 Voyager 2, after a ten-year odyssey outward from the Earth through the solar system, reached Neptune. The photographic record has shown us that this gas giant is more active than Uranus, with zonal bands and atmospheric storms that include a huge oval spot not unlike the similar disturbance on Jupiter. The spacecraft increased the known family of Neptune's moons from the previously discovered two to a new total of eight. Close-up photos of the largest, Triton, revealed an atmosphere surrounding that frigid little world, consisting of nitrogen and methane with a trace of hazy cloud that may be ice crystals.

Neptune will repay your efforts at telescopic observation to only a limited degree. You will need fine seeing conditions and a magnification well above 200× to resolve the planet as a disc. With an apparent diameter that ranges between 2 and 2.5 arc-seconds, Neptune at best appears as a greenish blob of light, never truly focusable to a sharp-edged disc. A moderate to large amateur telescope will permit a glimpse of the largest satellite Triton, whose 13th-magnitude point of light will barely be detected by a trained eye that knows where to look for it. Occasionally the monthly astronomical magazines publish finder-charts for the brightest satellites of both Uranus and Neptune, indicating their position-angles relative to the parent planets on dates of favourable separations.

Pluto

This remote little world is now known to be an intriguing double planet, its large satellite Charon being a substantial fraction of Pluto's own diameter and mass.

Pluto is too small and distant to be resolved as anything more than a single, faint point of light in even a very large telescope. Yet at magnitude 13.7 it can be at least marginally glimpsed with an instrument of 20 cm or greater, a challenge that every suitably equipped amateur will probably want to attempt at least once in his or her observational lifetime.

Locating Pluto is a matter of having a precise and detailed intimacy with the starfield through which the planet will be travelling at the time of observation. Your telescopic sketches of the field on a number of separate nights will reveal the moving object among the stars – the elusive little pinprick that is our solar system's most distant known planet.

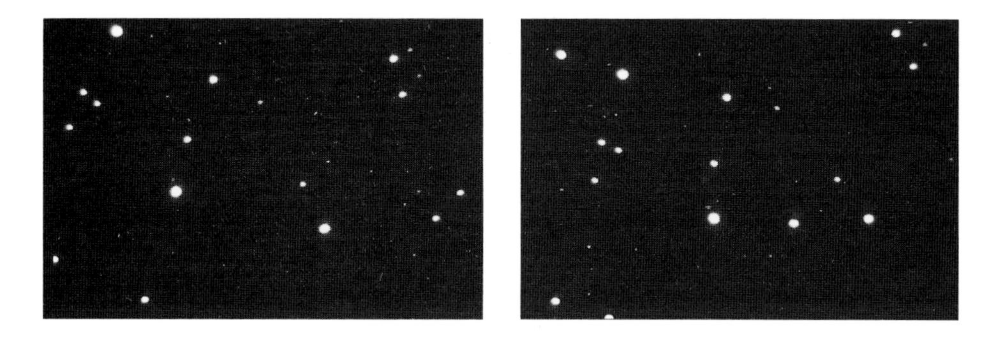

Plate 28. Pluto, showing the planet's movement through a starfield between May 20 (left) and May 22 (right), 1991.

Finding the outer planets

Tables of locations among the constellations, such as those provided in earlier chapters for the brighter planets, serve no useful purpose in the case of Uranus, Neptune and Pluto. Only precision finding-charts for each year's positions will serve to guide you to these relatively faint bodies among the mazelike backdrop of stars. These charts appear in annual publications such as *The Observer's Handbook*, and usually in the January issues of the major astronomical magazines.

2.9 The tiny planets: tracking the asteroids

The authors of the present book were both small boys when they read the story of the discovery of Ceres, the first of the minor planets. I still remember my own reaction: I conjured up a romantic vision of Father Giuseppe Piazzi at the telescope of his observatory at Palmermo on the eve of the 1801 New Year. I imagined his puzzlement when, in the telescopic field as the instrument was pointed toward the constellation Taurus, he suddenly noticed a 'star' that should not have been there. With what breathless excitement must he have discovered, when he peered through the eyepiece on a subsequent night, that the object had shifted its position. It was a solar system body – a little, hitherto unnoticed planet!

My own first attempt at locating Ceres in the night sky and identifying the tiny 1000-km-diameter planet by its movement relative to background stars was a deeply satisfying project of observation. Following the yellowish 6th-magnitude object over a period of several nights, I felt almost as if I were repeating Piazzi's historic experience of discovery.

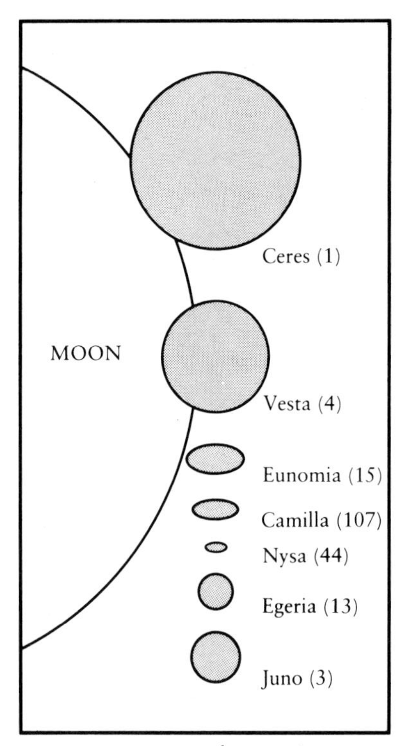

Figure 2.21. Some frequently observed asteroids, to scale. Those with extremely elongated forms may display a detectable variation in brightness as they rotate.

The population of asteroids

In the years since the finding of Ceres it has been learned that there are several thousand minor planets, most of them moving in fairly circular paths between the orbits of Mars and Jupiter. Some of these bodies, having more eccentric and less typical orbits, travel periodically into parts of the inner solar system much nearer to the Earth. A small number of them actually cross the Earth's orbit on what potentially may be collision courses.

The asteroids range in size from Ceres' 1000 km diameter, through a rather large population with diameters of the order of about 100–200 km, down to subminiature worldlets only a kilometre or so across. Among this swarm of small, rocky planets probably only the largest have an even approximately spherical form. Many of the medium- to small-sized specimens are thought (in several cases, known) to be of highly irregular shape. Figure 2.21 shows several named asteroids, all on the same scale, with an indication of their believed forms.

In any case, none of these tiny planets shows a resolvable disc when observed with even the greatest earthbound telescopes. Hence the term 'asteroid', which is based on two Greek words meaning 'stellar in appearance'.

The four minor planets that were discovered first are the most readily located and observed. Ceres, Pallas, Juno and Vesta are all very easy binocular objects, with magnitudes at opposition of 7.5, 8, 8.5 and 6 respectively. Vesta may be occasionally glimpsed with the unaided eye. In addition to these prominent four, however, many hundreds of smaller asteroids are bright enough, at least during a favourable opposition, to be tracked with modest telescopes.

An asteroid will readily identify itself by its movement within a star field. If it lies at a considerable distance from the Earth, out in the principal belt beyond the orbit of Mars, it will typically shift by about half a degree (a substantial part of the telescopic field) during a twenty-four hour period. Occasionally, a body is observed to shift at a very much greater rate – perhaps two degrees or more per day. Such an object is probably an 'earthgrazer', moving into the inner solar system on a path that brings it relatively near to the Earth.

Tracking asteroids

Finding and following these obscure little worlds can be fun.

Each year the *ephemerides* (daily or weekly positions) for several

Plate 29. Asteroid Eunomia near fringes of galaxy M33, September 1985.

asteroids that will be prominent at opposition appear in the annual astronomical handbooks. The monthly magazines also report apparitions of major asteroids from time to time. The observer's first step is to plot the predicted daily positions of a minor planet on the most detailed star chart available. Before observing on a given night, one must begin by studying the chart to memorize the principal stars that will appear in the telescopic field, and the position among them of the minor planet itself.

When the telescope is directed to the indicated part of the sky, the observer will eventually (after some minutes of search) recognize the stars that identify the correct field. It is with some excitement that one notices the object that does not seem to belong to the memorized pattern of fixed stars.

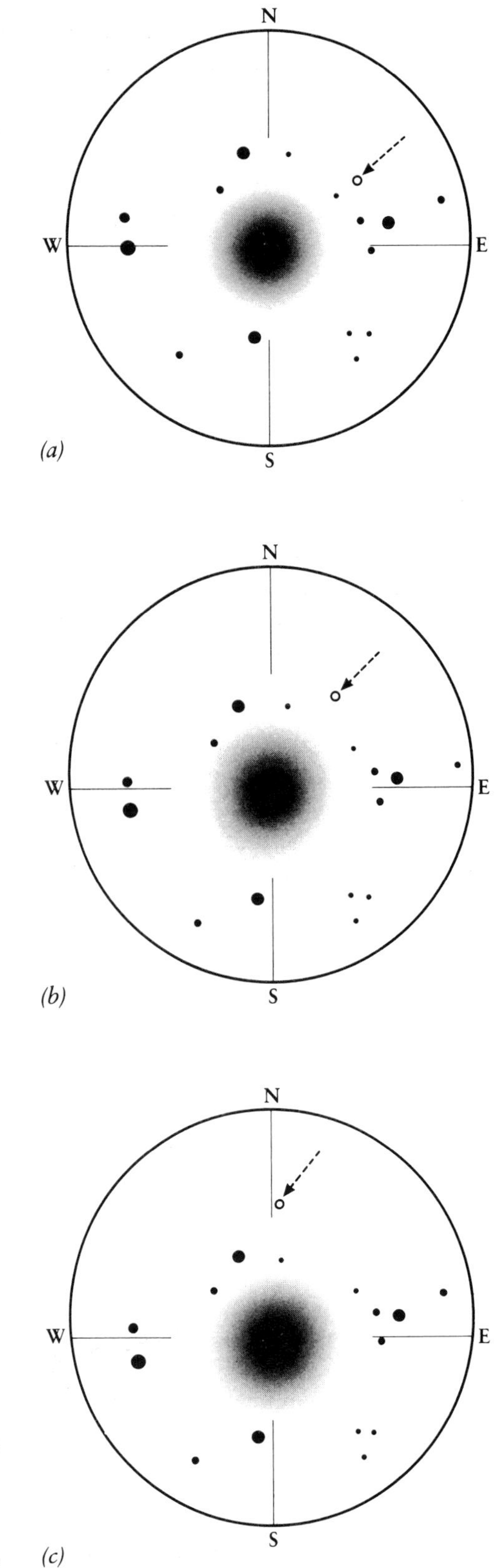

Figure 2.22. Asteroid Eunomia in a field of galaxy M33, on *(a)* 17 September, *(b)* 18 September, and *(c)* 19 September 1985 at 9.30 pm. Observed with Celestron-90 at 40×.

But an asteroid identification is not complete at this stage; a second confirmatory observation will be required. As in Piazzi's original discovery procedure, the modern amateur 'discovers' his or her minor planet by confirming, on a subsequent night, that it has moved.

Some people photograph the asteroid's field each night. My own preference is to draw the field very carefully; there is no observational practice that so thoroughly familiarizes one with a pattern of stars as this old-fashioned technique. The sequence of sketches shown in Figure 2.22 is typical.

Occultations by asteroids

Here is an area in which amateurs can contribute information of major scientific value.

When a minor planet occults a star, the precise timing of its passage in front of the star can yield instant information about the little object's size. Simultaneous observations of the event by a variety of observers at different latitudes can even provide a quite precise picture of the asteroid's actual shape. This work has assumed a new importance because of NASA proposals to send space probes to selected minor planets. If such a mission is to succeed, our currently rather imprecise knowledge about some of these bodies must be refined.

The first attempt at firsthand asteroid exploration, the successful approach of the Galileo spacecraft to asteroid 951 Gaspra in October 1991, has shown the potential of this kind of activity. The photographs returned by Galileo have given us a high-resolution view of a most intriguing minor planet, a potato-shaped worldlet whose largest dimension is only about 19 km. Its fascinating geography includes rocky plains, sharp-rimmed craters and gouged valleys. What was formerly a mere pinpoint of light in the night sky has now been seen as a world whose terrain we know in detail.

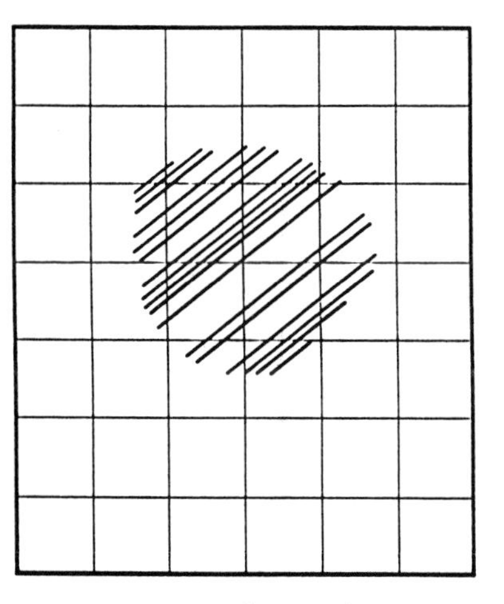

Figure 2.23. Occulation of star *1 Vulpeculae* by Pallas (29 May 1983). A plot of durations timed by observers at many separate latitudes provides an indication of the asteroid's shape.

In January of each year *Sky & Telescope* magazine usually publishes times and details of predicted occultations, with a map indicating which geographic locations will lie on the occultation paths. If this or a similar published source indicates that an asteroid's occultation path is likely to fall across your own location, you can plan to attempt a very precise timing of the event.

If you are confident that your technique has been sufficiently good to ensure an accuracy of about half a second (or better), it will be useful to make a formal report of your observation. One co-ordinator of such data is the International Occultation Timing Association at Lowell Observatory in Flagstaff, Arizona.

A particularly successful co-ordinated effort of this kind was the occultation of the star 1 Vulpeculae by Pallas on 29 May 1983. Professional and amateur observers from many locations across the occultation path submitted enough careful timings to produce a time–latitude graph which we (very roughly) simulate here (Figure 2.23). It will be seen that the varying durations of the event as observed at various locations provide a surprisingly direct picture of this minor planet's shape.

Light curves and rotation

Some asteroids show a noticeable variation of apparent brightness over a brief period of time. Such fluctuations are in some cases evidence of a highly eccentric shape rotating in such a way as to present greater or smaller reflecting surfaces to the Sun. In other cases the rotating body may have a large, dark albedo feature that periodically reduces the amount of reflected light.

An experienced amateur may be able to detect such variations and plot the fluctuations of a minor planet's brightness over a period of hours or days. If the variations are observed over a large number of complete cycles, a pattern may be recognized that can be a clue to the object's rotational period – the length, that is, of the tiny world's 'day'. The variations can be detected either by purely visual estimate or by the use of a photoelectric photometer linked to the telescope.

In the few cases where a variation of more than about a quarter-magnitude is present, it can be detected positively and beyond doubt by an experienced eye. Variable star observation is the best training for this kind of work. Among asteroids that have been found to exhibit large, short-period variations are Iris (variation of about 0.3 mag in seven hours), Metis (0.3 mag in five hours), Eunomia (0.5 mag in six hours) and Massalia (0.25 mag in eight hours).

Even more dramatic are the rotational variations of Kleopatra, an apparently rather extremely elongated asteroid whose 'day' is 5.4 h, and whose light fluctuates by an enormous 1.4 mag in that period. Another extreme case is the very small earthgrazer, Geographos, whose close approach to us in the summer of 1994 will bring it within the range of amateur telescopes. This 2 km or 3 km cosmic rock-splinter shows a periodic variation of more than a full magnitude.

If you succeed in monitoring the subtle light changes in a variable asteroid, think about the physical implications of what you are seeing. Kleopatra, for instance, can be visualized as an irregularly shaped miniature planet of perhaps 80 km diameter, rotating at a pace that brings each of its surface features to a new dawn once every five hours. Such a consideration allows us to think of this minor planet, not as a feeble point of light in our telescopic view, but as a little world.

2.10 Comets: the amateur's prize

Friday the thirteenth of November 1925 brought thrilling good fortune to Leslie Peltier. During the early evening of that day Peltier, who was not a triskaidekaphobe, discovered his first comet.

The delightful story of this classic amateur discovery of a comet is told in Peltier's autobiography, *Starlight Nights*. After three long years of careful sweeping of the skies in search of comets, Peltier on this fateful evening was guiding his telescope slowly across the northern reaches of the constellation Boötes. Suddenly he stopped. 'A small round fuzzy something was in the center of that sea of stars.'

The story continues with the young amateur's hurried efforts to establish the new object's movement across a telescopic field that was quickly sinking toward the western horizon. So rapidly was the comet moving that its position had considerably shifted after only fifteen minutes of observation. The final pages of Peltier's dramatic chapter describe his midnight bicycle ride into town, to telegraph details of his discovery to Harvard College Observatory.

Keeping a comet watch

For centuries comet discovery has been a field in which amateur astronomers have distinguished themselves. Even today, in an age when professional astronomy surveys the sky using sophisticated techniques of electronic detection, backyard stargazers continue to be the finders of a surprisingly large portion of each decade's new comets. Seki and Ikeya in Japan, Bradfield in Australia, Levy in the United States and Meier in Canada are among recent discoverers whose successes have fired our imaginations and inspired increasing numbers of us to sweep the skies for views of comets.

If your interest is aroused, bear this in mind: the fun of comets does not depend on your discovery of new ones. You will find much joy in observing the delicate, changing mystery of any comet that swings within range of your eye, binoculars or telescope.

Plate 30. Comet Kobayashi–Berger–Milon 1975 was memorable for its slender, spiky tail.

Most of the comets that appear in any given year (whether returning or freshly found) will be subtle objects, probably requiring a telescope for a satisfactory view. Yet comets occasionally deliver major surprises. I personally recall Comet West, in 1976, as one of the most stunning of such surprises, its tall bright tail fanning upward into the cold blackness above the predawn horizon on frigid March mornings.

Many comets that have been much less spectacular than West have been at least as fascinating to observe. Strong impressions remain of lovely telescopic comets such as slender-tailed Kobayashi–Berger–Milon 1975, dusty little Bradfield 1987 and swift-moving Levy 1990. Kohoutek 1973, very unfairly given a bad press when

Plate 31. Comet Swift–Tuttle 1992 was carefully observed in order to refine its orbital elements, which seemed potentially alarming.

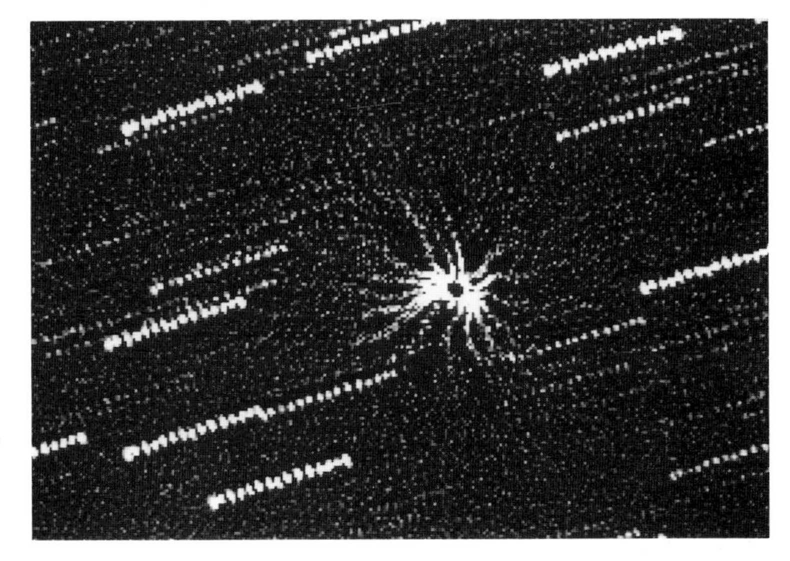

Plate 32. CCD image taken through Jack Newton's 25 in. reflector, processed by Dave Ballam, University of Victoria. Note the comet's (Swift–Tuttle) jet-structure, a detail never before photographed.

it fell rather short of its predicted brightness, was in fact a startlingly perfect gem for binoculars or low-power telescope.

A rare few comets possess a special dimension of excitement – in fact, an element of actual hazard. A case in point is the 'missing' comet Swift–Tuttle. When this object was recovered (by amateur observers) in 1992, comet expert Dr Brian Marsden noted that its orbital uncertainties included one very alarming possibility. On its predicted return on August 14, 2126, this body may actually collide with the Earth. With a diameter of about 8 km, this is the largest known 'earthgrazer', a body capable of inflicting a lethal environmental disaster. Even in the more likely eventuality of a near miss our Earth may at least be brushed by the tenuous fringes of the coma, producing dramatic atmospheric effects.

The first aim of your own comet watch should be to observe – and simply to enjoy – as wide a range of such comets as the night sky may present, each year.

Announcements of cometary positions appear frequently in the astronomical magazines. To locate, track and study large numbers of these will be your best apprenticeship; the practice of observing all such objects that lie within the light grasp of your telescope will quickly acquaint you with the characteristic appearance of comets. You will soon grow familiar with the subtle difference between the soft, diffuse glow of a comet and the more distinctive forms of deep-sky objects such as galaxies or globular clusters. Some comets will display obvious elements of cometary structure: a bright nucleus, diffuse coma and perhaps the hint of a tail. Many telescopic specimens will not be so co-operative; their appearance will be merely that of a dim, unfocused star.

It may be that each year's feeble and seemingly uninteresting 'minimalist' comets will prove to be your most valuable practice objects. The challenge of identifying them by eliminating other possibilities and of confirming the cometary nature of your object by detecting its motion among the background stars will sharpen your skills. Those skills may eventually be useful, should you wish to embark on a search for something not previously discovered.

Not for the fainthearted

What is involved in the search for a new comet?

The rather daunting first requirement is a program of *many hundreds of hours* (perhaps all the clear nights for many years!) of systematic observation. David Levy's advice in *The Observer's Handbook* comet section is almost frightenly stark, and yet it gives us a clue to the secret of his own remarkable record of six comets. 'The person who makes repeated discoveries', he tells us, 'has a definite plan of work that usually is placed above virtually everything else in importance'. We are reminded of Thoreau's remark that worthwhile achievements are made only by those who devote themselves to those ends, at the expense of all else.

Clearly such a programme is not for the lazy or fainthearted amateur.

Keeping a watch for new comets will involve regular sweeping of the sky using an eyepiece that will yield the widest field that your telescope can deliver.

Because the broad expanse of the whole night sky is vastly too large to be covered systematically, you will want to focus your attention on the regions most likely to yield success. Comets are brightest and most accessible to backyard-sized telescopes when they are in the solar neighbourhood. Thus, a large arc of sky that lies above the eastern horizon at dawn and the similar region above the western horizon at sunset are the hunting grounds on which amateur comet seekers normally concentrate their efforts. Observers' manuals have traditionally described these semicircular swatches of sky as 'the comet haystacks'.

Each evening's (or early morning's) search routine will be a matter of doing a series of slow telescopic sweeps executed *systematically*. Your aim will be to guarantee that a large arc of sky above the horizon is covered without omissions or gaps. Many comet hunters use a simple altazimuth mounting for the process. This mechanical configuration makes it easy to execute horizontal sweeps parallel to the horizon. Each successive sweep is just sufficiently higher than the preceding one to allow only a modicum of overlap.

On most occasions (perhaps on hundreds of nights) your routine of sky-sweeping will net only a catch of star clusters, pale galaxies and intriguing double stars. No enthusiast is disappointed by hours thus spent, for browsing among these deep-sky treasures is a meditative experience that brings peace and joy.

On one occasion, however, you may stumble on a faint, diffuse *something* at a position where your best charts show nothing. Perhaps, after an hour of careful monitoring, you will know the heart-stopping thrill of discovering that your mysterious object has shifted relative to the framework of background stars. You will then want to examine the most detailed star atlases and photographic sky surveys to which you have access, and to consult the latest journals and circulars for news of comets that may be known to be near the position of your object.

If no deep-sky object or solar system body can be identified with your find, perhaps you have observed an unknown comet.

Details of such a find – date/time, the object's position and direction of movement, its estimated magnitude – can be reported to the Central Bureau of Astronomical Telegrams in Cambridge, Massachusetts. But be cautious. Only after considerable observing experience will you be expert enough to recognize a truly likely new comet suspect. While you are developing your skills you will be wise to savour comet discovery as a purely private joy. In astronomy, as in most aspects of life, the private landmarks will always be treasured among your keenest pleasures.

2.11 Meteorwatch: fireflies in an atmospheric net

To astronomer and casual skywatcher alike the sudden, unexpected flash of a 'shooting star' is an exciting spectacle. The visual impression is rather eerie; it is that of an actual star coming adrift from its fixed location in the heavens.

Even the most uninformed observer nowadays is aware of the true nature of meteors. These particles of interplanetary debris, typically of sizes ranging from the scale of sandgrains to that of small stones, occasionally collide with our Earth, creating a brief glowing track as they are vapourized by their friction-heated passage through the upper atmosphere. More rarely the object is considerably larger than pebble-size. If so, it may blaze a startling track across the full breadth of the night sky without being utterly consumed; a charred remnant may even reach the ground.

Does amateur observation of these phenomena serve any purpose? On first consideration it will seem that such random and unpredictable occurrences do not lend themselves to any systematic programme of study.

Yet there are meteor projects that form a major part of some amateurs' work. One approach depends on the fact that there do happen to be predictable patterns of activity – the many annual meteor showers that are a product of the Earth's periodic collision with clusters of debris lying in known (or at least approximately known) orbits about the Sun. The famous Perseid shower in mid-August, for instance, can usually be depended upon to inject a dense scatter of 'falling stars' into the sky for several consecutive nights. Another approach focuses upon much rarer chance-sightings of individual fireballs – meteors of such brightness and duration that there is a possibility of surviving fragments being found and examined.

Monitoring a shower

The visual analysis of a meteor shower by counting or plotting the frequency of individual tracks per hour was once the chief means of studying the meteor groups. Nowadays it is less important; the use of radar and radio-counting methods has greatly reduced the need for amateur visual surveys.

Nevertheless, meteor counts are enjoyable projects that can still contribute usable data, especially when a shower happens to exhibit features that are more readily detected by the human eye than by electronic means. It turns out in some cases, for instance, that a shower includes numerous meteors that are prominent visually, but weak or undetectable electronically. Also, some showers are simply not monitored electronically.

As meteors stream into the Earth's atmosphere from space, the direction from which they approach is a function of the cluster's orbital geometry. Each shower characteristically breaks through into our atmosphere from a region of the sky located within a specific constellation, at a point described as the shower's *radiant*. A recurrent shower's name is derived from the name of its radiant constellation.

Although all the individual members of a cluster travel in parallel orbits and fall into our atmosphere on parallel trajectories, the terrestrial observer sees their paths as a starburst, with the flashes of the meteors streaking across the sky from their radiant spot toward all points of the compass. This is a perspective effect, like that which causes wind-driven snow in front of your face to appear to radiate off in various directions (see Figure 2.24).

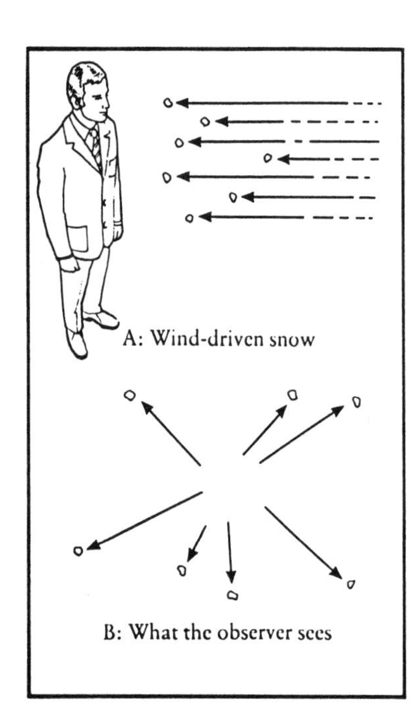

Figure 2.24. The apparently radial pattern of a meteor shower can be understood through the analogy of wind-driven snow.

Because of the radial nature of a large meteor fall, it is best monitored by a team of observers positioned in such a way that the whole visible sky is covered. A single observer, however, can produce a count that will indicate the shower's intensity and its variation over a period of time. To measure the level of activity in a meteor shower the observer should do a count that continues throughout the night (and preferably through all the nights of the shower), recording the number of tracks seen during each hour. *Sporadic*

Table 2.9. *Prominent meteor showers*

Name	Co-ordinates of radiant	Approx. dates	Characteristic OHR (single observer)
Quadrantids	15h 28m/50°	1–4 January	40
η Aquarids	22h 24m/00°	21 April–12 May	20
Perseids	03h 04m/58°	30 July–15 August	50
Leonids	10h 08m/22°	14–20 November	Highly variable
Geminids	07h 32m/32°	7–15 December	50

meteors should be noted as such; these are objects that originate from a sky-location other than the radiant point of the shower under observation. It is useful to record the presence within the shower of unusually bright members, or those of untypically great duration. The overall count of the shower's observed hourly rate (its 'OHR') is the main product of each observer's project. An estimate of the radiant's altitude in degrees above the observer's local horizon should be noted beside each reported hourly count. This is essential if scientific use is to be made of the data. In assessing a specific observer's OHR, a scientific analyst will need to apply a correction factor that depends on atmospheric obscuration and on the position of the horizon. Results of amateur meteor counts can be submitted to appropriate committees of the various national astronomical associations.

Table 2.9 provides information on a small sample of the most dramatic recurrent meteor showers. Large numbers of further showers usually monitored by amateurs can be found in the standard annual handbooks.

A thought that might have occurred to the reader is that a typical all-night meteor vigil must be one of the most exhausting (and chilling!) amateur astronomical projects. To minimize the physical hardship, most practitioners of amateur meteor science dress like high-altitude mountaineers, equip themselves with deck chairs for a comfortable supine observing position, and clutch thermos-flasks containing potent hot beverages.

A luxurious improvement on these basic measures can be seen in a special meteor-observers structure used by Royal Astronomical Society of Canada members in Ottawa, where the nights are very long and cold (Figure 2.25). Looking ominously like a radial arrangement of coffins, this low circular plywood shelter is divided into segments that face all points of the compass, and which are arranged so that one member of a meteor observing team reclines comfortably in each segment, with a lid snugly closed over most of his or her body. In this array of enclosures, each containing its own handy writing-tray, snack-shelf and other amenities, the team can monitor a shower during the long hours of a subfreezing night. A central seating position houses the team's co-ordinator and time-keeper.

Figure 2.25. Circular plywood shelter for cold-weather meteor watching.

Capturing a fireball

Extremely brilliant meteors, often not associated with known meteor showers, are occasionally seen. A *fireball* is an object of this kind that is more luminous than a bright planet, and may even rival the light of a full Moon. The largest of these cross a significant portion of the sky, perhaps surviving to strike the Earth's surface.

In the most promising cases, astronomers may wish to locate the impact site and examine a remnant of the space-object itself. Success in a search of this kind will depend on the number and quality of observations that have been recorded. The experienced amateur meteor tracker who chances to see a fireball should note the time and the object's estimated magnitude. A key feature of the report is a statement describing the point of origin, and the direction and length of the fireball's track. A technique used by some observers is as follows: immediately after the sighting, while the object's path across the sky is still fresh in mind, one holds a ruler (or more likely in the circumstances, any straight stick that comes to hand) at arm's length against the sky, along the remembered track of the fireball. Mentally one notes specific stars that lie near the ends, and along the length, of the track. Later, the track can be plotted on a star atlas, using these specific stars as a guide.

Detailed and accurate reports of a major fireball sighting should be sent at once to one of the centres that correlate and use such information. In Great Britain, the Meteor Section of the British Astronomical Association ('The Harepath', Mile End Lane, Apuldram, near Chichester, West Sussex, PO20 7DZ) receives these data. In North America, USA sightings are directed to the Scientific Event Alert Network (Mail Stop 129, 20560) and Canadian sightings to Meteor Centre, Herzberg Institute of Astrophysics (Ottawa K1A OR6).

Undoubtedly the luckiest of all fireball events is the capture of such an object on film. A photograph of the object's track across the sky can be of great value in establishing facts about its trajectory through our atmosphere. One of the most spectacular pieces of good luck with a fireball was a chance photographic record of the Great Northwestern Object in August 1972. Linda Baker, who happened to be holding an 8 mm zoom movie camera at the opportune moment, caught a total of 461 separate frames in a 26-second sequence of film, tracking this extraordinary daylight meteor across the Wyoming sky. Her record of the event appeared as a full-colour feature in *Sky & Telescope*, July 1974 (pp. 6–7).

Meteors and your radio

In recent years some amateur astronomers in Europe, Japan and North America have experimented with a simple electronic method of counting meteors. Perhaps your reaction to this news may be the thought: 'That is all very well, but I can't afford specialized electronic equipment'.

Yet the principal instrument used by many of these experimenters may already be in your possession.

The main tool for a meteor-count of this kind is an FM radio. The principle is as follows: if a radio receiver is tuned to an FM station that lies beyond the horizon (say, about 70 or 80 km distant) the station's signal will normally be exceedingly weak or completely unhearable. But when a meteor passes through the atmosphere between the station and the receiver, the ionized trail that it creates will reflect the upward-radiating signal downward, to be detected momentarily by the radio. The amateur observer 'watches' the sky by listening to the static noise on his selected FM band, and by counting the intermittent bursts of strong, intelligible signal.

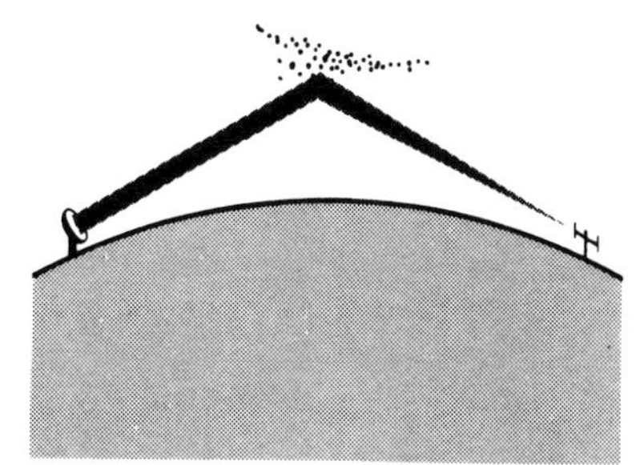

Figure 2.26. Signals from an FM radio station beyond the horizon bounce earthward, briefly deflected by the ionized trail of a meteor in the atmosphere.

In the signals received there may even be an indication of the meteors' relative sizes. The smallest particles, as they burn up in our atmosphere, trail ion-clouds that survive for only a matter of seconds; large objects' trails may persist for a few minutes.

In practice, a workable system based on this principle may require one or two refinements. Some amateurs modify their radio equipment to receive 24-h-per-day noncommercial sources, such as aeronautical beacons. Others link the receiver to a home computer, letting the computer record voltage-rises caused by increases in signal strength, while the 'observer' is sleeping or at work elsewhere. Full details of amateur experiments of this kind appear regularly nowadays in the popular journals. See, for example, *Sky & Telescope*, July 1983, pp. 61–2 and *Popular Science*, May 1984, p. 80.

PART 3 A deep-sky guide

3.1 Deep space

Solar system astronomy, which we described in the previous chapter, seems to encompass a vastness of space. The orbit of distant Pluto, after all, lies so far from our Earth that light reflected from the cold little planet's surface takes about five hours to reach us.

Nevertheless, the amateur deep-sky observer's Universe becomes so unimaginably large that the solar system shrinks to the status of a mere pinpoint within it. An ordinary backyard telescope can reveal galaxy clusters half a billion light-years distant, and may permit a glimpse of a quasar so remote in space (and in time) that its light has travelled several billion years to arrive at the Earth.

To the imaginative observer, deep-sky astronomy can be an awesome, even an unsettling, experience!

Deep-sky objects fall into two broad groups: features of our own Milky Way Galaxy and objects that lie elsewhere in the Universe, beyond our 100-billion-star local system. Perhaps surprisingly, scores of examples of even the very distant latter group are easily located with optical equipment no more sophisticated than good binoculars. The following chapters include a detailed catalogue of such objects.

Here, in brief summary, is a descriptive census of the deep-sky-object types that will be encountered by every amateur who explores the night sky with a telescope of any size.

Star clusters

Open clusters (Plate 33) are found in the spiral arms of our Galaxy. They are aggregations of several dozen (or several hundreds) of suns that move together in space, gravitationally bound into a loose knot, or family. Typically the clusters are not very distant from us – a few hundred light-years or so.

Globular clusters (Plate 34). Enormously larger and more dense star-swarms, comprising as many as a million individual stars in a compact spherical clump. Lying at distances of tens of thousands of light-years, they are arranged in a halo above and below the plane

Plate 33. M35 in Gemini.

of our galaxy. While the open clusters are often families of very young stars, the stellar populations of the globulars are exceedingly ancient.

Nebulae

Diffuse nebulae (Plate 35) are interstellar clouds of gas and dust – principally hydrogen excited to luminosity by the energy of hot stars

Plate 34. M22 in Sagittarius.

embedded within (reddish areas) or glowing with reflected light of foreground stars (bluish areas). Many of these spectacular objects lie nearby in our Milky Way Galaxy. Similar knots of glowing hydrogen gas are seen in the spiral arms of other galaxies as well. Typically, these are objects of quite large mass – as much as tens of thousands of solar masses.

Plate 35. M16 in Serpens.

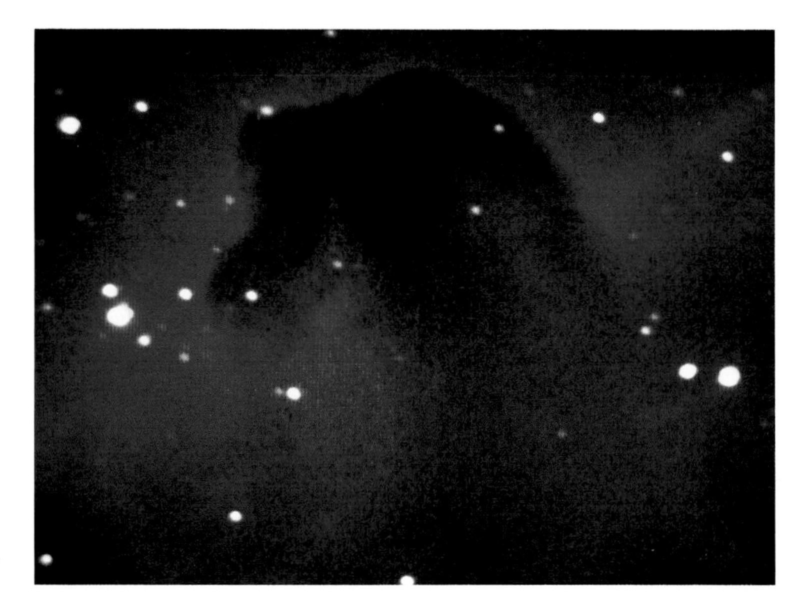

Plate 36. Horsehead Nebula, Orion. ST6-CCD.

Obscuring nebulae (Plate 36) are similar to the objects described above, but not illuminated. Minute dust-particles in these clouds are highly effective absorbers of light; thus, stars and bright nebulosity in regions of space behind are blocked, to leave the obscuring cloud in dramatic silhouette.

Planetary nebulae (Plate 37). The later stages of some stars' evolution includes the release of an expanding envelope of atmo-

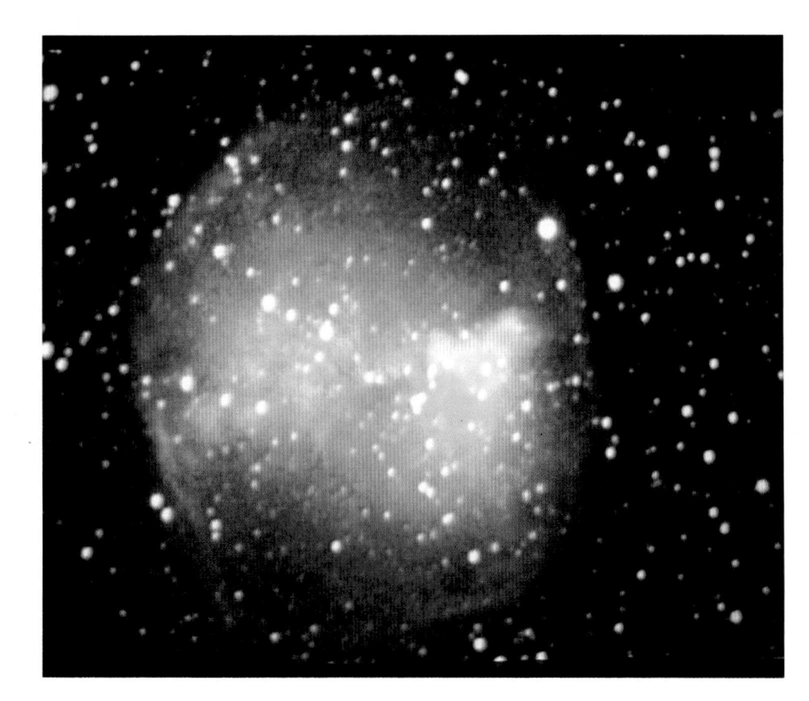

Plate 37. M27 in Cygnus: a planetary nebula. ST6-CCD.

Plate 38. Veil Nebula, Cygnus: a supernova remnant.

spheric gas, presenting the appearance of a compact 'bubble' easily discerned with the telescope.

Supernova remnants (Plate 38). Some scattered threads of gas and irregular splashes of interstellar material are the debris left behind after the destruction of a star in an event described as a supernova-explosion.

Galaxies

Spiral galaxies (Plate 39). Characterized by a small bright nucleus, a surrounding disc and a number of more or less loosely wound 'arms', these systems of hundreds of billions of stars are distant copies of our own Milky Way Galaxy.

Barred spirals (Plate 40). The structure of this class of galaxies differs from that of the more common spirals in that the arms attach to the ends of a central 'bar', rather than directly to the galaxy's central disc.

Elliptical galaxies (Plate 41). This class, which includes both supermassive giant galaxies and dwarfs, is devoid of structural details (arms, disc, lanes of dust) that characterize the spirals. This form of galaxy is an elliptical mass of chiefly very old stars, among which apparently little new star formation occurs.

Irregular galaxies (Plate 42). Usually small in both mass and linear dimensions, these systems present a rather vague and scattered appearance, with little structure. In some cases, vestiges of unformed spiral arms may be suspected.

In addition to galaxies of the types described above, there are 'peculiar' systems, so described by astronomers because they exhibit features not usually seen, or because they seem to be morphologically halfway between two normal classes.

Plate 39. M51: a typical spiral galaxy. CCD-ST6.

The celestial haystack

If you have never attempted to find and observe a small nebula or faint galaxy with a telescope, your first try will reveal a surprising difficulty. The night sky is a cosmic haystack in which an individual object may seem as elusive as a lost needle.

Most astronomical telescopes have an exceedingly narrow field of view. While typical binoculars may show as much as a seven- or eight-degree circle of sky at a glance, the telescope will cover only

Plate 40. M109: a barred spiral. ST6-CCD.

Plate 41. M110: an elliptical galaxy.

about one degree. To aim with sufficient precision so that you are viewing the tiny, degree-wide field that contains a specific deep-sky target requires some technique.

As described earlier, telescopes used by amateur astronomers have mountings that may be of two kinds. The *altazimuth* style is simplest; it permits the telescope to move vertically and horizontally. The *equatorial* mounting is more sophisticated; its principal axis can be tilted so that it is parallel to the Earth's own axis of rotation. When the telescope is moved about this polar axis, its sweep across

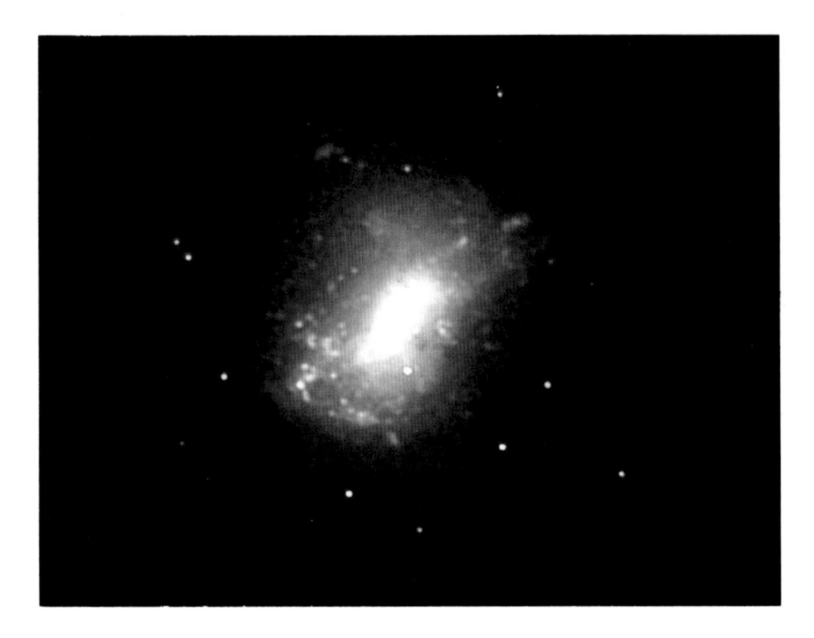

Plate 42. Irregular galaxy NGC 4449 (CCD image).

the sky is parallel to the curving lines of declination found of star-charts. This is a useful finding aid.

The most basic finding method will work on either type of mounting. This 'step' method is illustrated in Figure 3.1. After noting the position of the desired object on a chart, locate on the chart the nearest star that can easily be identified with the unaided eye, and also a series of two or three other stars that lie in easy steps leading in the direction of the faint target-object. The telescope (or rather, its wide-field auxiliary 'finderscope') is first pointed at the bright star. It is then moved, one short step at a time, from one 'stepping stone' to the next, until the target is reached. If the apparent distance between the selected stars is less than the width of the finder-telescope's field, then the method is virtually foolproof; as each star is found, the next is already in view near the edge of the field.

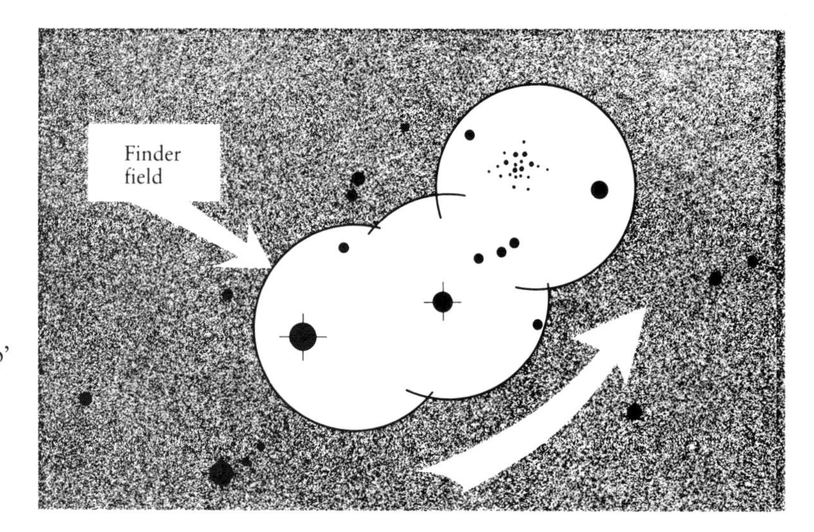

Figure 3.1. Finding M35 by the 'step' method. Naked-eye star (left) is located in finder. Sweeping upward along 'stepping stone' stars, the finder telescope comes upon the cluster.

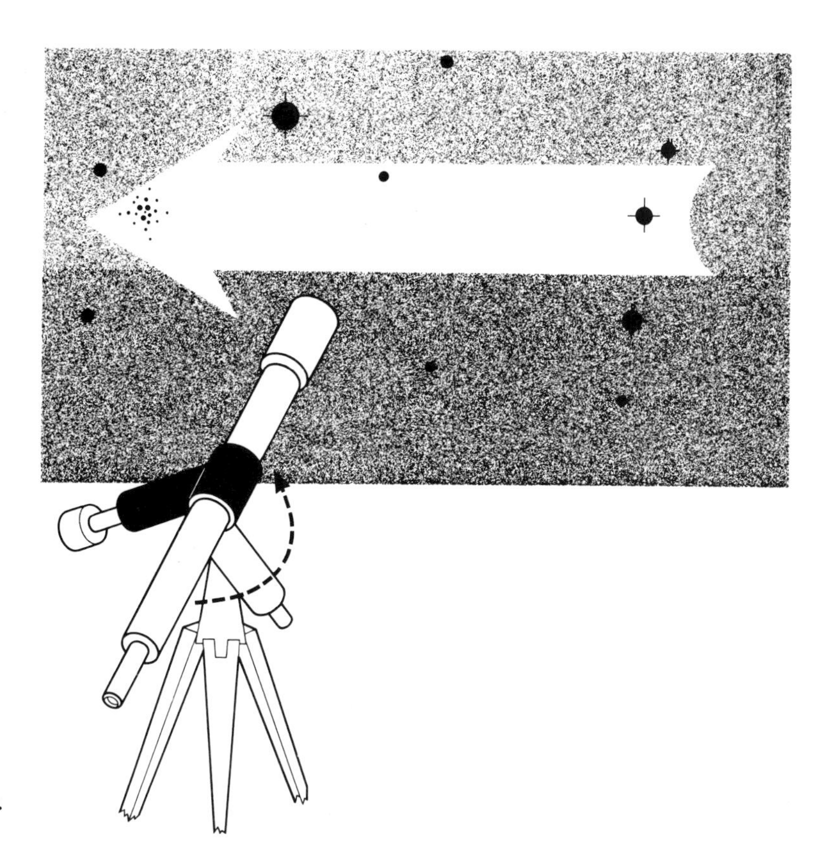

Figure 3.2. Finding M44 by 'declination sweeping'. Star υ *Geminorum* (right) is located. Turning the 'scope eastward about the polar axis brings M44 into view.

A second finding technique, for telescopes on equatorial mountings, is the 'declination-sweeping' method. On an instrument whose polar axis has been carefully adjusted to point toward the celestial pole (as indicated in Figure 3.2) this finding method is a very surefire system.

The declination-sweeping method works as follows: an observer begins, once again, by locating his or her target-object on a chart. He or she then searches westward or eastward on the chart for the nearest naked-eye star on, or very nearly on, the same declination (the east–west lines running across the star-charts) as the target. The telescope is then set on the bright guide star, its declination axis is locked, and the instrument is turned slowly east or west, as required, around the polar axis. By doing this, one sweeps along the target-object's line of declination in the sky and will automatically run the object down.

It should be noted that with this second method, unless the deep-sky object is fairly bright, the sweeping will have to be done while the observer looks through the telescope itself, rather than its finder. Care must be taken not to sweep so quickly that the object slips past the field of view unrecognized.

There are, of course, more sophisticated means of locating faint objects in the night sky. An even more advanced offering of some instrument firms these days is the home-computer finding system, which allows the user simply to punch digital co-ordinates into a terminal, for automatic adjustment of the telescope to a given object's sky position

We personally make a rather reactionary suggestion: for the greatest enjoyment of the night sky, stay with the simplest finding techniques. If you use the crudest method of them all – the stepping stone approach – you will learn most about the constellations and their deep-sky contents. A few months of this basic style of sweeping will result in a surprisingly intimate knowledge of the visible Cosmos. At star parties it is easy to identify those observers who normally use the old fashioned sky-hunting methods; they are the telescope users who impress everyone with their ability to hop from object to object, as if by some intuitive gift, without consulting charts or tables of co-ordinates.

3.2 Stars: the classic doubles

Amateur astronomy, like most other aspects of human life, is highly subject to changes of fashion. Until a generation ago, double stars were greatly in vogue. The closely spaced, subtly coloured twin jewel of a telescopic double was considered by many to be the loveliest of all celestial objects.

Nowadays the trend has shifted. With the advent of huge-aperture backyard reflectors, dim nebulae and remote galaxies have become the popular challenge.

For some of us, however, double (and especially *binary*) stars continue to hold a powerful fascination. The visual appearance of the giant orange sun α *Herculis* with its small greenish companion almost in contact alongside is unforgettably fine. In addition, the double star enthusiast is struck by the intriguing physical nature of such a system: the Hercules pair is a true binary, an enormous star and a smaller attendant circling each other in a gravitational embrace, like a pair of massive dancers.

Perhaps the shift of interest away from wonders such as this is due in part to something that modern amateurs have forgotten: the specialized excellence of a good refractor for binary star observation. Large Newtonian reflectors and the more complex catadioptrics that are superb for extended deep-sky objects often yield an indifferent stellar image when compared with the classic refractor. The latter, with its tiny, clear star image, permits crisp resolution of binary companions only an arc-second (or less) apart.

Among the binaries of smallest separation are examples of something especially interesting: systems whose revolution can actually be detected, if the observer is patient enough to monitor them over a sufficient period of time. Indeed, for the impatient observer there is a way to cheat at this game. A comparison of the relative separation and orientation of a binary that you have measured tonight can be compared with the measurements recorded by an early enthusiast such as Thomas Webb, in his famous old *Celestial Objects*

Table 3.1. *A: Showpiece doubles*

Albireo (β *Cygni*)	Cygnus	Bright double separated by an easily resolved 34 arc-seconds (34") and displaying dramatic gold/sapphire colour contrast. Recently discovered to be not a true binary, but only a chance optical pair.
Almach (γ *Andromedae*)	Andromeda	A similarly colourful gem, perhaps even more beautiful than Albireo, and probably a true binary system. Apparent separation is 10".
Mesarthim (γ *Arietis*)	Aries	Stunning, equal white pair, like twin diamonds. Only 8" apart; use fairly high magnification.
Castor (α *Geminorum*)	Gemini	Closely spaced (2.8") binary involving only slightly unequal whitish stars; splendid at high magnification.
Algieba (γ *Leonis*)	Leo	Close (4") yellow/orange binary whose orbital period seems to be about seven centuries.
Cor Caroli (α *Canum Venaticorum*)	Canes Venatici	Easily resolved (20") binary whose subtle colour-contrast is variously reported by different observers: 'white/lilac', 'pale yellow/copper', 'yellow/blue'.
Epsilon (ϵ) *Lyrae*	Lyra	The famous 'Double Double'. Very keen-eyed observers can split this wide (208") pair with the unaided eye. A good telescope at high magnification shows each component itself as a close binary (separations: 2.7" and 2.3").

B: 'Rapid' binaries. These show a measurable change of angle or separation within a few years

Designation	RA/Dec.	Magnitudes	Position angle and separation (1986) (degrees(°))	(arc-seconds ("))	Period (y)
Σ 186 (Cetus)	01h 55m/01° 45'	7/7	56	1.3	170
ξ Ursae Majoris	11h 17m/31° 39'	4/5	87	2.1	60
ζ Boötis	14h 40m/13° 49'	5/5	304	1.0	125
ξ Boötis	14h 50m/19° 12'	5/7	329	7.0	150
ζ Herculis	16h 41m/31° 38'	3/6	105	1.5*	35
70 Ophiuchi	18h 05m/02° 32'	4/6	278	2.0	88

* The considerable magnitude difference between components makes this binary more difficult to split than separation indicates.

for Common Telescopes. If the positions have changed, then you have discovered (vicariously at least) that your system is in motion.

A selection of binaries worth studying is shown in Table 3.1. In the first part of the table, stars have been selected chiefly for their sheer aesthetics. The beauty of colour and arrangement will stun you! The second part focuses on binaries worth measuring, from time to time, with a view to observing their orbital motion. For a comprehensive guide, consult *Burnham's Celestial Handbook* (Dover).

Measuring a double

As in many kinds of amateur projects, the monitoring of double stars can be undertaken either in a technically sophisticated manner that yields precise scientific results, or in a more casual fashion, just for fun. The latter approach, incidentally, should not be despised. Although it may produce data of too imprecise a sort to have much practical value, it can still give the amateur observer a truly eye-opening insight into the workings of the Universe – one aspect of it, at least.

To judge whether, and in what manner, a double star is changing, one must be able to measure or estimate two simple features of the pair. One of these features is the separation, in arc-seconds, of the two components. The other is their position angle, the position in the telescopic field of the secondary star relative to the primary (the brighter).

Estimation of a position angle lends itself more easily to simple, visual methods than does the measurement of separation. For uniformity, all double stars are described in terms of their orientation in a classic refractor, with *north* at the bottom of the field of view and *west* at the left-hand side.

While observing a binary, one imagines the grid illustrated in Figure 3.3 superimposed on the telescopic field. The correct orientation of the east–west axis can be established by noting the path of the star across the field when the telescope is slewed in the east–west direction around its polar axis. The intersection of the two grid lines is visualized as lying on the primary star. The position angle of the companion is measured or estimated in terms of its displacement *counterclockwise from north*, which is designated zero degrees.

The position angle and separation of ξ Bootis (from the list above) are plotted on the grid here. After a few nights' practice, most observers can begin to estimate a binary's position angle with at least very rough accuracy. One can test oneself by observing pairs that show no movement over a period of centuries, and comparing the results with handbook-data on these pairs. In the case of rapidly moving systems, even a very crude estimate will suffice to show the stars' shift of orientation, when compared with a measurement made some years earlier.

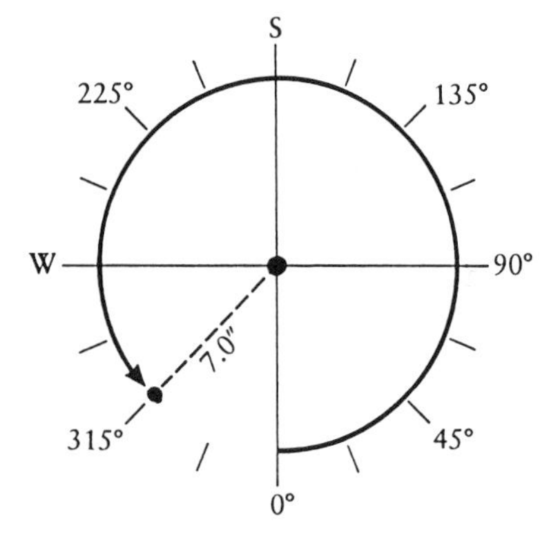

Figure 3.3. Position angle of companion star is measured in degrees, counterclockwise from North (0°). Separation of components is measured in arc-seconds.

A tool for binaries

Information about a binary star can be accurate enough to provide a basis for scientific work such as the discovery of the pair's true orbital circumstances. If this degree of precision is wanted, then a simple but accurate instrument must be employed to measure the binary.

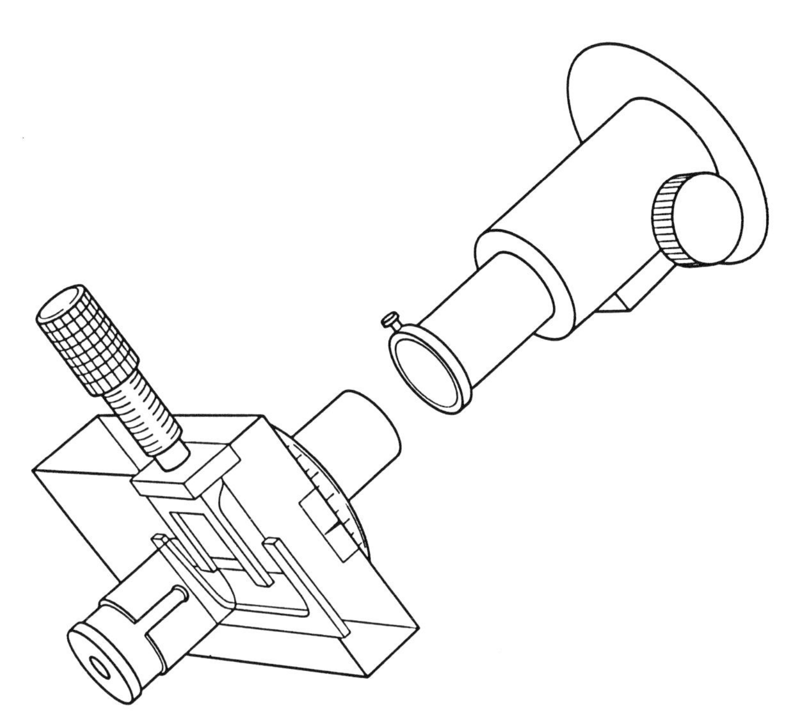

Figure 3.4. Simplified cutaway view of a filar micrometer.

The double star observer's special tool is the *filar micrometer* (Figure 3.4), which attaches to the telescope at its eyepiece end. This device is a frame that holds a superfine hairline fixed at the telescope's focal plane, and permits a second filament to be moved relative to the first. As the observer looks at a binary star through the eyepiece, he or she places the fixed filament across the primary star; the second filament is then moved, by means of a micrometer-screw, to the position of the companion star. A measurement of their separation is read from the micrometer-screw's graduated scale. The remaining part of the measurement – the pair's position angle – can be read from a graduated circle within which the instrument is rotated to bring a fixed transverse filament through the images of both stars.

Although some expert amateur machinists have produced good homebuilt micrometers, the filar micrometer is a tool that most of us will prefer to buy from a commercial source.

Dividing superclose pairs

Doubles whose components lie very close together present a tantalizing challenge to the observer. A large aperture will, of course, resolve more tightly knit pairs than a smaller aperture.

The so-called 'Dawes Limit' (see p. 34) indicates that a 12 cm telescope should be just capable of dividing a pair of stars whose apparent separation is one arc-second. The same empirical formula leads us to expect that a 20 cm instrument will split a 0.6-second pair. Yet in practice binaries at the telescope's theoretical threshold of resolution may prove frustratingly difficult to separate, especially if the two components are highly dissimilar in brightness.

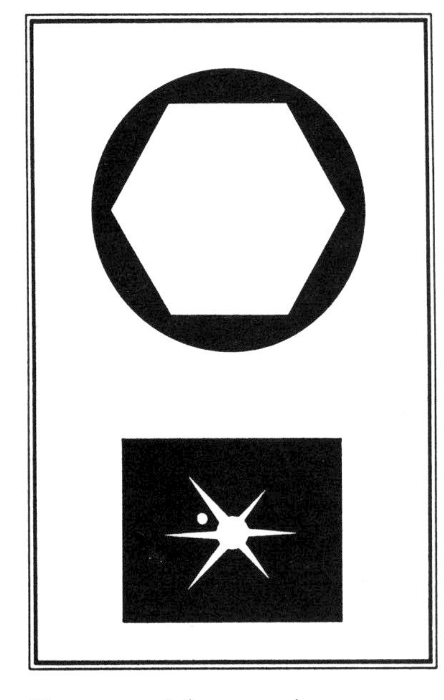

Figure 3.5. A hexagonal aperture mask draws light from the central disc of the star-image into diffraction spikes, perhaps revealing a superclose companion between the spikes.

Refractors and long-focus reflecting instruments of good optical quality show a star-image that is not a simple blob or point of light. As one of the consequences of the wave nature of light, the focused starlight produces a small central disc (the 'Airy Disc') surrounded by a number of fine concentric circles of light called *diffraction rings*. Often a bright primary star's faint companion remains unseen because the tiny companion happens to be masked by one of the diffraction rings. (Experienced observers usually suspect the presence of the small, close star because of a localized thickening of the diffraction ring itself.)

Hidden companions of this kind are extremely challenging; the dedicated binary star enthusiast may spend many evenings attempting to crack the secret code of such a pair. One simple, but powerful technique involves the use of a precisely cut metal or cardboard mask over the objective lens (or over the open tube-end of a reflecting telescope). The mask is cut to alter the shape of the telescope's aperture from circular to hexagonal (see Figure 3.5). This will have a surprising effect on the star-image: the central disc of the bright primary star is minimized in diameter, throwing much of its light out into radial spikes, instead of into bright diffraction rings. If the mask is slowly rotated (causing the spikes to rotate in the telescopic field), a small very close companion star may be revealed when its position is brought into the gap between the bright spikes. Thus, a superclose pair at the telescope's very threshold of resolution may successfully be divided and perhaps measured.

The end result

From the astrophysicist's point of view, what is it that makes binary stars so interesting?

The fact is that these gravitationally bound systems are keys to much fundamental knowledge about the Universe. A classic use to which binaries have been put is the actual weighing of stars. Where two bodies in space are found to be circling each other in a mutual tidal embrace, it is possible to apply the equipment of Newtonian orbital mechanics, and to solve the problem of what each component actually 'weighs'. Thus we find stars whose mass surprisingly turns out to be many times that of our Sun. Similarly, we discover celestial featherweights, a tiny fraction of the Sun's weight.

These stellar systems abound in physical peculiarities that challenge and reward scientific study. In one or two of the most extreme cases (where a star is apparently orbiting an empty space, rather than a visible companion), astronomers have detected the best evidence yet for the presence of that theoretical monster, the black hole.

3.3 Stars: the challenging variables

There is one way in which our Sun is different from a very large number of other stars in our Galaxy: its thermal output is relatively steady and unvarying.

In the steadiness of the Sun, we are lucky. What would the terrestrial environment be like if the source of its energy were a star that was subject to declines and outbursts representing a variation in luminosity by a factor of, say, 10 000? If such were the case the Earth's surface would be alternately incinerated and deep-frozen!

Yet suns characterized by this degree of variability are common within the galaxy. For instance, watch χ *Cygni* over a long period. In some months this star is easily located with the unaided eye in the constellation Cygnus (The Swan), at an energetic 3rd magnitude. On other occasions it will be found to be declining in brightness. Sometimes it will have sunk to the limit of visibility in a 10 cm telescope; occasionally it will have disappeared entirely from view in even a 15 cm instrument. With a total luminosity-range of about 10 magnitudes over a period of 407 days, χ *Cygni* is a sun that is 10 000 times brighter at maximum than at minimum.

The amateur's opportunity

Few scientific projects lend themselves so ideally to amateur work as the careful recording of variable star observations. The several types of variables all offer clues to the detailed physical nature of stars, but only if significant numbers of them are systematically monitored. A precise knowledge of each variable's behaviour (its range and period of fluctuation, the characteristic shape of its light curve) depends on very large numbers of observations, preferably by a variety of different observers.

In general the amateur's procedure is as follows: he or she observes a selected star as frequently as possible, always with the same telescope and magnification to obtain a uniform impression of stellar brightnesses. The magnitude of the target star is estimated very carefully, by techniques described later in this chapter. The basis of the estimate is comparison of the variable with other stars in the telescopic field whose magnitudes are known (and trusted as invariable). Dates, times and magnitudes are recorded, and may eventually be plotted to show a graphic profile of the variable's behaviour.

The important final step is, of course, to report this information to researchers who can make use of it. In North America a centre that co-ordinates such observations is The American Association of Variable Star Observers (AAVSO – 25 Birch St., Cambridge, Massachusetts 02138, USA). This association instructs new contributors on the details of a standard reporting form, and will also assess the quality of the beginning observer's contributions, to assist him or her in achieving a useful level of technique. In the United Kingdom, the Variable Star Section of the British Astronomical Association (BAA) correlates amateurs' observations in the same way.

Types of variable star

Variables behave in their bizarre fashion for a number of different reasons, and their variations are of several characteristic types. Some of the light-fluctuations are *intrinsic*; the increase and decrease of brightness that we see is actually happening, for some reason. Other variations are only *apparent*; the light of a star is periodically blocked from our view because it has an orbiting companion that passes in front of it. Systems of this type, called *eclipsing binaries* (Figure 3.6) have their own special contributions to make to astrophysics. Details of the orbits of these close-knit pairs, as revealed by their light curves, can reveal much about the masses and structures of the two individual stars.

Suns that actually pulse or waver in their output of energy include the following principal classes.

(a) *Cepheid variables* These stars pulse with clockwork regularity, at rates that are proportional to their true luminosity. Because their periods are in this way a clue to their brightness, the cepheids have been recognized as important distance-indicators. Comparison of apparent and true luminosities of cepheids observed in nearby gal-

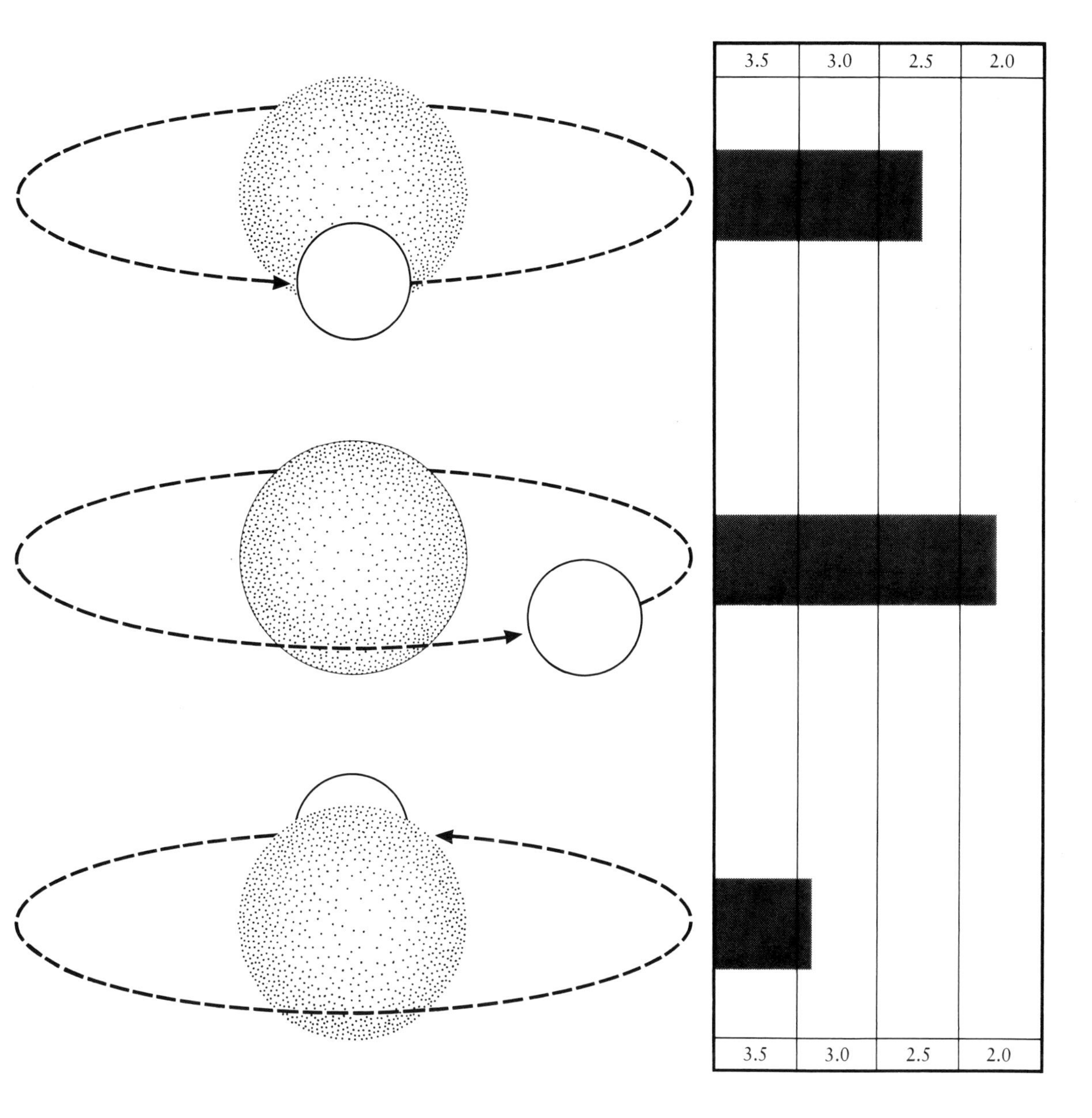

Figure 3.6. A typical eclipsing binary star. The graph at the right shows the varying effect of orbital movement on light level as seen from the Earth.

axies, for instance, is a basis for estimating those galaxies' distances from us. The periods of typical cepheids range from $1\frac{1}{2}$ days to several weeks. The prototype of this class, *Delta (δ) Cephei*, is an easy example on which to practice observations; it varies from mag 4.4 to 3.5 in 5.36 days. Conveniently, two other stars in the same binocular field (ε and ξ) have useful comparison magnitudes of 4.2 and 3.6. If you are a notice at variable work, try estimating *Delta's* brightness relative to its comparison stars, over a period of several nights. This is an easy one!

(b) *RR Lyrae variables* Super-rapid pulsating stars, these stars fluctuate with periods of only a few hours. They can be highly dramatic in that, in spite of their rapid pulsation, some of them show a variation of as much as two magnitudes. Very old and large (giant) stars, the *RR Lyrae* types are often found within globular clusters.

(c) *RV Tauri variables* These are also giant stars, but with long periods that range from about one month to several months. The *RV Tauris* bear a distinctive signature by which they are characterized: a light curve that shows alternating minima that are deep and shallow. One highly interesting example, *R Scuti*, which varies from mag 5 to deep minima of about mag 8, can easily be monitored with binoculars.

(d) *Long-period variables* The famous *Mira* in Cetus is an example. These giant stars vary over a predictable period, and usually have a quite dramatic magnitude-span, but on a very protracted time scale (several months to several years). A simple 6 cm refractor is capable of following *Mira's* full 11-month variation from about 3rd magnitude to about 9th.

(e) *Novae* are an exciting, but unpredictable phenomenon. A nova is a sudden, explosive outburst that may brighten an obscure (perhaps hitherto unnoticed) star by a factor of 10–15 magnitudes. The 'victim' then settles very slowly to its original obscurity, but may still be worth monitoring over a period of many decades for signs of further activity. The initial discovery of a nova is often an amateur achievement.

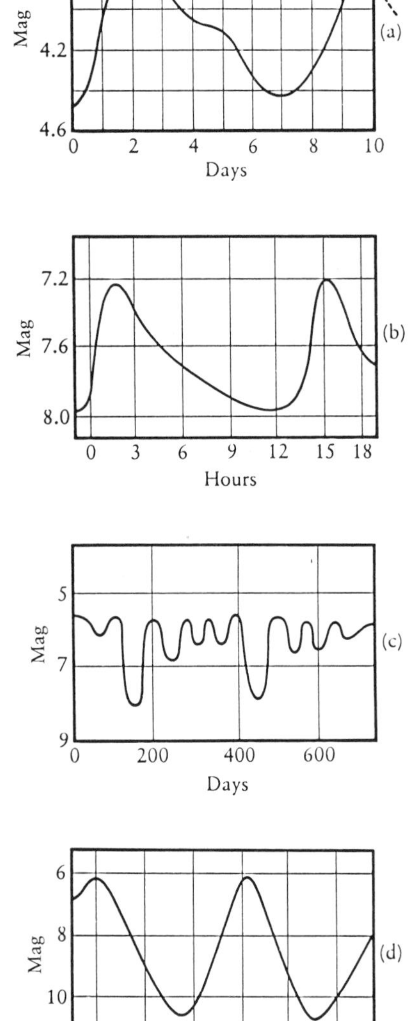

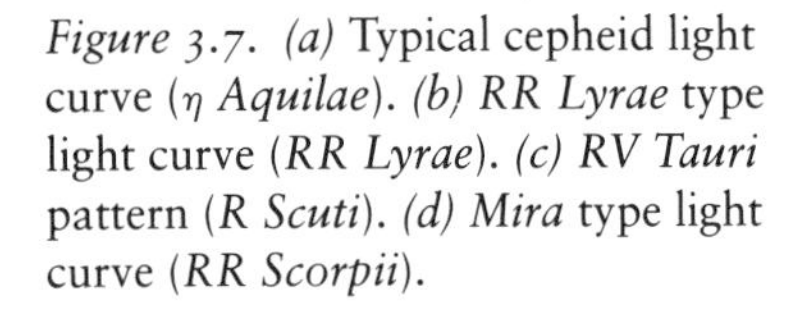

Figure 3.7. (a) Typical cepheid light curve (η *Aquilae*). *(b) RR Lyrae* type light curve (*RR Lyrae*). *(c) RV Tauri* pattern (*R Scuti*). *(d) Mira* type light curve (*RR Scorpii*).

Estimating brightness

Although a few amateurs apply a photoelectric photometer to variable star work, the vast majority do their brightness-estimates with a much more elegant and simple instrument – the eye.

It may at first sound incredible that any observer can accurately judge a star's brightness to within a small fraction of one magnitude by a means so crude as a mere visual estimate. Yet it can be done. The secret of success is the conscious adoption of a specific method.

The *fractional method* is the simplest and probably the best. Basically, it involves judging a variable star's brightness in terms of a fraction of the interval between two comparison stars. Suppose, for instance, that it is δ *Cephei* whose brightness we are trying to estimate on a given night. A finder chart for this variable is shown in Figure 3.8. Magnitudes of two useful comparison stars, ζ and ϵ, are 3.6 and 4.2 (note that on variable star-charts the decimal point is omitted in magnitudes, so that it cannot be mistaken for a star position). Let us say that on a given night the variable δ *Cephei* appears fainter than the nearby star ζ, but brighter than ϵ. Using the fractional method, we must judge by what fraction of the interval our variable is fainter than the former, and brighter than the latter.

There is a standard way to record this estimate. If the brightness of δ is judged to be exactly halfway between the magnitudes of the two comparison stars, we note that fact by writing $\zeta 1 \delta 1 \epsilon$. This indicates that the brightness difference is the same on each side. (δ

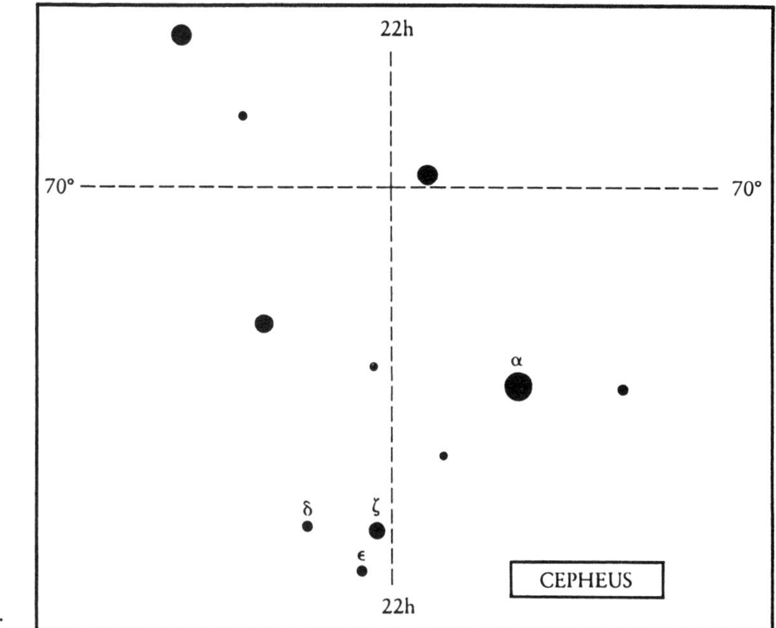

Figure 3.8. Finder chart for δ *Cephei.*

is thus exactly halfway between the two known magnitudes, that is, 3.9.)

On another occasion the brightness of δ may be judged to be $\frac{2}{3}$ fainter than ζ, and only 1/3 brighter than ε. This is jotted down as ζ2δ1ε. δ is thus brighter than ε by only $\frac{1}{3}$ of the interval, and is mag 4.1.

The key to it all, of course, is a chart for the variable star that includes precise magnitudes for at least a few nearby comparison stars. Such charts appear in some popular star catalogues – notably the superb three-volume set, *Burnham's Celestial Handbook*. Participants in BAA or AAVSO variable star observing programmes can only obtain charts from those organizations for specific variables under study.

In principle, variable star estimates are easy to make. In practice, however, acceptable results depend on much training of the eye. The beginner may find it a challenge to recognize brightness differences even as great as half a magnitude. Yet, with experience, a quarter-magnitude can be judged fairly easily, and some very well-trained observers can detect visually a mere $\frac{1}{10}$ magnitude difference between the variable and a comparison star.

Accuracy depends also on good observing conditions. In a night sky that is degraded by haze, light pollution or bright moonlight the relative brightnesses of stars of different colours may become difficult to determine correctly. Like most other kinds of astronomical work, variable star observation should ideally be practised at a site as far from urban lighting and atmospheric pollution as possible.

Table 3.2. *The variables: some examples*

Star	Magnitude (max.)	Magnitude (min.)	Period (days)	Comments
β Persei	2.1	3.3	2.87	'Algol', the most famous eclipsing binary.
λ Tauri	3.3	4.2	3.95	Bright eclipsing binary, easily observed with unaided eye. Use nearby *γ Tauri* (mag 3.68) and *ξ Tauri* (mag 3.74) for comparison.
η Aquilae	3.5	4.3	7.18	*Cepheid*: easy with binoculars. *β Aquilae* (mag 3.7) is useful for comparison.
R Scuti	5.0	7 or 8	144 (approx.)	*RV Tauri* class variable with complex, irregular light curve. Comparison stars include *ε Scuti* (mag 5.2) and several 7th-mag stars adjacent to *RV*.
o Ceti	3.4	9	332	*Mira*, the best known long-period variable. Good project for small telescope.
χ Cygni	3.5	14	407	*Mira*-type variable, but more dramatic even than that prototype. See opening remarks, this chapter.
R Cassiopeiae	7.0	14	431	Long-period variable. Exciting, but requires careful chart work for location in crowded star field.
SS Cygni	8.0	12	Irregular	Peculiar dwarf variable of highly irregular habits. This star is normally mag 12, but displays frequent nova-like outbursts.
BM Orionis	8.0	8.7	6.47	Eclipsing binary: this is the faintest of the four closely knit components in θ_1 (The Trapezium) within the Orion Nebula.
Y Ophiuchi	6.5	7.3	17.12	*Cepheid*: the $2\frac{1}{2}$ week period is long, compared with most well-known specimens of this class.
R Serpentis	5.7	14 (approx.)	357	Long-period type, very easily located because of proximity to bright *β Serpentis* (*R* lies about 1° south-east of *β*).

A binary eclipse method

In some research projects that involve eclipsing binary stars, it is not only the general pattern of the light curve that is of interest. Of prime importance is a timing of the precise interval between minima: it is crucial to discover the times of fullest eclipse. For the professional astrophysicist these timings can be the key to detecting *apsidal motion* (a gradual precession of the binary's orbital axes), and to insights into the physical structure of the two companion stars.

Typically, it is known that the minimum in an eclipsing binary will occur within a period of a few hours on a specific day. What is required is a refinement of this general time. During what minutes does the eclipse reach its deepest point?

Canadian amateur George Bell has used a highly effective photographic approach to this kind of timing. The excellence of his method lies in its simplicity.

Measurement of the binary's eclipse minimum involves photo-

Plate 43. Time lapse photo of the variable star *U Cephei*, by George Ball. Each 2 minute exposure was taken 5 minutes after preceding one; the telescope was moved 0.8 arc-minutes between exposures.

graphing it repeatedly over a period of a few hours. During this period the telescope is driven at a speed slightly faster than the sidereal rate, so that the target-star retrogresses slowly across the field, rather than remaining precisely centred. The camera's shutter is opened for short exposures continually throughout the period, at precisely timed intervals of, say, five minutes. When the film is processed, the star's image appears in the form of a trail of evenly spaced dots, a few of which will be identifiably the dimmest of the series. The time of this dimmest moment within the eclipse can easily be determined, because the starting time of the series has been noted, and the interval between each image and the next is also precisely known. To find the time, one simply counts the dots!

As a refinement, George Ball has developed an electronic device that opens his camera shutter at five-minute intervals. The same device also slightly advances the telescope in right ascension after each brief exposure, thus shifting the star image.

The lure of the variables

The mysterious, faltering light of variable stars is a phenomenon of special significance in amateur astronomy. No other celestial puzzle has attracted so many non-professional observers into true scientific work.

The joy of investigating variable stars has been persuasively described in one of astronomy's most delightful autobiographies, Leslie C. Peltier's *Starlight Nights* (Harper and Row, 1965). This famous amateur variable star addict has aptly identified the source of the variables' great attraction: 'It is their unpredictable behaviour, more than any other factor, which has so long sustained my interest and has made the watching of them a literal *Ten Thousand and One Nights* of entertainment.'

3.4 Stars: close solar neighbours

An interesting small project that few amateurs do is the visual observation of stars that occupy our Sun's very nearby neighbourhood in space. What gives this little study its special fascination? The interest lies in the odd character of most of the stars that we find, if we observe our solar vicinity. The Sun, as it happens, lies in a region of space whose stellar population is in many ways surprising.

The three nearest stars beyond our own solar system are all components of the 4.3 light-years-distant *Alpha Centauri* system. Two of the companions in this triple star are not dissimilar to our Sun in size and luminosity. Their very faint attendant *Proxima Centauri* is a distinctly unsunlike red dwarf star only a twentieth the diameter of our Sun, and about 13 000 times less luminous.

Although (for southern hemisphere observers) *Alpha* is a spectacular naked-eye object and a splendid bright binary in the small telescope, the 11th-magnitude dwarf companion is so obscure as to be almost unfindable by the casual telescope user.

It turns out that *Proxima's* dwarfness is typical of stars in the Sun's immediate family.

For northern hemisphere observers, the nearest star upon which we can turn our instruments is *Barnard's Star*, in the summer constellation Ophiuchus. A finder chart for this dim little sun is provided here (Figure 3.9*a*), for it is worthwhile to locate and view this intriguing nearby neighbour. That this star lies very close by was early recognized because of its unusually large proper motion; it has been nicknamed 'The Runaway Star'. (The chart shows its rapidly shifting position relative to other stars in the telescopic field.) Yet although it is only 6 light-years distant from us, *Barnard* appears telescopically as a feeble reddish light-point of about magnitude 9.5. It is a red dwarf one-sixth the Sun's diameter and 2500 times less luminous than the Sun. This tiny object is the nearest sun that has been suspected of having a planet-sized body (or bodies) in orbit about it.

Another very nearby star, at only 7.7 light-years, is the exceedingly dim red dwarf named *Wolf 359* (Figure 3.9*b*). With a visual magnitude of 13, this tiny sun requires a 15 or 20 cm telescope, but is worth observing for this reason: it is instructive to look at its feeble red image and to realize that we are viewing an extremely nearby star. Interstellar space may be thickly populated by such miniatures, but obviously most will remain undiscovered unless they are in our immediate vicinity, like this example.

Wolf 359 is about ten per cent the diameter, and 1/60 000 the luminosity of our Sun. It is located in Leo, about $1\frac{1}{2}$ degrees northwest of the star 59 in that constellation (about 14 degrees south-eastward from *Regulus*).

Two very much easier telescopic objects in our region of space are *Epsilon (ϵ) Eridani* and *Tau (τ) Ceti*. At magnitudes 3.5 and 3.7 these neighbours are more like our Sun than many of the nearest stars, although both are dwarfs. ϵ *Eridani* is approximately nine-tenths the solar diameter, and a third the Sun's luminosity, τ *Ceti* is a sunlike G-class star of about nine-tenths and one-half the Sun's diameter and luminosity. The special interest in these two stars lies in the fact that they are the nearest of all stars that are sufficiently sunlike to have possible planetary systems capable of harbouring life similar to our own. ϵ *Eridani* is 10.8 light-years distant, and τ *Ceti* 11.8 light-years.

Krüger 60, in Cepheus (near the famous variable δ *Cephei*, Figure 3.9*c*), is 12.9 light-years from us – still, on the interstellar distance scale, part of the Sun's immediate coterie. At magnitude 9.8, it is not inordinately difficult to locate. Once again, we will find ourselves looking at a dwarf object, this time a miniature binary system consisting of two very dim companion stars one-third and one-fifth the diameter of the Sun. This little orbiting pair has a short period

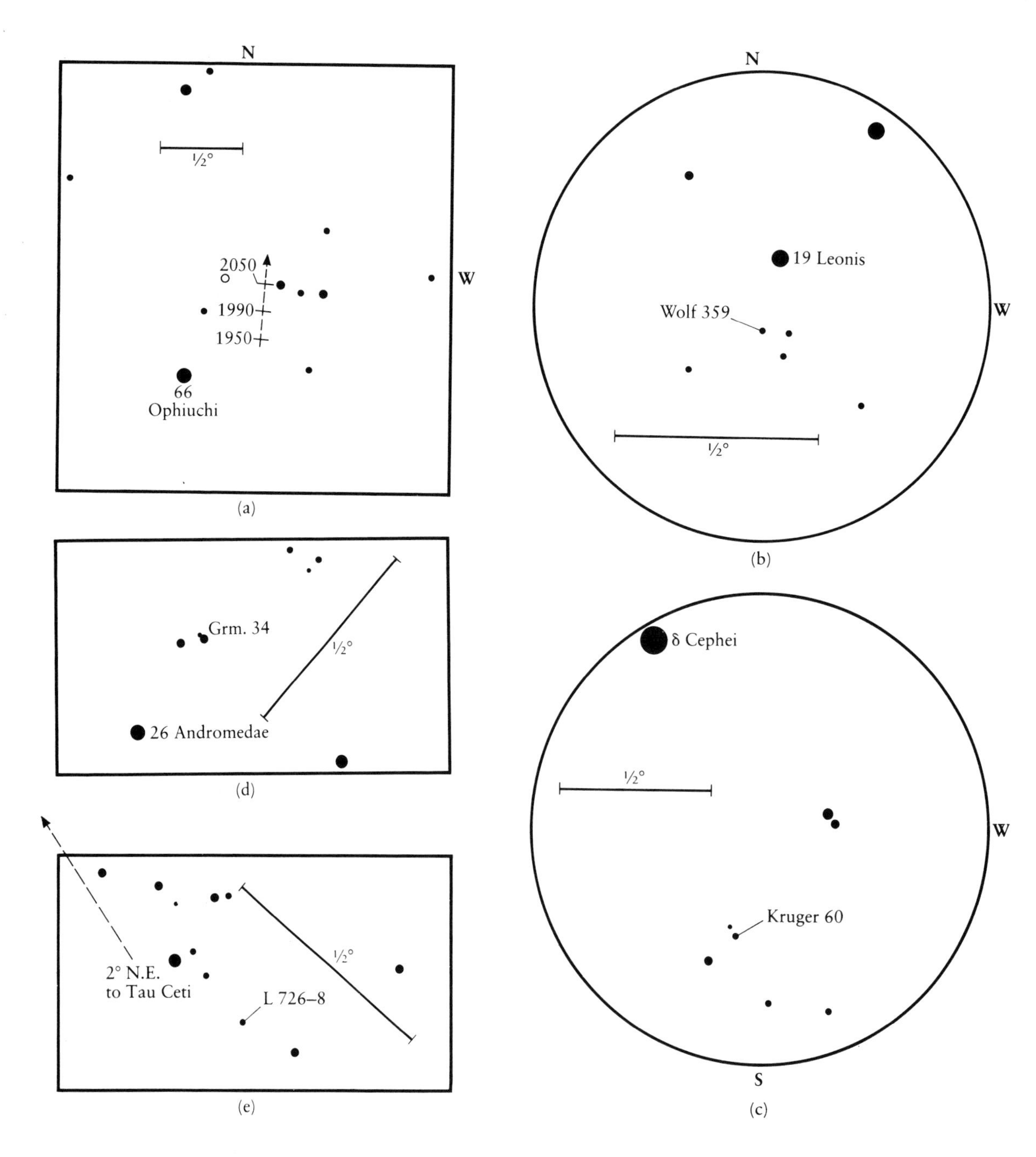

Figure 3.9. (a) Field of *Barnard's Star. (b)* Field of *Wolf 359. (c)* Field of *Krüger 60. (d)* Field of *Groombridge 34. (e)* Field of *Luyten 726-8.*

of revolution (44 y); their separation will be found on some occasions to be too small for resolution. In other years their maximum distance of three arc-seconds may permit them to be split with a moderate- to large-aperture instrument.

Among the large number of inconspicuous dwarf stars that lie near to us, an easier binary pair is *Groombridge 34*, in Andromeda (Figure 3.9*d*); this is another orbiting pair of red dwarfs. It lies 11 light-years distant, and its two little components are 8th- and 11th-magnitude objects – not an undue challenge to even a quite small telescope. The larger of the companions is less than a hundredth as luminous as our Sun, and the smaller is a feeble 0.00045 times the solar brightness. With a separation of 39 arc-seconds, *Groombridge 34* can easily be resolved by any telescope large enough to detect the faint 11th-magnitude component. The characteristic reddish colour of an M-class dwarf can be detected visually by an experienced observer, at least in the brighter star of the pair.

Perhaps the most extreme member of the local stellar association is a bizarre little system called *Luyten 726-8*, in the constellation Cetus (Figure 3.9*e*). This binary star is a mere 8.4 light-years distant from us, and yet it is an extremely feeble telescopic object of only twelfth magnitude. (Use an aperture of 15–20 cm to locate it.) The strange thing about this system is the tiny mass of each component star. Each of the red dwarfs that make up the pair has a mass only about 40 times that of the planet Jupiter – a mass that must lie near the defining boundary between a planetary body and a minimal sort of star!

Pondering our local association

The locations of several very nearby stars are given in Table 3.3. Although all are dwarfs – some of them extreme miniatures – they are typical of the stellar population that surrounds our Sun within a radius of about 15 light-years. Of the 48 stars that lie within that heliocentric sphere, only three are larger and more luminous than our Sun (*Alpha Centauri A, Sirius, Procyon*). All the rest are dwarfs. We seem to inhabit a sparse and ill-lit neighbourhood!

When observing the nearest stars, the amateur astronomer should use a telescope of the largest aperture available; most of these pale

Table 3.3. *Selected dwarfs in the solar vicinity*

Name	RA/Dec.	Distance (light-years)	Comments
Barnard's Star	17h 58m/04° 34'	6.0	9.54 mag. Easily located with 6 cm telescope.
Wolf 359	10h 56m/07° 01'	7.7	13.5 mag (very small, non-luminous dwarf). Needs a 20 cm aperture.
Luyten 726-8	01h 39m/ – 17° 57'	8.4	12.5 mag star of planetary mass. Needs 20 cm.
ϵ Eridani	03h 33m/ – 09° 28'	10.8	3.7 mag; easy naked-eye object.
Groombridge 34	00h 18m/44° 01'	11.2	8 mag binary (companion 11th). 10 cm telescope will resolve.
Krüger 60	22h 28m/57° 42'	12.9	9.8 mag fast-moving binary. Use large telescope to resolve.
Van Maanen's Star	00h 49m/05° 23'	14.1	12.4 mag. White dwarf visible in 10 or 15 cm telescope. (Of sunlike mass, but only Earth diameter!)
Ed + 50°1725	10h 11m/49° 27'	14.7	6.6 mag orange K-class dwarf.

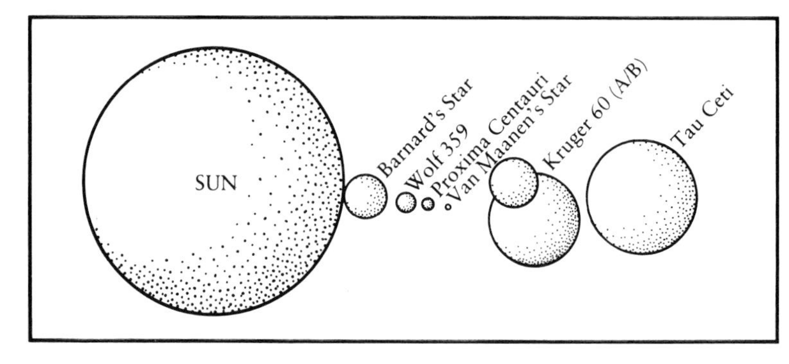

Figure 3.10. The Sun and some nearby stars, to scale.

little luminaries are rather elusive. But more important than the telescope is the quality of the observer's imagination. That is the most critical tool for an appreciation of these dim reddish pinpricks of light within the telescopic field.

3.5 The deep sky: summer

This chapter, along with the three that follow, comprise a deep-sky guide to the night sky in each of the four seasons.

Although our selection of seasonal offerings is necessarily brief,* it has not been limited to only the best-known or most spectacular objects usually listed for each constellation. Instead, we have attempted to mix some of our own more obscure favourites with the easily recognized items that form the basis of every published catalogue. Thus, while the most prominent telescopic targets will provide material for the beginner's first observations, at least a few objects in each list may perhaps surprise and challenge even the reader who has enjoyed many years of deep-sky exploration.

We begin with the wonders of the summer sky because it is in this season that many people are inspired to make their first attempts to recognize the constellations, and to direct a telescope towards the heavens. During the summer months, the Earth's position in its

* For the observer who needs a more exhaustive catalogue, the obvious choice continues to be *Burnham's Celestial Handbook* (Dover, 1978), unanimously recognized by all amateur astronomers as the most definitive descriptive list of every constellation's telescopic highlights.

orbit about the Sun is such that we gaze in the hours of darkness towards the dense plane of our galaxy (along the Milky Way in Cygnus, Aquila, Scutum) and towards the galactic centre itself (in the thronging starclouds of Sagittarius). It is a rich sky in which to begin exploration.

It should perhaps be noted that the seasonal layout that we have adopted necessarily involves an element of hemispheric chauvinism. The authors (both northerners) apologize to their southern-hemisphere friends, for our 'summer' selections appear in their winter sky.

Aquila		
NGC 6709	18h 49m/10° 17'	Open cluster; faint and loose, but arranged in neat diamond formation with bright star at each apex.
Pi (π)	19h 48m/11° 45'	Close (1.4") binary star with 6th mag white/blue components. Easily divided with 12 cm refractor; test for 8 cm.
Eta (η)	19h 50m/00° 52'	Cepheid variable star, with nearly 1 magnitude change in brightness over a seven-day period.
Ara		
RW	17h 31m/ – 57° 07'	Unusually dramatic eclipsing binary: mag 8.7 to 12 in 4.4 days.
NGC 6397	17h 37m/ – 53° 39'	Globular cluster, loose enough for easy resolution into individual stars by 15 or 20 cm telescope.
Capricornus		
Alpha (α)	20h 53m/ – 12° 42'	Very wide double, divided by unaided eye. Both components have faint (9th, 11th mag) companions.
M30	21h 39m/ – 23° 15'	8th mag globular cluster. Not one of best in its class, but a 20 cm telescope shows partial resolution into stars.
Cygnus		
		This is the constellation that for most people epitomizes the summer season. Although its glowing starclouds abound with fine telescopic objects, it is a region that should be viewed and contemplated first with the unaided eye. Gazing along the central 'spine' of Cygnus, we find ourselves peering into the crowded plane of the Milky Way Galaxy. More precisely, we are looking at the relatively nearby Cygnus Spiral Arm, in the fringes of which our own solar system lies embedded. Perhaps the most exciting aspect of this region is the non-telescopic view. To appreciate its star-swarms and dramatic dark rifts (lanes of obscuring material), however, one must find a sky that is far darker and more free of pollution than is typical nowadays in most suburban settings.
Beta (β)	19h 30m/27° 56'	'Albireo', the famous colour-contrasted double star, a stunning pair in a small telescope. Recently found to be not a true binary, but only an optical double whose components lie at radically different distances from us.
NGC 6918	19h 40m/40° 06'	Open cluster. Small, compact and circular, like an easily resolved globular.

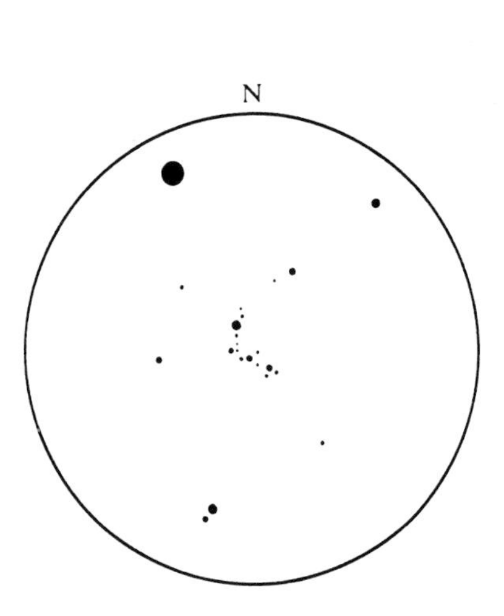

Figure 3.11. Open cluster NGC 6910 in Cygnus.

Plate 44. A seldom observed nebulosity in the field of γ *Cygni*.

Delta (δ)	19h 44m/45° 04'	Challenging binary (mags 2.9, 6.3; sep: 2.4"). More difficult than distance between components might indicate.
Chi (χ)	19h 49m/32° 47'	Regarded as a standard 'training' variable. Long-period fluctuation from about mag 4 to below mag 12 in 406 days.
b_1, b_2, b_3	20h 07m/35° 55'	Sweep this region slowly and carefully at low magnification. Rich, colourful field: many contrasting doubles and tight groups.
Omicron$_1$ (o_1)	20h 12m/46° 35'	Intensely yellow star with two turquoise stars adjacent; a stunning low-magnification field.
Omicron$_2$(o_2)	20h 14m/47° 35'	Note, approximately 50 arc-min eastward of this star, a wide blue/red pair. The red component is *U Cygni*, a dramatic 460-day variable (mag 6 to mag 12).
Gamma (γ)	20h 22m/40° 13'	Centred in dense Milky Way starcloud; view this region with large binoculars or very low-power telescope.
NGC 6910	20h 22m/40° 37'	A rare little cluster, like a delicate string of tiny pearls.
M29	20h 23m/38° 27'	Tiny 7th mag open cluster that appears like a miniature 'Pleiades'. If 10–15 cm telescope is used, two minute stars within central rectangle are test of atmospheric transparency.
NGC 6990/ 6992	20h 45m/30° 38'	'The Veil Nebula': very large, but elusive, thread of nebulosity. Usually considered only a large-telescope object, but we have viewed parts of it with a 9 cm aperture.
NGC 7000	20h 57m/44° 08'	'The North America Nebula'. Too vast (2° in width) for most telescopic field, but can be glimpsed with binoculars if sky is very dark. Try this experiment: view just with eye plus hand-held nebular filter.
61	21h 06m/38° 34'	Binary star of subtle gold/bronze colour contrast (mags 5, 6; sep: 30"). This nearby system is now suspected of having associated planets.
M39	21h 32m/48° 21'	Huge, bright galactic cluster; fine in binoculars.

Plate 45. The Veil Nebula in Cygnus.

Delphinus

NGC 6934	20h 32m/07° 14'	Globular cluster; only mag 9, and relatively small.
Σ2725	20h 44m/15° 43'	Gemlike little binary (mags 7.5, 8; sep: 6"), usually overlooked, although in same field with the following object:
Gamma (γ)	20h 44m/15° 57'	A showpiece binary star comprising 4.5 and 5.0 mag components separated by 10". Superb in the smallest telescopes, showing slight (but controversial) contrast of colours.

Draco

Nu (ν)	17h 31m/55° 13'	Famous wide double (mags 5, 5; sep: 62"), resolvable in binoculars, and very striking through low-magnification telescope.

Hercules

M13	16h 41m/36° 30'	Most popular of all northern hemisphere globular clusters – large, bright, easily resolved by telescopes over 10 cm. A 20 cm instrument shows individual member stars even near the dense core.
NGC 6210	16h 42m/23° 53'	A neat, intensely blue little planetary nebula. Use high magnification to resolve its disc shape.
NGC 6207	16h 42m/36° 56'	11th mag galaxy, often missed although in field with M13. We have glimpsed it with a 12 cm refractor.
NGC 6229	16h 46m/47° 37'	Distant 8.7 mag globular cluster. Observe this as a comparison object to the similar, but much nearer, M13.
Alpha (α)	17h 14m/14° 24'	Dramatic orange/green binary needing moderately high power to divide close (4.6") components. Primary star is an irregular variable, fluctuating by nearly 1 mag.
M92	17h 17m/43° 10'	6th mag globular cluster, only slightly less impressive than M13, and almost as easily resolved into component stars by small telescopes. (12 cm shows individual stars at edges of cluster.)
95	17h 59m/21° 36'	The famous double that the pioneer nineteenth-century amateur Smythe described as 'apple green, cherry red'. These colours are debatable; how do you see this intriguing pair?
100	18h 58m/26° 06'	Binary with perfectly matched 6th mag components – a splendid pair of diamonds easily divided (sep: 14") by even a 5 cm telescope.

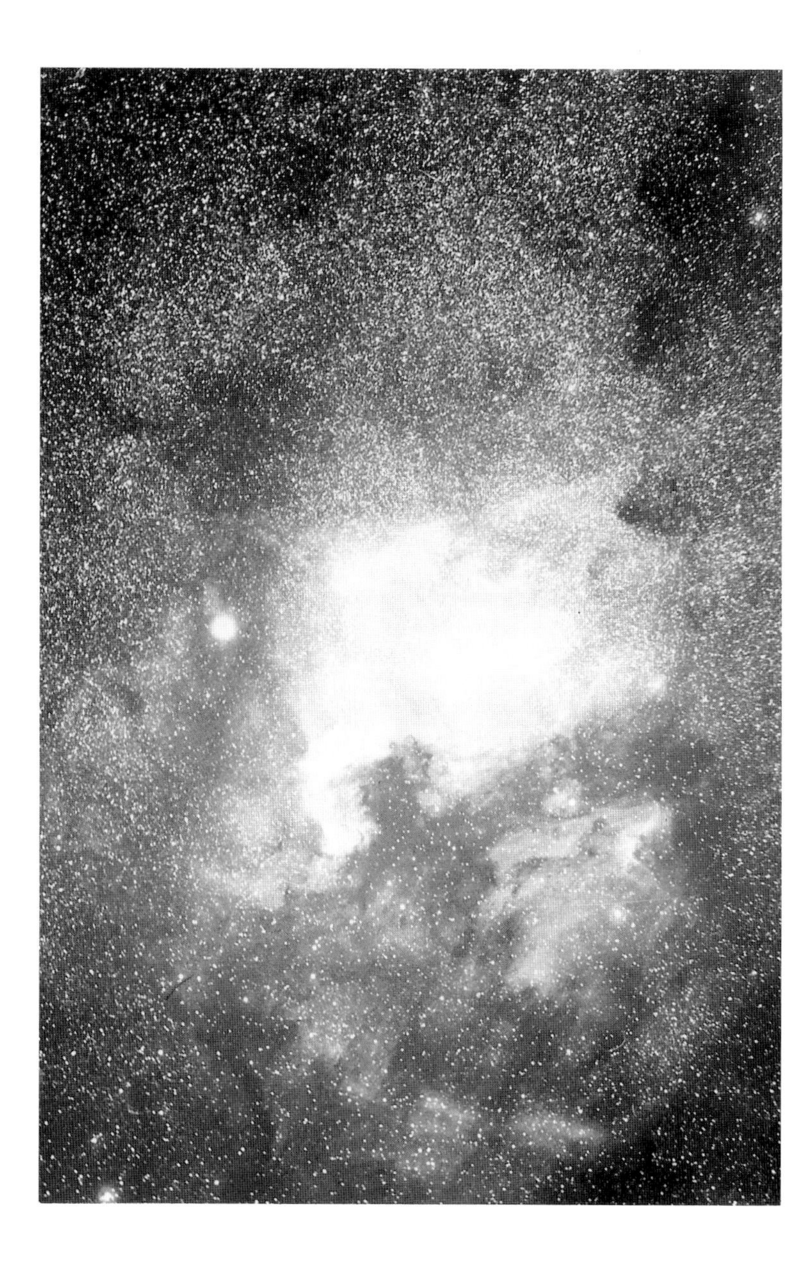

Plate 46. NGC 7000: The North American Nebula in Cygnus.

Lyra

Epsilon (ϵ)	18h 44m/39° 38'	Wide double that, as a naked-eye object, is a test of eyesight. Each component is itself a close binary, divided at high magnification (not difficult for a 6 cm refractor). Called 'The Double Double'.
M57	18h 54m/33° 02'	'The Ring Nebula': even a very small telescope reveals the intriguing 'smoke ring' shown in photos of this famous object.
Σ2470/Σ2474	19h 07m/34° 40'	These two superb little binary stars, lying close together in field at low magnification, form Lyra's charming second 'double double'.
M56	19h 16m/30° 08'	8th mag globular cluster. Although small, it is resolvable into stars by a 20 cm instrument.

Plate 47. Crescent Nebula (NGC 6888) in Cygnus (CCD image).

Ophiuchus

M12	h 46m/ − 01° 55'	6.6 mag globular cluster. Low surface brightness, but relatively easy to resolve.
M10	h 50m/ − 04° 05'	6.7 mag globular. Noticeably more dense and compact than M12.
IC4665	h 46m/05° 44'	Very large open cluster, spoiled by high telescopic power, but excellent in binoculars.
Y	17h 50m/06° 08'	Cepheid variable star that varies from mag 6.5 to 7.3 over a 17-day period.
70	18h 05m/02° 32'	Close (2") binary; interesting because of short period and detectable change of position-angle from year to year. Components mags 4.2, 6.0.
NGC 6633	18h 25m/06° 32'	A neglected open cluster. Bright and dramatic at low to medium power; shows some yellow/blue colour contrasts.

Plate 48. M13, a globular cluster in Hercules (CCD image).

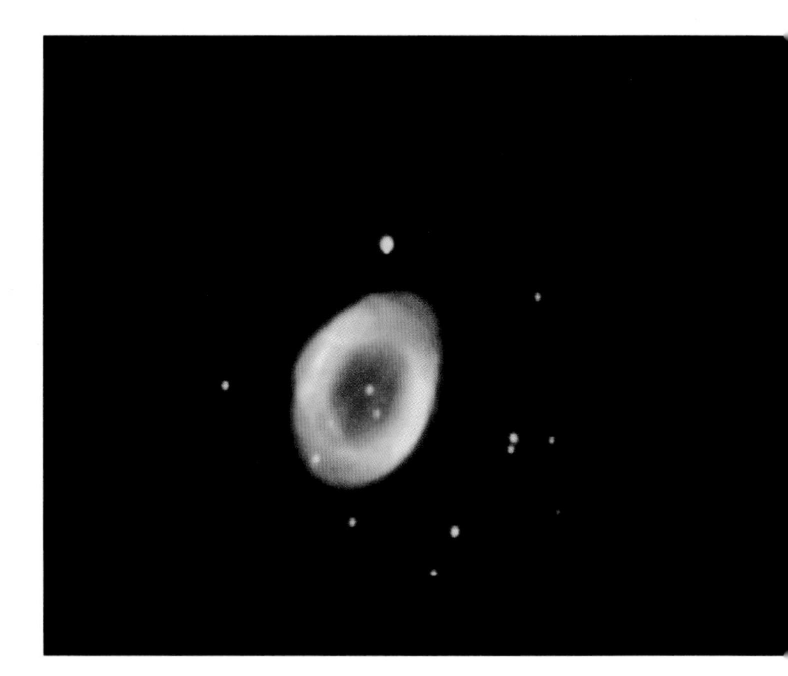

Plate 49. Lyra's finest object, The Ring Nebula (CCD image).

Pavo		
NGC 6744	19h 05m/ – 63° 56'	Barred spiral galaxy of about 10th magnitude. Very large, dim; best in medium-sized telescope at very low magnification.
NGC 6752	19h 07m/ – 60° 04'	Fine globular cluster. Quite condensed, but small aperture resolves scatter of bright fringe stars.
Sagitta		
M71	19h 53m/18° 44'	Globular cluster: peculiarly featureless.
Sagittarius		
		In this part of the summer sky the Milky Way appears to climax, reaching a brilliance and density unrivalled elsewhere. This impression is an accurate one, for in this constellation we are gazing towards our vast spiral galaxy's centre, which lies in the direction of *Chi (χ) Sagittarii*. So rich is this segment of the sky in spectacular telescopic wonders that our little selection here must exclude dozens of major objects. Sweep the region with binoculars or a rich-field telescope (low magnification, wide field) to discover an almost inexhaustible goldmine of clusters, nebulae, small dark 'rifts', etc.
M20	18h 01m/ – 23° 02'	'The Trifid Nebula'. A 15 cm telescope shows a hint of internal dark rifts; 50 cm reveals red/blue colours of emission and reflection components.
M8	18h 02m/ – 24° 23'	'The Lagoon Nebula', one of the most easily observed with small optical aids. Principal bright lobe surrounds star cluster NGC 6530; fainter mass lies adjacent, to west. Binoculars show cluster and nebulosity as a sizeable glowing patch.
RS	18h 14m/ – 34° 08'	Visually a double star, but one component is also an eclipsing binary that dims from 6.0 to 6.5 mag at 2.4-day intervals.

Plate 50. Globular cluster M12 in Ophiuchus.

M17	18h 20m/ – 16° 12'	'The Omega Nebula'. Bright, easy for smallest telescope. Even a 6 cm refractor shows characteristic 'checkmark' shape.
M24	18h 31m/ – 18° 27'	Compact, superdense galactic starcloud best viewed with binoculars. A small globular cluster (NGC 6603) embedded in the northern end of this cloud is often identified as M24, but Messier clearly describes his object's huge 2° extent. We have seen the large aggregation to good advantage with large binoculars, and the tiny globular with a 10 cm telescope.
M22	18h 35m/ – 23° 55'	Very large globular cluster, finer even than the more famous M13. Resolved into stellar components by 10 cm aperture.
RY	19h 13m/ – 33° 37'	Intriguing irregular variable star: usually about mag 6.5, but suddenly fades to below 12 at unpredictable intervals.
M55	19h 39m/ – 31° 00'	Huge, sparse globular cluster.

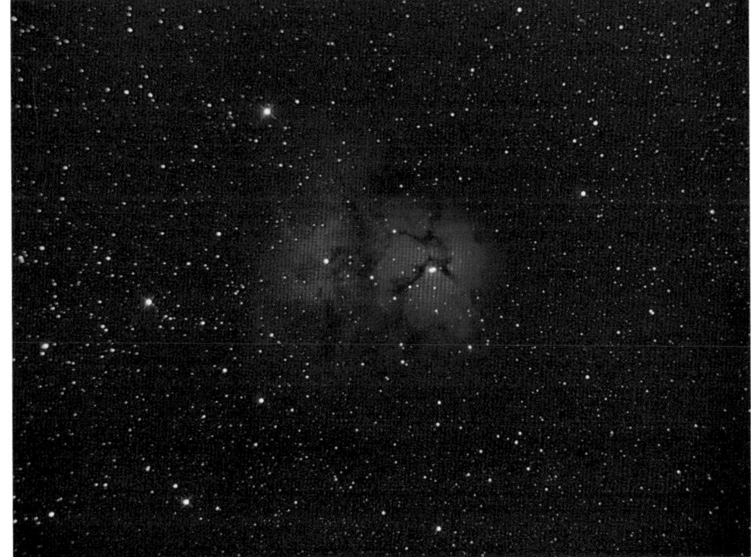

Plate 51. M20, The Trifid Nebula in Sagittarius.

Plate 52. M8, The Lagoon Nebula in Sagittarius.

Scorpius

Beta (β)	16h 05m/ – 19° 46'	Splendid golden double (mags $2\frac{1}{2}$, 5; sep: 14") for 6 to 8 cm telescopes.
M4	16h 22m/ – 26° 27'	This bright, loose globular cluster lies only slightly more than one degree west of *Antares*, which is described as follows:
Alpha (α)	16h 29m/ – 26° 24'	*Antares*, the blood-red giant star. It is a very challenging binary (mag 6.5; sep: 3") whose tiny greenish companion is lost in the primary's first-magnitude glare, except in a very good telescope of at least moderate aperture.

Plate 53. M17, The Omega Nebula in Sagittarius.

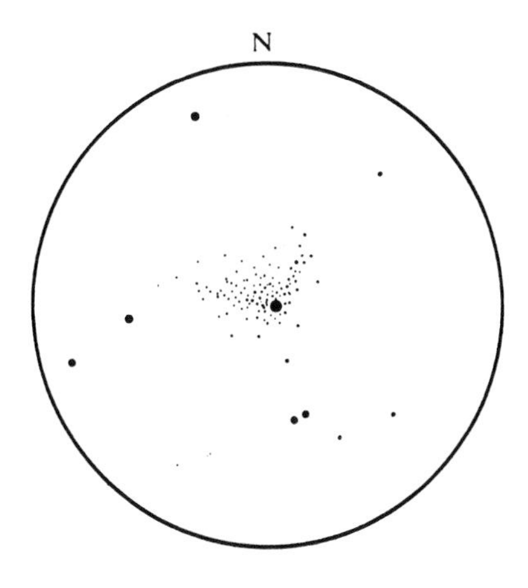

Figure 3.12. Open cluster M11 in Scutum.

Plate 54. M11, The Wild Duck Cluster in Scutum.

NGC 6231	16h 51m/ − 41° 43′	One of the most outstanding galactic star clusters in the southern sky, a source of envy to us in northern latitudes from which it is never seen. Easy and dramatic binocular object.
M6	17h 39m/ − 32° 11′	Another fine binocular cluster.
M7	17h 53m/ − 34° 48′	Large, magnificent open cluster, the most southerly of the Messier objects. Visible to the unaided eye as a luminous swathe, it is superb in a telescope at only about 20×, but spoiled at higher magnifications.

Scutum

Here once again we encounter a region whose special character is best appreciated if the telescope is laid aside, and the sky surveyed with the eye alone. The dominant feature of the constellation is the *Scutum Starcloud*, a dense, bright knot of Milky

		Way stars covering a substantial part of Scutum. In a dark, transparent sky this phenomenon often appears at first like an actual atmospheric cloud; on closer examination its true nature dawns upon the observer.
M26	18h 44m/–09° 25'	Open cluster; appears as a grainy clot of stars against the Milky Way background.
M11	18h 50m/–06° 18'	'The Wild Duck Cluster', a misty spray of minute stars. Visually this lovely, dense group has a distinct V-shape; oddly, photographs show a more typically spherical array. Needs about 60x for good resolution.
Serpens		
M5	15h 18m/02° 11'	Excellent, large globular cluster. Note star *5 Serpentis*, 20' south of the cluster: it is a binary with a close, faint companion.
Delta (δ)	15h 32m/10° 42'	Bright, close jewel of a binary: 3rd and 4th mag components separated by 4".
R	15h 48m/15° 17'	Long-period variable, easily followed with a small telescope: 6th to 13th mag in approximately 357 days.
M16	18h 18m/–13° 48'	'The Eagle Nebula': smallest apertures show only a bright, coarse star cluster; 20 cm reveals the haze of surrounding nebulosity. Photographic exposures bring out this object's complex structure of emitting and obscuring material.
Theta (θ)	18h 54m/04° 08'	Stunning, equal white pair, shown to best advantage in 5 or 6 cm telescope (mags 4, 4; sep: 22").
Vulpecula		
C399	19h 20m/20° 00'	Very large, loose cluster that appears in binoculars like an inverted coathanger. Contains a challenge-object: the very faint little cluster NGC 6802 (at eastern end of the 'coathanger'), which our 12 cm refractor shows as a minute, milky patch.
M27	19h 59m/22° 40'	The famous 'Dumbbell', easiest of all planetary nebulae for a small instrument. 6 cm shows elongated, narrow-waisted shape.

3.6 The deep sky: autumn

By mid-autumn the character of the night sky has altered quite radically. The broad Milky Way clouds and bold patterns of first-magnitude stars that made August skies so dense and rich have given way to something different. By comparison, we look up into a dark void that seems relatively empty.

The apparent vacancy of the fall Cosmos is not a false impression. At this season the Earth's position in its solar orbit is such that in the hours of darkness we no longer face directly into the plane of the Galaxy, but gaze outward through a thin layer of Milky Way stars toward the starless vastness of intergalactic space. There is a fair exchange involved in this situation: in place of summer's abundance of nearby clusters and nebulae we gain a new prospect, that of numerous galaxies that lie far beyond our own.

When the beginning observer has enjoyed the impressive binocular view of his or her first alien galaxy (the bright, easy Andromeda system, M31) he or she usually feels the challenge of locating other remote star systems, millions or even tens of millions of light-years distant from our Milky Way. Thus invariably begins the long-term galaxy-hunting pastime that may eventually take the observer out into regions of the Universe as remote as 500 million light-years from the Earth. A telescopic observation of that kind is a sort of time travel, carrying the astronomer half a billion years into the past.

Autumn is a fruitful season in which to begin the exploration of the galaxies. The following constellation notes include examples at various levels of difficulty.

Andromeda

NGC 7662	23h 24m/42° 14'	Planetary nebula: a small bluish ellipsoid, well shown by 6 or 7 cm telescope. (Use high magnification.)
M31	00h 42m/41° 09'	Huge spiral galaxy: in a dark sky good binoculars show the full 3 degree extent of this great spiral's outlying regions. A 20 cm telescope can reveal some hint of dark dustlanes. Small, bright elliptical satellite galaxy M32 lies immediately adjacent, to the south; fainter companion NGC 205 is about half a degree northward.

Plate 55. Galaxy NGC 891 in Andromeda (CCD image).

NGC 752	01h 55m/37° 26'	Fine galactic cluster, best viewed at very low magnification.
Gamma (γ)	02h 02m/42° 16'	Perhaps the most beautiful of all double stars; strong blue/orange colour contrast (mags 2, 5; sep: 10").
NGC 891	02h 19m/42° 07'	Spiral galaxy viewed edgewise. Faint, but a 20 cm telescope shows dramatic spindle shape.

Aquarius

NGC 7009	21h 01m/ – 11° 34'	Easy planetary nebula for smallest aperture. Narrow east–west projections (visible in a 20 cm telescope) appear rather like planetary ring viewed edgewise; hence the name 'Saturn Nebula'.
RY	21h 18m/ – 11° 01'	Neat little eclipsing binary (mag 8.8 to 10 in 1.97 days).
M2	21h 32m/ – 00° 54'	Globular cluster, partially resolved by 15 cm instrument.

Plate 56. The Helix Nebula in Aquarius.

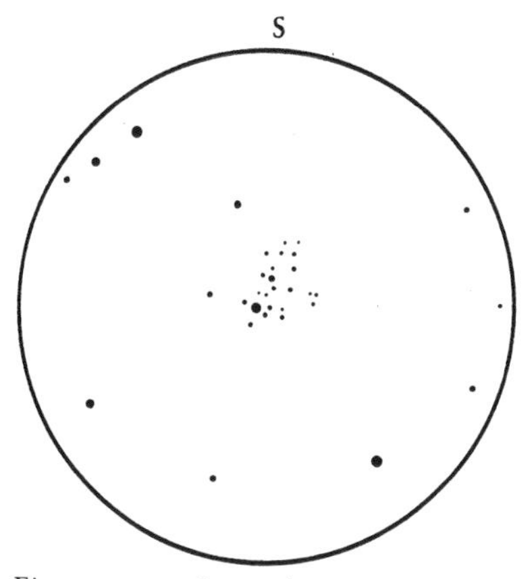

Figure 3.13. Open cluster M52 in Cassiopeia.

Plate 57. Open cluster M103 in Cassiopeia.

41	22h 12m/ – 21° 19'	Striking bronze/azure double star (mags 6, 7; sep: 5").
NGC 7293	22h 27m/ – 21° 19'	'The Helix Nebula'. Huge, dim; needs low magnification and wide field.

Aries

Gamma (γ)	01h 51m/19° 03'	White, equal pair separated by 8" (mags 4.8, 4.8).

Cassiopeia

This well-known constellation is one of the highlights of the autumn sky. When we direct our gaze (or binoculars) high overhead to survey Cassiopeia we are again looking into the galactic plane; it is for this reason that the constellation bristles with fine open clusters, a few of which are listed here:

Plate 58. Stephan's Quintet in Pegasus (CCD image).

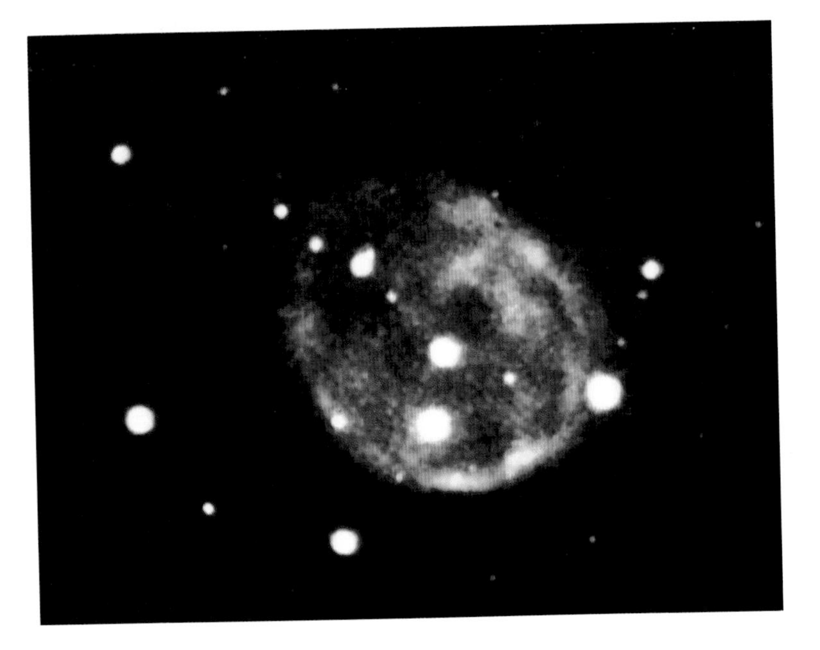

Plate 59. NGC 246 in Cetus (CCD image).

M52	23h 23m/61° 29'	Intriguing wisp of light at 20×. Higher magnification resolves it into a dense, compact star cluster.
NGC 7789	23h 55m/56° 26'	A faint, but unusually rich cluster.
NGC 457	01h 16m/58° 04'	Coarse, dramatic cluster with a colourful double *Phi* (ϕ) at south-east edge. A smaller, but richer cluster NGC 436 appears in the same field at low magnification, about a degree northwest.
M103	01h 32m/60° 35'	Bright triangular group. 6 cm aperture resolves double star *Σ131*, at cluster's apex.
NGC 663	01h 43m/61° 01'	Large cluster; nearly fills a 1 degree field.
Cepheus		
Delta (δ)	22h 28m/58° 10'	Wide double, and brighter component is the classic Cepheid variable (mag 3.5–4.4 in 5.4 days). Easily followed with binoculars.
U	00h 58m/81° 37'	Eclipsing binary: mag 6.8–9.2 in 2.5 days.
Cetus		
NGC 246	00h 45m/−12° 09'	Planetary nebula. Very large, but challengingly dim. Requires 20 cm.
Omicron (o)	02h 17m/−03° 12'	'*Mira*', the famous long-period variable (mag 3.4–10 in 332 days).
NGC 936	02h 25m/−01° 22'	Faint barred spiral galaxy, with more difficult NGC 941 closely adjacent in field. Use largest aperture available.
M77	02h 42m/00° 04'	A much easier galaxy, bright and compact through 14 cm telescope.
Pegasus		
M15	21h 29m/00° 04'	Bright globular cluster, impressive even in a 6 cm telescope.
Σ2841	21h 52m/19° 29'	Yellow/blue double star, in coarse cluster at low power.
NGC 7331	22h 35m/34° 10'	Fine spiral galaxy, just visible with 7 or 8 cm instrument.

Plate 60. Galaxy NGC 7331 in Pegasus (CCD image).

NGC 7332	22h 35m/23° 32'	Like a smaller, dimmer version of the above. Needs 20 cm.
NGC 7479	23h 02m/12° 03'	Barred spiral galaxy; 40 cm telescope shows bar structure.
Σ2978	23h 05m/32° 33'	Delicate binary star; close white/blue pair (mags 6, 7; sep: 8").
NGC 7814	00h 01m/15° 51'	Faint spiral galaxy viewed edgewise. Use 20 cm or larger telescope.

Perseus

Probably the best constellation of the season, this glowing chain of first- and second-magnitude stars overlies a background mist of fainters suns. Perseus is another of those sky regions that reveal their true character best if observed with the unaided eye. The dominant feature, a sprawling 5-degree star throng surrounding *Alpha* (α), is a true association of young, hot stars.

Plate 61. Messier 76 (CCD image).

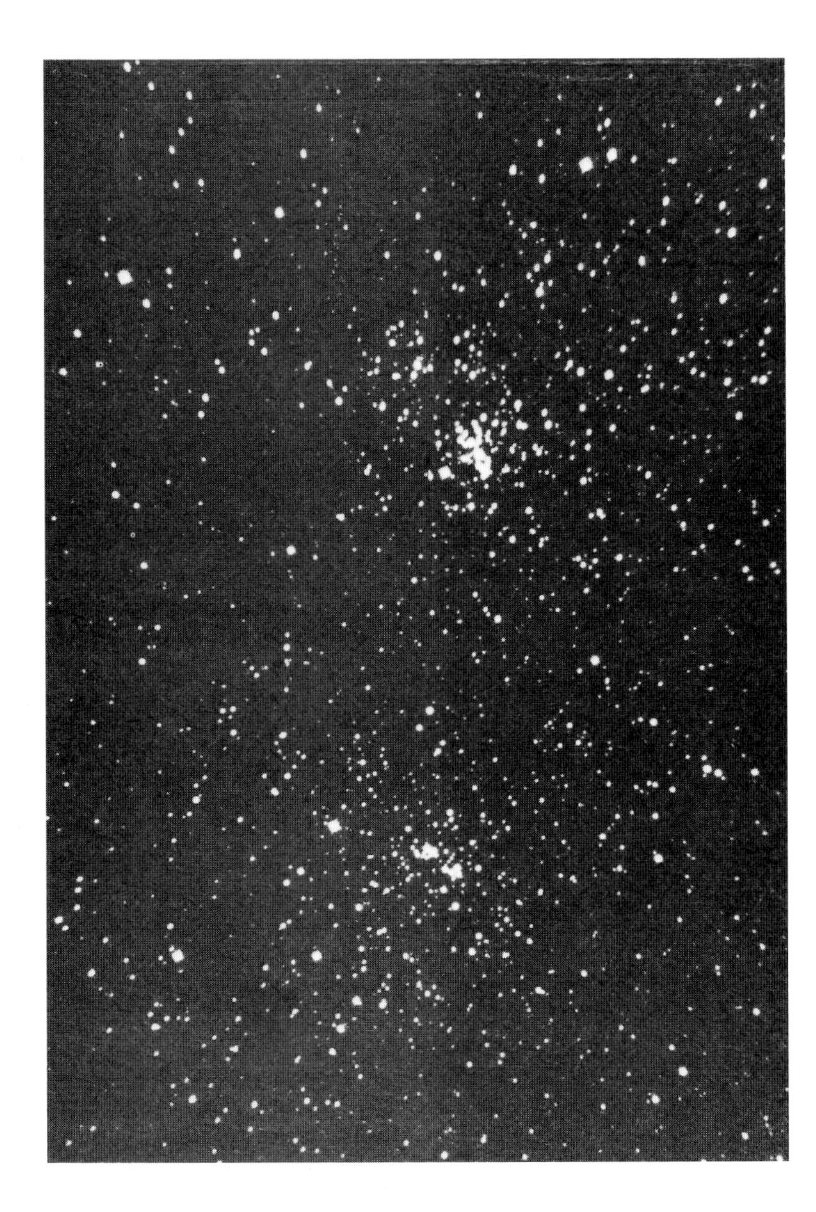

Plate 62. The Perseus Double Cluster (NGC 869/884).

Northward, the sky rises to a second climax of interest near the border of the neighbouring constellation Cassiopeia, where clustered knots of stars appear to the naked eye as tantalizing patches of light.

M76	01h 41m/51° 28'	Tiny planetary nebula with shape similar to the Dumbbell Nebula. Visible in a 10 cm instrument.
NGC 869/884	02h 16m/56° 55'	'The Double Cluster', one of the finest of all deep-sky objects. The two brilliant, coarse star groups are easily seen with binoculars, and are visible to unaided eye. Ineffably glorious through a rich-field telescope at low magnification. Two prominent red giants stars between the clusters show their colours well in medium apertures.
NGC 1023	02h 37m/38° 52'	Elliptical galaxy: a dim oval patch in 15 cm instrument.

M34	02h 41m/42° 43'	Naked-eye cluster, fine through binoculars.
Beta (β)	03h 05m/40° 46'	'*Algol*', the most famous eclipsing binary (mag 2.1–3.3 at 2.9-day intervals).
Epsilon (ϵ)	03h 55m/39° 52'	Nice little binary (mags 3, 8; sep: 9"), with white/grey contrast.
*Σ*552	04h 28m/39° 54'	White, equal pair (mags 6, 6; sep: 9") – lovely!
Sculptor		
NGC 55	00h 13m/ – 39° 50'	Bright edgewise-oriented galaxy. Can be located with the smallest telescopes, but is inaccessible in northern latitudes.
NGC 253	00h 45m/ – 25° 34'	Large, dramatic spiral galaxy, viewed nearly edgewise. One of the best for small instruments, although rather low in sky for northern observers.
Taurus		
		The principal naked-eye feature of this constellation, the great V of bright stars that delineate the head of the bull Taurus, is not a chance arrangement of objects at various distances in space. The Hyades, as this four-degree-long asterism is named, is a true physical association. It is a star cluster like the Pleiades, but three times nearer to us than that famous stellar family. (The Hyades Cluster lies about 130 light-years distant from our Sun.)
M45	03h 46m/24° 03'	'The Pleiades': famous naked-eye cluster, stunning in binoculars. 15 cm telescope shows hint of the enveloping nebulosity, if sky is suitably dark and transparent.
Lambda (λ)	03h 58m/12° 21'	Prominent eclipsing binary: mag 3.5–4.0 in 4 days.
T	04h 19m/19° 25'	Irregular variable, much studied. Mag 9 to 13 at frequent but unpredictable intervals.
Chi (χ)	04h 20m/25° 31'	Easy and lovely double with white/blue contrast (mags 5, 7; sep: 20").
NGC 1647	04h 43m/18° 59'	Open cluster: a dim splash of stars 4 degrees eastward of ϵ *Tauri*.

Plate 63. The Pleiades.

Plate 64. M1, The Crab Nebula in Taurus (CCD image).

NGC 1807/ 1817	05h 08m/16° 28'	Pair of clusters that appear together in view of a rich-field telescope. Both quite sparse, but interesting at low magnification.
M1	05h 33m/22° 01'	'The Crab Nebula'. A 10 cm telescope shows bright oval patch of light; 20 cm begins to give some hint of the supernova remnant's irregular structure.
Triangulum		
M33	01h 33m/30° 33'	Large, nearby galaxy viewed face-on. Needs low magnification and wide field for satisfactory presentation; often best in binoculars. 20 cm telescope shows an intriguing internal feature, NGC 604, an emission nebula on the galaxy's northern fringe.
Tucana		
NGC 104	00h 22m/−72° 21'	47 Tucanae, very large, bright globular cluster. Vastly outranks the famous Hercules Cluster, but cannot be seen by northern-latitude observers.
NGC 292 (SMC)	00h 50m/−73° 30'	'The Small Magellanic Cloud'. This irregular galaxy, a satellite of our Milky Way system, is a huge, dim naked-eye feature, sometimes mistaken by casual viewers for a small atmospheric cloud. A small telescope reveals many internal knots and clusters of stars.

3.7 The deep sky: winter

Although winter observing can be both uncomfortable (because of the cold) and frustrating (because of the frequency of cloudy weather), the amateur astronomer should be aware of another salient fact: winter skies can be glorious! Personally, if we were to cast a vote for the best seasonal deep-sky show, the constellations of winter would be our choice.

In December our night-time perspective is upon the galactic plane once again, but in an outward direction (that is, away from the general direction of the galactic centre). Among the large-scale features of the Milky Way's structure overhead on these frosty nights is the 'Orion Spur', a nearby branch of the spiral arm in which our Sun is embedded. It is a crowded treasure-store of clusters and bright nebulae for the small telescope, as is the dense galactic belt just eastward of Orion, angling obliquely across the heavens through Auriga, Gemini and Monoceros.

Surveying the crowded winter star fields the imaginative amateur develops a strong perception of the depth of the cosmic scene above his or her head. Nearby *Sirius*, blazing on our very doorstep a mere few light-years distant, is viewed against background clouds of stars that lie a thousand light-years more remote. The bright stars of Orion, a few hundred light-years from us, form a windowframe through which we gaze at The Great Nebula, three times further away across a small rift in the galactic structure.

Auriga

NGC 1893	05h 23m/33° 30'	A sparse cluster or knot of stars. If seeing is perfect, a moderate telescope shows a haze of nebulosity surrounding this group. (We have glimpsed this with a 12 cm refractor.)
M38	05h 27m/35° 48'	Open cluster, loose and misty in appearance. Note the much smaller, fainter cluster NGC 1907 about half a degree southwest of M38, and visible in same field at low magnification.
M36	05h 35m/34° 05'	Bright, loose cluster. Fine in 5 or 6 cm refractor.
M37	05h 52m/32° 33'	Magnificent, dense open cluster; appears like a globular in small telescope. Resolved into a mist of tiny stars at mid-power.
Theta (θ)	05h 56m/37° 13'	Close binary (mags 3, 7; sep: 3.4"). Much more challenging than separation might indicate; difficult for a good 10 cm refractor.

Plate 65. Open Cluster M36 in Auriga.

SS	06h 10m/47° 46'	A strange and exciting variable. Normally 15th mag (visible only in large amateur instruments), but explodes at approximately 55-day intervals to 10th mag brightness.
*Σ*872	06h 12m/36° 10'	Colourful double, with gold/blue components (mags 6, 7; sep: 11").

Canis Major

Alpha (α)	06h 45m/ – 16° 43'	'*Sirius*': famous challenge binary. The dwarf companion has been observed with 20–30 cm amateur instruments, but is commonly much more difficult (mags 1.4, 8; sep: 8").
M41	06h 46m/ – 20° 43'	Fine open cluster just 4 degrees south of *Alpha*.
*h*3945	07h 15m/ – 23° 13'	Double star (mags 5, 7; sep: 27"). One of the most dramatically coloured – orange/green.

NGC 2362 — 07h 17m/ – 24° 52' — Bright, compact open cluster surrounding *Tau* (τ) *Canis Majoris* – a very satisfying object at low magnification.

Dorado

LMC — 05h 24m/ – 69° 46' — 'The Large Magellanic Cloud' (LMC), a very nearby external galaxy that is a satellite of our own. To the unaided eye it is a huge and obvious nebulosity. A 15 cm telescope resolves this galaxy into numerous component clusters and nebulae, including the following:

NGC 2070 — 05h 39m/ – 69° 06' — 'The Tarantula Nebula', an enormous and bright emission region that appears in binoculars as a detached portion of the LMC. Telescopically the nebula and several associated clusters of stars fill a 1 degree field.

Gemini

The situation of this constellation against the Milky Way backdrop is interesting. If you visualize Gemini in the traditional way as The Twins (with the bright pair of stars *Alpha* and *Beta* marking the two standing men's heads), then the feet of the twins rest among the dense clusters on the galactic plane, while their heads and shoulders rise into the sparser and more starless regions above the central plane. At an observing site where the sky is sufficiently free of urban light pollution, this gradation of density is easily observed.

M35 — 06h 08m/24° 21' — Bright, loose cluster, best observed at relatively low magnification. Much fainter (more distant) cluster NGC 2158 can be glimpsed near the southwest edge of M35, as a dim misty spot.

20 — 06h 33m/17° 50' — Colourful double star for the smallest telescope (mags 6, 7; sep: 20"). Yellow, blue.

38 — 06h 52m/13° 15' — Nice yellow/grey pair (mags 5, 7; sep: 7").

NGC 2392 — 07h 26m/21° 01' — 'The Eskimo Nebula'. This intense, compact planetary is easy

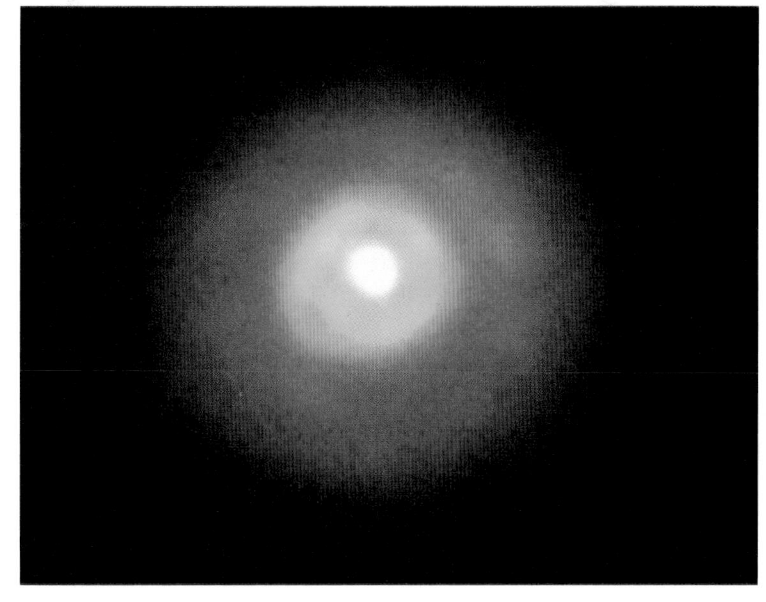

Plate 66. The Eskimo Nebula in Gemini.

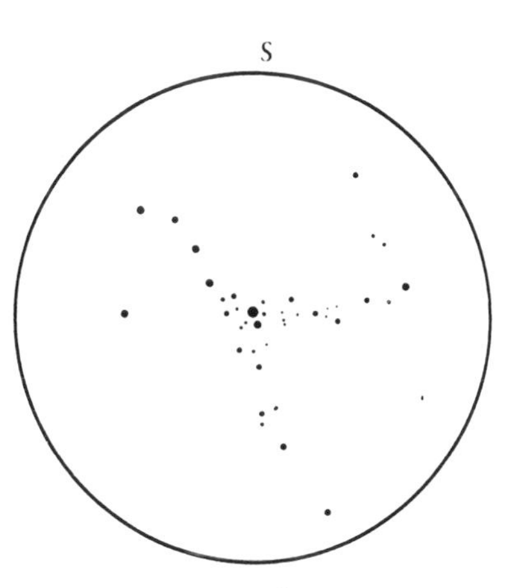

Figure 3.14. Open cluster NGC 2301 in Monoceros (CCD image).

		even in a little 6 cm telescope. 25 cm shows pale outer fringe that surrounds the more prominent inner disc.
Alpha (α)	07h 33m/31° 55'	'*Castor*': a superb binary star (mags 2, 2.8; sep: 2.8") that can just be divided by 6 cm refractor at high power. Very striking in larger instruments. The 9th mag star about 72" southwest is also a component of this multiple system.
U	07h 52m/22° 08'	Eruptive variable, worth monitoring. Usually 14th mag, but explodes to 9th mag at unpredictable intervals.

Monoceros

Beta (β)	06h 26m/−07° 00'	Outstandingly fine triple star. Small telescope at about 50× shows equal pair (mags 5, 5; sep: 7"). At 150×, one of the components itself divides (companion 6th mag; sep: 2.8").
NGC 2244	06h 30m/04° 54'	Coarse, brilliant cluster. Largest amateur instruments show hint of the surrounding gas envelope ('The Rosette Nebula', NGC 2237).
NGC 2261	06h 36m/08° 46'	'Hubble's Variable Nebula', a small, bright spot of diffuse nebulosity that can be glimpsed with a 15 cm telescope.
NGC 2264	06h 38m/09° 56'	Very large, loose triangular cluster. Impressive in very low-power rich-field telescope, but lost at higher magnification. A denser and more compact star-swarm surrounds the prominent 4th mag star at the cluster's base.
NGC 2301	06h 49m/00° 31'	Subtle, fascinating open cluster – a straggling chain of stars against a branching mist of fainter stars.

Orion

This most prominent of all constellations glitters in the transparent, frosty skies of December, marked by its easily recognized 'belt' and 'sword', and blazing with first- and second-magnitude stars. The stars themselves overlie a rather extraordinary background. Optical telescopes show a high concentration of bright and dark nebulae within the region; several are described below. Infrared studies by the orbiting observatory IRAS revealed the true nature of these objects: they are visible outcroppings of a single enormous nebula that spans virtually the full breadth of Orion. Parts of this vast galactic cloud are dark, while other parts (such as the glowing swatch that we know as Messier 42) are excited to luminosity by young embedded stars – some of them, in fact, very recently condensed from the surrounding material.

NGC 1662	04h 49m/10° 45'	Sparse, but pretty little cluster for low magnification. 12 cm refractor shows some colour contrasts among the member stars.
Beta (β)	05h 14m/−08° 13'	'*Rigel*'. Binary (mags 0, 7; sep: 9.4"), whose extreme difference of brightnesses makes resolution much more difficult than one might expect from the separation. Nevertheless, we have spotted the tiny companion with a 6 cm telescope.
NGC 1973,	05h 33m/−04° 48'	Faint complex of nebulae surrounding star 42 *Orionis*, half a degree north of θ_1 in M42. Can be glimpsed with a 10 cm telescope.
Lambda (λ)	05h 34m/09° 56'	Delicate binary (mags 4, 6; sep: 4.4").
M42	05h 34m/−05° 24'	'The Great Nebula'. What can we say? Everyone agrees that it is the noblest of all the deep-sky objects! Visible to naked eye as a

Plate 67. M42, The Great Nebula in Orion.

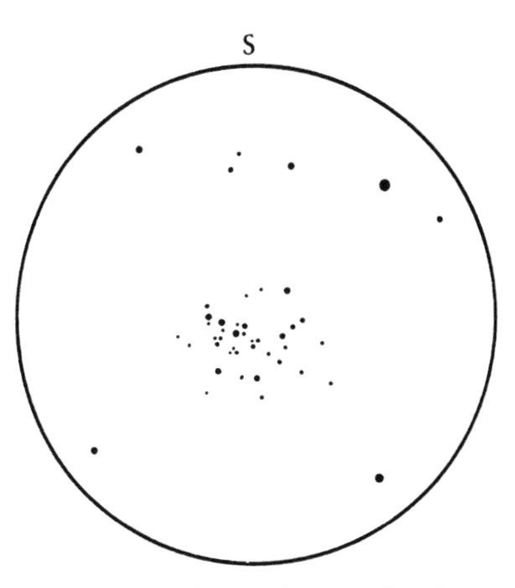

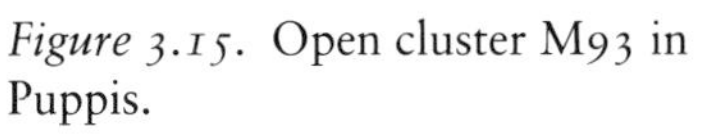

Figure 3.15. Open cluster M93 in Puppis.

		fuzziness surrounding the sword's central star, the nebula is glorious in the smallest telescope. 15 cm aperture clearly shows the distinctive greenish tint of this interstellar cloud. Note the multiple star *Theta*$_1$ (θ_1) at the head of the multiple principal dark rift. Detached portion of the nebula immediately adjacent, to the north, is M43.
Sigma (σ)	05h 39m/ – 02° 40'	Interesting multiple star (mags 3.7, 10, 7, 6; separations: 11", 13", 42"). Needs at least medium power.
B33/I434	05h 39m/ – 02° 26'	'The Horsehead Nebula' and its luminous background cloud. Visually extremely difficult, but a favourite target for astro-photographers.
Alpha (α)	05h 54m/07° 24'	'*Betelgeuse*'. Irregularly variable red supergiant with a period several years in length. At maximum the star may closely match *Rigel* in brightness; in some years it is about 1½ magnitudes dimmer.
NGC 2169	06h 08m/14° 00'	Pleasing quadrangular knot of stars, best viewed at low power. Higher magnification (100×) splits brightest member into a close, equal pair.
NGC 2194	06h 11m/12° 50'	Open cluster. Faint, small, but very rich – a dim spray of tiny stars.

Puppis

M47	07h 36m/ – 14° 27'	Open cluster (loose). Second brightest member is a colourful double.
k	07h 57m/ – 26° 41'	Binary star – a matched gem! (Mags 4.5, 4.5; sep: 10".)
M46	07h 41m/ – 14° 46'	Dense, populous cluster that looks like a circular haze of small stars, at low magnification. The tiny planetary nebula NGC 2438, which lies in the cluster's northern fringes, usually requires at least a 15 cm telescope, but has been glimpsed with a 9 cm instrument.
M93	07h 44m/ – 23° 49'	A fancy little cluster whose shape is like that of a butterfly with wings spread.

3.8 The deep sky: spring

Spring, like autumn, finds the Earth on a segment of its orbit that orients our planet's dark side outward from the Milky Way plane. Again, we look through a relatively thin veil of stars in the local system toward intergalactic space. The dominant feature overhead, on the warming and shortening nights of this stimulating season, is a broad swathe of comparatively starless sky that astronomers sometimes call the *Realm of the Galaxies*.

From Ursa Major southward through Canes Venatici, Coma Berenices and Virgo there extends a window through which we glimpse the swarming star systems of the Universe beyond our own galaxy. The hundreds of these galaxies that can be discovered with very modest-sized telescopes are all impressively remote objects. The easily observed M94, for instance, lies about 14 million light-years distant from us. The distance of the famous Whirlpool Galaxy (M51) is considerably more than twice as great, at something like 37 million light-years. The thronging galaxies of the Virgo Cluster (most of them feasible targets for small apertures) are situated in a region of the Cosmos from which light takes about 40 million years to reach the Earth. It is worthwhile to realize that fact, while viewing the pale, elusive images of these objects.

While we are on the subject of elusiveness: even the noblest of the great galaxies are exceedingly subtle telescopic objects. Do not expect success in observing or photographing them from a site that is flooded with city lights. Yet in a dark rural location you will be pleasantly surprised by the clarity (and perhaps even detail!) of the best of the great spirals. In fact the number of spring galaxies that are commonly observed by amateurs vastly exceeds the page-space available here. Our selection is therefore only a very small representative sample.

Boötes

Kappa (χ)	14h 12m/52° 01'	Spectacular gold/blue pair (mags 4, 6; sep: 13").
Epsilon (ϵ)	14h 43m/27° 17'	Close orange/greenish binary (mags $2\frac{1}{2}$, 5; sep: 2.9").
Xi (ξ)	14h 50m/19° 12'	Fairly rapid binary in which positional changes can be observed over several years (mags 5, 7; sep: 7").
R	14h 50m/20° 57'	Long-period variable: mag 8 to approximately mag 12 in 258 days.

Cancer		
Zeta (ζ)	08h 11m/17° 43'	A favourite triple star (mags $5\frac{1}{2}$, 6, 6; sep: 6", 1"). The close components are a rapidly moving binary, with noticeable changes from year to year. Use high power.
Iota (ι)	08h 37m/28° 57'	Wide, but dramatic double star. Yellow/blue pair (mags 4, 6; sep: 31").
M44	08h 39m/20° 04'	'The Beehive Cluster', visible to the unaided eye as an unresolved cloud of starlight. Exciting at lowest telescopic power; spoiled at higher magnification.
M67	08h 50m/11° 54'	Compact open cluster: a nebulous patch in smallest apertures, splendid in 15 cm.
Canes Venatici		
M106	12h 18m/47° 25'	Bright, elongated galaxy, an easy object for 10 cm telescope.
NGC 4631	12h 40m/32° 49'	Also viewed edgewise. About 1 mag fainter than M106.
M94	12h 50m/41° 14'	Luminous face-on spiral, prominent enough to be located easily with 6 cm refractor.
Alpha (α)	12h 54m/38° 35'	The famous binary star *Cor Caroli* (mags 3, 5; sep: 20"), in which some observers see a contrast of colours – pale yellow/blue-grey.
M63	13h 15m/42° 08'	Obliquely viewed spiral, easy for small apertures. 20 cm shows bright core and much fainter peripheral disc.
M51	13h 29m/47° 18'	'The Whirlpool Galaxy'. Visible, with its adjacent smaller companion galaxy, in 8 cm telescope. 25 cm shows hint of spiral arms and dark rifts.
M3	13h 41m/28° 29'	Globular cluster. Large, intense.
Carina		
NGC 2808	09h 11m/– 64° 39'	6th mag globular cluster, unresolved with very small apertures, but rich in 20 cm or larger.
Eta (η)	10h 43m/– 59° 25'	Irregular variable star, embedded in the great nebula described below (NGC 3372).

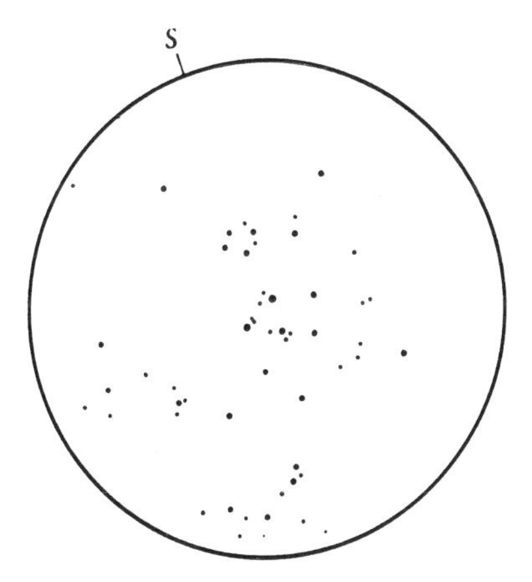

Figure 3.16. Open cluster M44, The Beehive, in Cancer.

Plate 68. Galaxy NGC 4631 in Canes Venatici (CCD image).

Plate 69. M51, The Whirlpool Galaxy in Canes Venatici. This photograph shows the 1994 supernova (detected April 1 1994) – a bright 'star' near the galaxy's nucleus.

NGC 3372	10h 43m/ – 59° 25'	Huge, diffuse nebula, a full degree in width. Visual impression includes bright lobes and dark lanes.

Centaurus

NGC 3766	11h 34m/ – 61° 19'	Open cluster: large, rich, intriguing small-telescope object.
Gamma (γ)	12h 38m/ – 48° 41'	Binary star: 3rd mag components separated by a close 1.6" – a stringent test for an 8 cm aperture.
Omega (ω)	13h 28m/ – 47° 12'	Enormous, bright globular cluster – the finest visible from the Earth. Observable with unaided eye and resolvable into individual stars with 6 cm telescope. Alas, never visible to observers in northern latitudes!.

Plate 70. (*Left*) Globular cluster M3 in Canes Venatici.

Plate 71. (*Right*) The *Eta Carinae* Nebula.

Alpha (α) 14h 39m/ – 60° 47' Superb binary star (mags 0.3, 1.7; sep: 21"). Only about 4 light-years from our Sun.

Coma Berenices

The most noticeable feature of this constellation is the grouping of ten or twelve stars that form a compact two-degree triangular mass. This aggregation is a true cluster, lying only 250 light-years distant (little more than half the distance of the Pleiades). In a dark sky it is readily seen with the unaided eye; in binoculars it is spectacular.

The chief glory of Coma Bernenices lies far beyond the visible star group. It is the throng of galaxies that speckle the telescopic field in this region. The majority of these systems, which can be

Plate 72. Globular cluster *Omega Centauri.*

Plate 73. Centaurus A.

picked up in moderately sized amateur instruments, are members of an enormous physical association, the so-called *Coma–Virgo Galaxy Cluster.* Like stars within our galaxy, galaxies themselves tend to occur in condensed knots and clumps throughout the observable Universe. The galaxies of this constellation and its neighbour Virgo are a knot of this kind, situated on average about 40 million light-years from us.

M98	12h 13m/15° 01'	11th-mag edgewise spiral. Not easy for small apertures.
M99	12h 18m/14° 32'	Spiral galaxy viewed face-on. Brighter than M98.
Σ1633	12h 18m/27° 20'	Very nice binary with equal, pale yellow components (mags 7, 7; sep: 9").
M85	12h 24m/18° 18'	9th-mag galaxy, well seen in 10 cm telescope. Larger instruments show NGC 4394 in same field.

Plate 74. The Coal Sack, a nebula in Crux.

M88	12h 31m/14° 32'	Spiral galaxy. 15 cm telescope shows the northwest–southeast elongation.
24	12h 33m/18° 39'	Coma's best double star, a colourful gem with yellow/turquoise contrast (mags 5, 6; sep: 20").
NGC 4565	12h 34m/26° 16'	Classic edge-on spiral. 10 cm aperture shows it as an elusive, highly elongated sliver of light. 30 cm reveals hint of dark equatorial lane. Handily found by sweeping $1\frac{1}{2}$ degrees east from the southeast corner of Coma Star Cluster.
M64	12h 56m/21° 48'	'The Black Eye Galaxy'. At a bright 8.8 mag, it is nicely seen with 6 cm refractor. 20 cm shows trace of the central black dust-lane that gives this object its name.
M53	13h 12m/18° 17'	Globular cluster. Small, distant, relatively difficult to resolve.

Crux

Alpha (α)	12h 24m/ – 62° 49'	Outstanding binary whose bright, diamond-like components (mags 1.5, 2) are separated by 4".
NGC 4755	12h 51m/ – 60° 05'	Some consider this the most superb of all open clusters. Medium-sized telescope shows bluish-white/reddish contrasts among its brightest stars. Named 'The Jewel Box'.

Hydra

M48	08h 13m/ – 05° 43'	Luminous, easy binocular cluster. Small telescope shows some hint of colour contrasts.
NGC 3242	10h 22m/ – 18° 23'	Planetary nebula, slightly brighter than the more famous Ring Nebula in Lyra.
M68	12h 38m/ – 26° 38'	Globular cluster. Small, but resolvable in 20 cm aperture.
M83	13h 36m/ – 29° 46'	Quite large, but rather dim face-on spiral galaxy.

Leo

NGC 2903	09h 29m/21° 44'	A fine, prominent spiral galaxy, obliquely oriented. A little 9 cm catadioptric easily shows elongated form; 12 cm refractor gives hint of bright core, fainter region of spiral arms.

Plate 75. NGC 2903, a spiral galaxy in Leo (CCD image).

Plate 76. M82, a disruptive galaxy that can be seen in same field with M81 in Ursa Major (CCD image).

R	09h 50m/11° 40'	Variable (mag 5–10 in 312 days) popular with beginning observers.
Gamma γ	10h 19m/19° 57'	Superb binary for all apertures (mags 2, $3\frac{1}{4}$; sep: 4").
M95	10h 43m/11° 49'	Somewhat challenging galaxy; appears in telescopic field with M96.
M96	10h 46m/11° 56'	Spiral galaxy that forms close pair with M95.
M105	10h 47m/12° 42'	Pale elliptical galaxy. If low magnification and wide field are used, M105 with M95 and M96 form a dramatic trio viewed together.
M65/M66	11h 18m/13° 13'	One of the most famous sets of 'paired' galaxies. Smallest telescope shows them both as elongated shapes, M65 being the narrower (that is, the most obliquely oriented).

Plate 77. M108, an obliquely oriented spiral galaxy in Ursa Major (CCD images).

Plate 78. M101, spiral galaxy in Ursa Major (CCD image).

NGC 3628	11h 18m/13° 53'	Dim spiral, viewed edgewise within same field as M65/66. Well seen through 12 cm aperture.
90	11h 32m/17° 04'	Intriguing multiple star: low power shows wide white/blue pair (mags 6, 9; sep: 63"). High magnification splits the brighter component into a close binary (mags 6, 7; sep: 3").

Ursa Major

The most widely known of all the constellations, Ursa Major contains the large, easily recognized asterism named variously 'The Plough' or 'The Big Dipper'. One reason for people's familiarity with this star group is its circumpolar location. For northern observers at least, the 'Dipper' never sets; it is prominent at every season. Spring is the best time to discover and observe Ursa Major's deep-sky objects, because at this time of the year the constellation lies high overhead. The opposite extreme – Ursa's poorest showing – occurs during autumn, when the Dipper swings to a position low above the northern horizon.

Sigma$_1$ (σ_1)	09h 06m/69° 09'	Close, challenging binary (mags 5, 8; sep: $2\frac{1}{2}$"). In finderscope, σ_1 forms dramatic colour-contrast with adjacent σ_2.
W	09h 40m/56° 11'	Super-rapid eclipsing binary (mag 7.9–8.6 in one-third day).
M81/M82	09h 54m/69° 09'	Pair of bright, easy galaxies in same telescopic field. Can be observed with 6 cm refractor or even large binoculars.
M108	11h 11m/55° 47'	Spindle-like edgewise galaxy; form clearly evident in 12 cm telescope.
M97	11h 14m/55° 08'	Planetary nebula, 'The Owl'. In one respect, a challenging object. Although not difficult to locate, it has details (the 'owl-eyes') that can prove elusive in even large telescopes.
Xi (ξ)	11h 17m/31° 39'	Fast-moving binary that shows movement of components over short period of years (mags $4\frac{1}{2}$, 5; sep: 2")
NGC 3941	11h 50m/37° 16'	Nice little spiral; bright enough for 10 or 15 cm instruments.

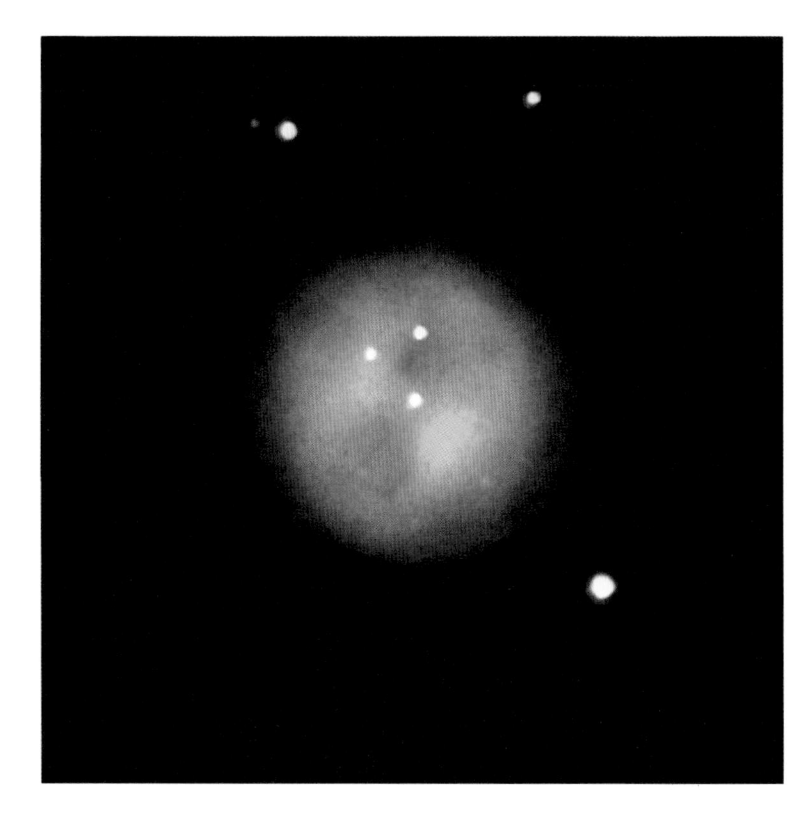

Plate 79. M97, The Owl Nebula in Ursa Major (CCD image).

Plate 80. Quasar 3C 273 in Virgo with jet (CCD image).

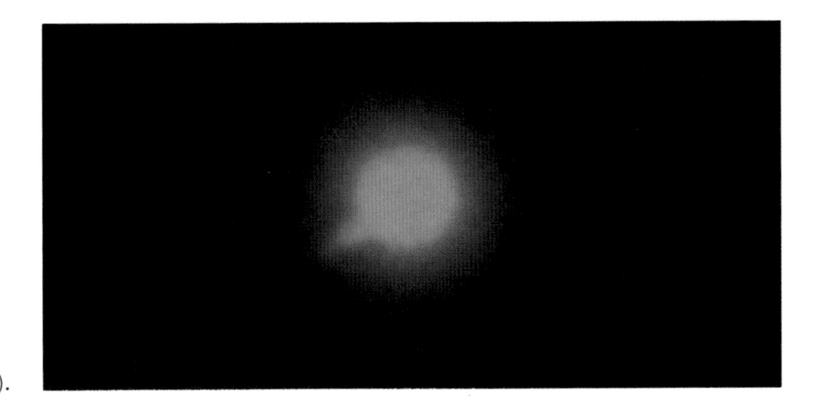

Plate 81. M87, a giant elliptical galaxy in Virgo with jet (CCD image).

M109	11h 57m/53° 29'	Very pale barred spiral galaxy.
Zeta (ζ)	13h 22m/55° 11'	'Mizar'. With Alcor, 12 arc-minutes distant, this is an easy naked-eye pair. In spite of great separation, these stars are believed to be true physical associates. Small telescope splits Mizar itself into a classic binary (mags $2\frac{1}{2}$, 4; sep: 14").
M101	14h 03m/54° 27'	Large, face-on spiral galaxy. Photographs show incomparable splendour of arms, dustlanes. Amateur telescopes give disappointing visual impression, because of this galaxy's low surface brightness. Use low power in dark sky; can be difficult to distinguish from sky background.

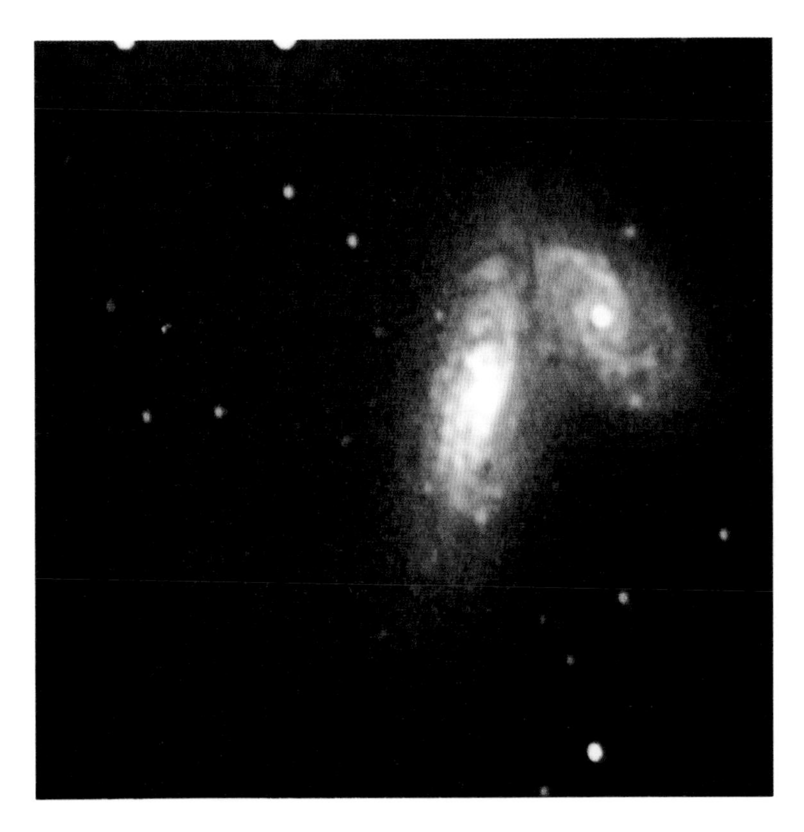

Plate 82. NGC 4567/8, the 'Colliding' Galaxies in Virgo (CCD image).

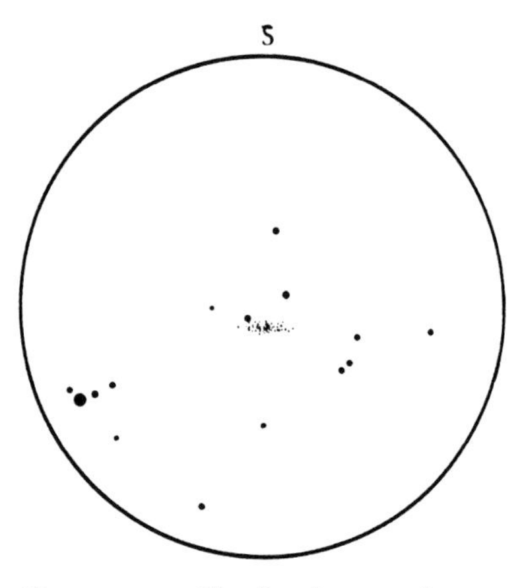

Figure 3.17. The Sombrero galaxy (M104).

Plate 83. M104, The Sombrero Galaxy in Virgo (CCD image).

Virgo

M61	12h 21m/04° 36'	Barred spiral galaxy. Rather dim.
M84/M86	12h 24m/13° 00'	Two bright elliptical galaxies close together in an extremely interesting field, crowded with fainter galaxies, of which two or three can be glimpsed with 15 cm aperture.
3C 273	12h 29m/02° 03'	Quasar, 3 billion light-years distant, but visible (as bright point of light) in a 15 cm telescope.
M49	12h 30m/08° 07'	Large elliptical galaxy, one of the easiest for very small apertures.
M87	12h 30m/12° 30'	Huge elliptical, but not as bright as M49.
NGC 4526	12h 32m/07° 58'	Elongated elliptical galaxy, about 1 degree southeast of M49. Can be found with 15 cm instrument.
NGC 4535	12h 32m/08° 28'	Spiral galaxy in same low power field as NGC 4526. Difficult!
NGC 4567/ 4568	12h 34m/11° 32'	Extremely challenging pair of objects (use largest available aperture). Visually, a single smudge of light; photographs show pair of galaxies in apparent collision.
M90	12h 36m/13° 16'	Oblique, elongated spiral galaxy. Not difficult for 12 cm refractor.
M58	12h 37m/11° 47'	Face-on spiral; easy for 10 or 15 cm telescope.
M104	12h 39m/ – 11° 31'	'The Sombrero Galaxy', an unusually intriguing object for small or medium-sized instruments. Edgewise orientation is obvious through even a 10 cm telescope. 25 cm gives unmistakable glimpse of dark equatorial dustlane.
W	12h 24m/ – 03° 07'	Cepheid variable star (mags 9.5–10.7 in 17 days).
Gamma (γ)	12h 41m/ – 01° 21'	Binary star, a lovely equal pair (mags $3\frac{1}{2}$, $3\frac{1}{2}$; sep: 3.4").
M59/M60	12h 41m/11° 47'	Easily observed elliptical galaxies, within same field at low magnification.
AL	14h 08m/ – 13° 05'	Cepheid (mags 9.1–10.0 in $10\frac{1}{3}$ days).

PART 4 The night sky on film: astrophotography

4.1 What is involved: an overview

Ever since people first gazed upward into the night sky, they have desired to record what they have seen. Astrophotography is the art of photographing those celestial objects.

Evidence of early interest are the dots chipped into the walls of caves by Neanderthal man to record the Sun, Moon and stars. Today we send rockets high into the atmosphere, and even spacecraft above it, to photograph the heavens.

Now that reasonably priced single-lens-reflex cameras, fast films and good commercially built telescopes are available, astrophotography is within reach of any amateur astronomer.

We shall start with the simple camera and tripod set-up for the beginner, and then deal with time exposures, eyepiece projection and, for the more advanced, the art of guided deep-sky photography with a 500 mm telescope and dry-gas cold camera. The subject matter for such photography is virtually endless, and each object takes on a new look when photographed at a larger aperture, or with longer exposure in the case of deep-sky objects.

As a beginner, I set out to map the sky with nothing more than a camera and tripod, and then went on eventually to photograph with my guided telescope all of Charles Messier's 110 deep-sky objects. These include beautiful nebulae, galaxies and star-clusters.

Nowadays, the rims of lunar craters can be photographed with eyepiece projection, and the belts and swirls in Jupiter's atmosphere, as well as Saturn's rings, can be recorded as part of your observations.

Who knows, you may one day be viewing the right part of the sky at the right moment to catch on your photograph a comet or nova – the shot of a lifetime! Most comets are discovered by amateur astronomers, as the professionals are generally too busy with specific projects to have the time to spend on random observing.

Filters are becoming now more than ever an important part of photographing the night sky. Deep red, and even light-pollution filters, help in capturing that last bit of contrast in deep-sky photography.

Darkroom procedures are briefly discussed in this section: dodging, unsharp masking, types of developer, etc.

The four most common systems for astrophotography are:

(a) Prime focus, or direct objective;
(b) Eyepiece projection;
(c) Negative lens projection; and
(d) Eyepiece/camera lens.

Prime focus is used to telephoto pictures of stars and nebulae. This system provides the brightest field but the smallest image scale. The standard 30 mm camera adapter cuts off some of the light cone and restricts the field to a circle of about 20–25 mm on the film. This circle can then be enlarged and printed to produce a preferred size. A 50 mm camera adapter will totally illuminate a 35 mm film format, eliminating this problem altogether.

Eyepiece projection is the use of the telescope and its eyepiece to project an image directly onto the film. This method is very popular for lunar and planetary photography, as it gives spectacular results. Image size is determined by the distance of the camera from the eyepiece: the closer they are, the smaller the image. Extension rings of varied sizes are used as required between the telescope eyepiece and the camera. Some eyepieces project a flatter field than others.

Negative lens projection is similar to positive eyepiece projection. It uses a Barlow lens to project an image onto the film, and provides a fairly flat field. This method is also used for planetary and lunar photography and gives full-frame illumination.

Of the four systems, the *eyepiece/camera lens* system is probably most suitable for the beginner. The camera is hand-held behind the eyepiece and the picture is snapped. A variation of this has the camera set on a tripod adjusted to receive the image from the telescope eyepiece. Brackets for this purpose can be purchased that will clamp the camera to the telescope.

All four techniques are discussed in greater detail in the relevant chapters, including information on making your own cold cameras and solar prominence spectroscopes.

Solar photography is introduced, and the author's experience – and adventures – on solar eclipse expeditions provides an introduction to another aspect of astrophotography.

4.2 Trying your hand: simple astrophotos

The simple camera with tripod is the starting point for most astrophotographers. Virtually any camera having either bulb or time setting, to allow the shutter to remain open, can be used to take fabulous photographs of the constellations and to map out the sky.

The basic camera, except the instamatic type, has several settings. The adjustment for focusing the lens is usually located on the lens itself, and calibrated in both metres and feet. The farthest focus, the one we use most, is generally indicated by the symbol '∞' to represent infinity. A separate dial indicates the shutter speed; this should range from the time setting to as little as 1/1000th of a second. These time settings are best used with a cable release, as the slightest vibration of the camera during the shot will show up on the film.

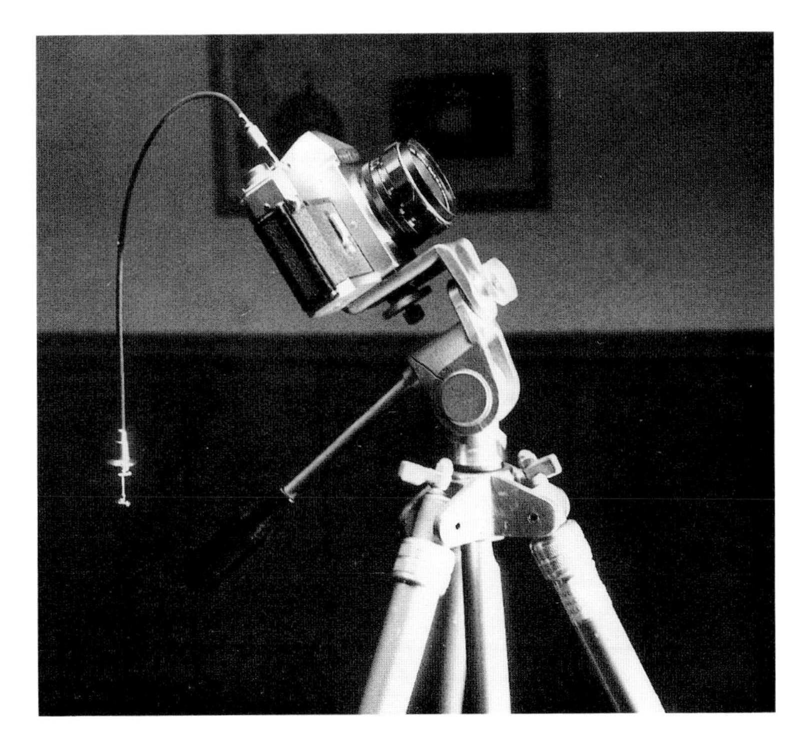

Plate 84. Camera and tripod: the simplest astrophotographic setup.

The 'f-stop' opens and closes the iris of the shutter. It allows you to vary the amount of light entering the camera, and when set at the lowest number, f1.4 or f2.8, the iris is wide open. When set at the higher numbers, f11 or f22, it is said to be stopped down for its minimum exposure. Each model of camera may be slightly different, so be sure to study the manual provided by the manufacturer.

The focus should be set at infinity, the lens wide open at f1.4, and the shutter speed at 'bulb' or 'time'. Then you can open the shutter for as long as you wish by locking the cable release. Meteor showers, star trails, satellites, aurorae, comets, and even lightning, can be photographed in this way.

The film you use should be fairly fast; I recommend Tri-X for black and white, Ektachrome 400 for colour slides, and Konica 3200 for colour prints, as good ones to start with. Be sure always to record the film type, its speed, and your camera settings – make it a habit.

When starting astrophotography, don't be afraid to experiment. Locate the constellation you wish to photograph, point your camera, and expose the film for 10 seconds. Next try the same shot for 20 s, at 30 s, and so on to 60 s. Then try a 2 min. exposure, and one at 5 min. These test shots enable you to find the longest exposure that will still give you good star images, with sharp stars and minimal trailing. The optimum exposure time will determined by analysing *all* the data from the experimental photographs, so a record should always be kept of camera settings and the subject in a log specially kept for this purpose (Table 4.1). It is important to note whether there was snow on the ground, noticeable wind, or any haze, and what the temperature was, as any of these can affect the quality of pictures you obtain.

Table 4.1. *Astrophotography log*

Subject____________________	Date__________
Time__________pm/am Sky conditions____________________	
Film type____________________	Speed__________
Picture no.________________	Exposure time__________s/min.
Comments ____	

Sky light or sky fog, which shows on the film as a lightening of the dark sky, is due to light pollution from a city or the bright Moon. Obviously you should try to avoid street and house lights when choosing your location for doing astrophotography.

Star trails

A five-minute unguided exposure will show star trails, an effect caused by the Earth's rotation. A spectacular photograph can be made by centring the camera on Polaris, the north pole star, with the lens stopped down to f8. Expose your film for about 20 min., and the stars on your picture will appear as beautiful circles of light scribing arcs around Polaris. A church steeple or other high building, or a tall tree, can be at one side to give perspective to the photograph. If sky fog is a major problem in your area, try using a K-2 yellow filter (with black and white film only) on your camera lens.

Plate 85. Star trails.

Some camera lenses will reveal serious distortion near the edge of the field when left wide open, at f3.5 or less. This may be corrected by stopping the lens to f4.5 or f5.6. At this setting only the centre portion of the lens is used and it gives a much sharper image, although less light is admitted.

Aurorae

The aurora is one of Nature's firework displays that can be photographed with stunning results. Set up the camera and tripod pointing to the area of greatest auroral intensity, or just generally towards the north. Colour film will capture the beautiful pageant most successfully. Ektachrome is best used wide open for one or two minutes, depending on the intensity of the display that particular night. Black and white film may also be used, but for a shorter exposure time.

Plate 86. The Aurora Borealis.

Satellites

Man-made satellites constantly circle our planet, ranging from expended rocket casings which blink on and off as they tumble through space, to round weather satellites that photograph the Earth and which reflect a steady light. The best time to locate satellites is just after dark or just before sunrise, when we are in the Earth's shadow and the sky is dark. Then, since they are so high and still illuminated by the Sun, we can clearly see and perhaps photograph their trails. High-speed black and white film such as Tri-X at f4.5 is recommended for this type of work, but sky conditions will dictate your 'f' setting. A two to four minute exposure

Plate 87. Satellite trail.

Plate 88. Apollo 14 liquid oxygen dump (30 s on Tri-X, with 450 mm f8 lens).

is recommended, dependent on the speed of the satellite; the idea is to shoot just ahead of the satellite so that its own brightness exposes a path across your film during this time.

With spacecraft going to the Moon and into deep space, we have been introduced to 'space spectaculars' such as dumps of fuel and releases of barium. Plate 88 shows an oxygen dump from Apollo

Plate 89. Apollo 14 liquid oxygen dump, showing expansion of the oxygen cloud (cf. Plate 88).

Plate 90. Lightning captured on film.

14 at about 16 000 km altitude. The photograph was taken with a 450 mm f8 lens and was an unguided thirty second exposure. Plate 89 shows a guided three minute shot with a 450 mm f8 telephoto lens.

As Apollo 14 headed for the Moon, the command module detached from the S4-B third-stage rocket, turned round and attempted to dock with the lunar excursion module (LEM), but failed at first attempt. The flight plan had called for venting of liquid oxygen (LOX) from the S4-B rocket after the lunar module had been secured to the command ship, and moved safely a few kilometres away. Because of the docking failure, the LOX dump was done with the LEM still secured inside the S4-B rocket. At 6.40 in the evening of 31 January 1971, we spotted the light ball in the sky, caused by the chemical dump, with 7 × 50 binoculars. We hurriedly trained our telescopes on the centre of the expanding ball of gas, and saw the S4-B rocket and the command module racing along. The S4-B was quite bright beside the command module, but as the module manoeuvered to dock with the LEM, its manoeuvering jets flashed even brighter than the S4-B. It was thrilling to watch this aerial display enacted for our eyes by men on their way to the Moon.

Lightning

I do not normally recommend shooting through windows, but to photograph lightning during a violent electrical storm, discretion is

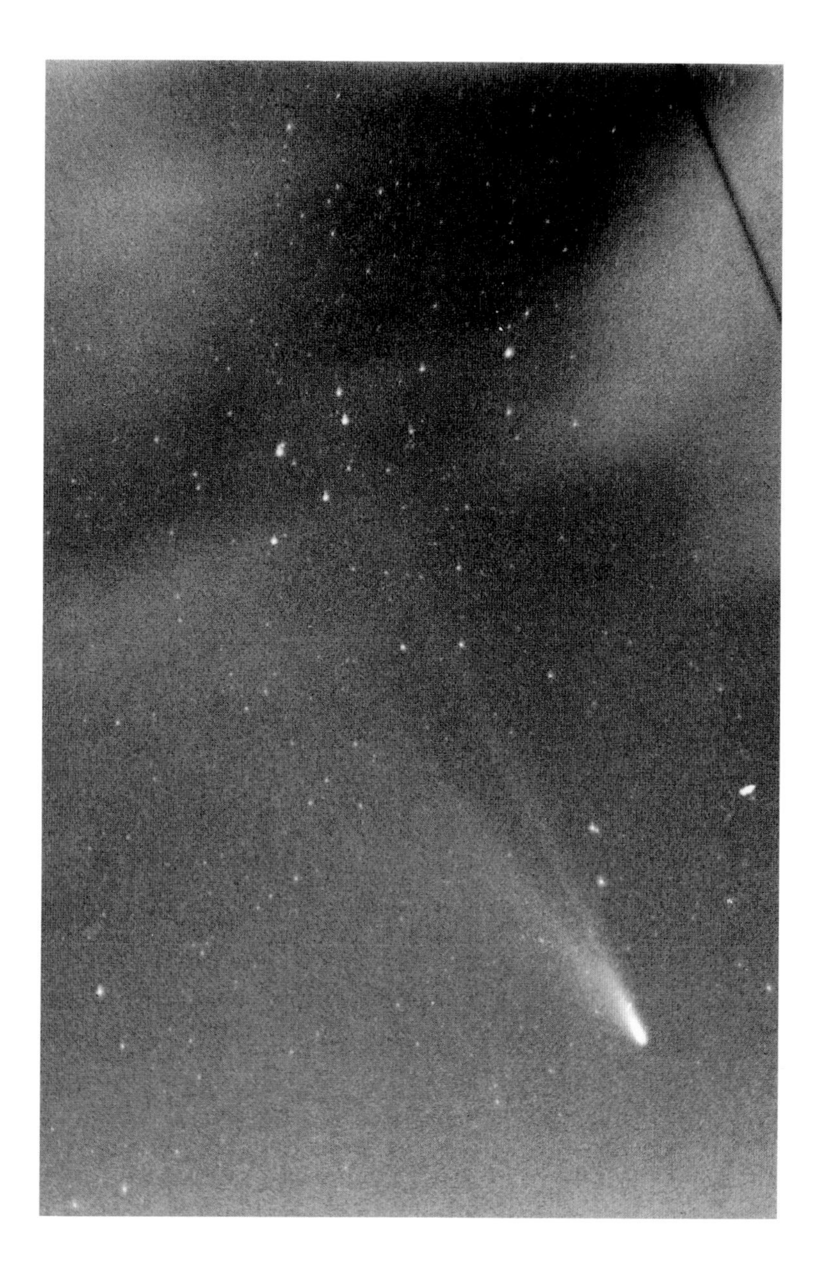

Plate 91. Comet West, 1976.

the better part of valour. Set up your camera and tripod indoors, loaded with medium-speed black and white film such as Plus-X at f16 or Ektachrome 400 at f11. Lock the shutter open and wait for the lightning flash. Then release the shutter. Make a mental note of the area of sky your camera is photographing. This will help you determine when a flash has actually been recorded. Remember not to leave your shutter open too long, because several near-misses will act like a flash bulb and overexpose your film. If possible, aim your camera at a water tower or high building, as you may very well record a strike. Many breathtaking pictures have been produced in this way.

Plate 92. Comet Halley, 1985.

Meteors

Meteor showers are great fun to observe. On a warm summer night, take a lawn chair and sit gazing at the sky, waiting for the flash of a meteor. This one of the most relaxing hobbies I know! How often have you seen a beautiful fireball lighting up the sky with fire and smoke, wishing you could photograph it? Well now you can, and it's quite easy.

Make a note of the 'radiant' of the particular shower you wish to photograph; the list in Table 2.9 will get you started. The radiant is the area, or constellation, from which a particular shower appears to originate. For instance, if you are looking for the Leonids, the meteors will appear to radiate from the constellation Leo.

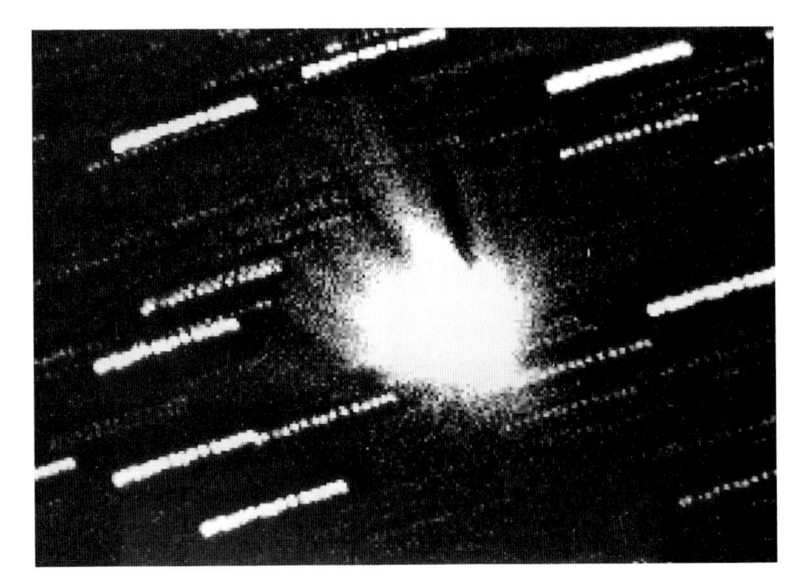

Plate 93. Comet Swift–Tuttle, 1992 (CCD image).

Plate 94. Comet Bennett, 1970.

Aim your camera/tripod combination at the radiant. Sky conditions, the phase of the Moon, and so on, will govern your 'f' setting and exposure time, as explained earlier. Slow film such as Plus-X at about f8 for 30 min. might be tried. Several cameras can be set up to photograph more of the sky. Unfortunately, there is little hope of photographing, or for that matter even seeing, any meteors when the Moon is in the sky.

Comets

For centuries, according to historical records, spectacular comets have caused great excitement. Some thought they foretold disasters.

Many amateur astronomers spend countless hours searching the evening and morning skies with binoculars or small telescopes for a new comet or to recover a returning one. The dream of many is to be the first to discover a new comet so that they earn the right to have it named after them.

One of the best methods of discovering comets is photographically. You may not even be aware of a discovery until you develop your film. Try shooting the western sky just after sunset, or the eastern sky just before dawn. By repeat photographs of the same area at one- or two-day intervals, you may find your own comet. Back in the warmth of your home, simply superimpose the two negatives to find whether any object has moved during the interval between the two shots.

When a comet has been located – usually those that other people

Plate 95. Comet Halley, 1985, during the early approach period.

have discovered – take a picture as described for constellations. The tail of a comet can become quite prominent as it nears the Sun, but it is very difficult to make out the tail of the average comet visually, though photographically you can see it. Most comets do not show much detail, as they are washed out in the twilight. However, you may find that two different negatives, when superimposed, bring out more of the faint detail in the comet's tail.

4.3 Aligning your telescope for astrophotography

Most astrophotography projects require exposures of at least several minutes. During this period, of course, the Earth's rotation will cause an apparent movement of the target object across the camera's field of view; the result will be a blurred or trailed image on the film. One solution to this problem might be (if it were possible) to stop the Earth from turning while the camera's shutter is open. An easier approach is to have the camera on a properly aligned, clock-driven equatorial mount.

Figure 4.1. Astrophotographic alignment of equatorial mounting. The polar axis laid precisely parallel to the Earth's rotational axis permits rotation of the telescope during a photographic exposure in a manner that exactly counteracts the Earth's movement.

The equatorial mount for telescopes was designed so that the right ascension axis of the instrument would be parallel to the Earth's polar axis (Figure 4.1). This enables you, with motion only in one direction, to focus the telescope on a star on the eastern horizon and follow it across the sky until it sets in the west. In fact, if the telescope had a clock drive that was left running, it would move along pointing at the ground all day and would again pick up that same star in the east the next evening.

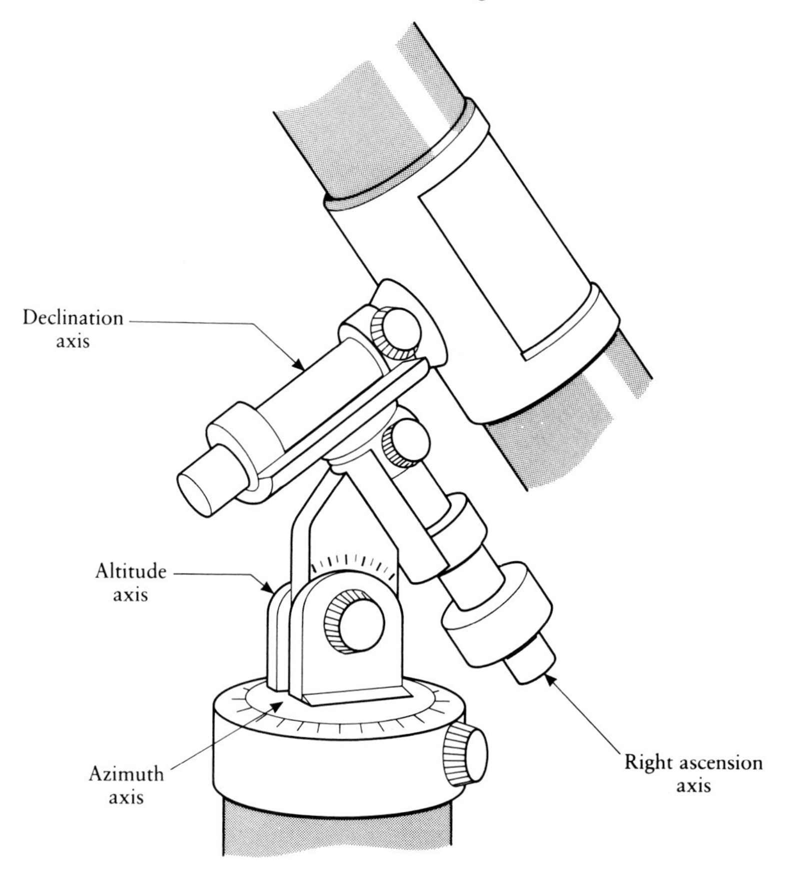

Figure 4.2. The astrophotographic mounting 'German style': its four axes of adjustment.

The German equatorial mount is made up of four major axes or adjustable dimensions: azimuth, altitude, right ascension, and declination (Figure 4.2). The *azimuth* axis, or horizontal motion of the base, permits the telescope to be swung 360 degrees around the horizon. This motion is the same as panning a full circle with a movie camera mounted on a standard tripod. Azimuth motion is used to position the telescope on Polaris or on the true celestial pole nearby. The German mount would not ordinarily have any mechanical azimuth, as the tripod can be moved around.

The *altitude* axis for vertical motion allows the telescope to be moved vertically from the horizontal, usually from 0 to 90 degrees. This adjustment is used to align the mount upward on the celestial pole; the angle formed by the mount and the pole is equal to your latitude. Some telescopes, such as the fork-mounted Questar, use the adjustable front leg of their tripods to accomplish this.

The *right ascension* (or 'polar') axis motion tracks the star with one motion. That is, it follows the star completely across the sky. The right ascension axis has setting circles mounted so that the observer using Universal Time may locate the celestial object from its co-ordinates. This axis often has a mechanically or electrically driven motor, the clock drive, that will move the telescope at the same rate as the Earth rotates, so keeping the object centred in the telescope without adjustment.

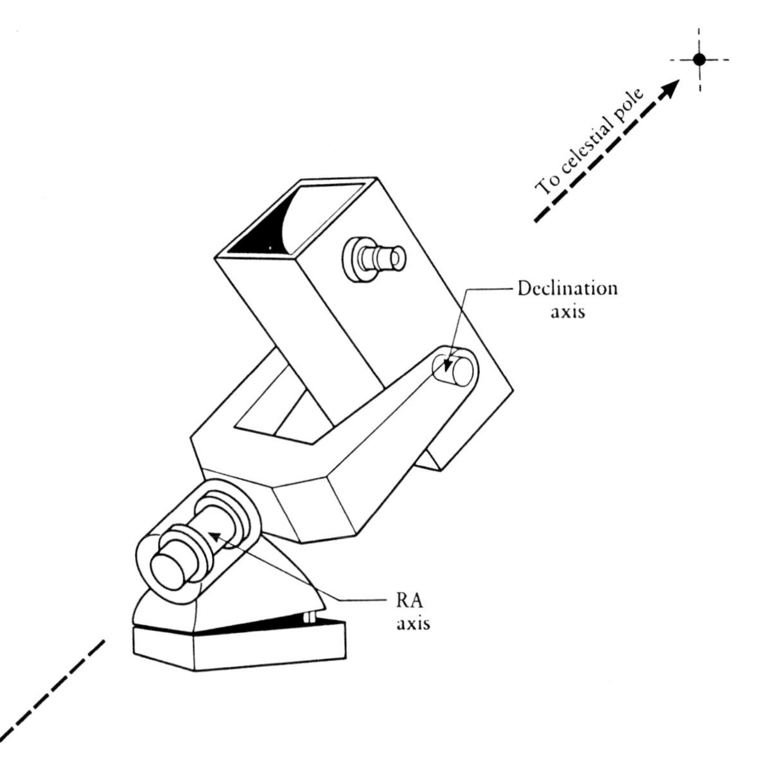

Figure 4.3. The polar aligned fork mounting.

The *declination* axis pivots at right angles to the right ascension axis. This motion is the vertical adjustment, and it follows the vertical lines drawn on your starcharts. It allows the telescope to move up or down in the sky and locate a star anywhere from the south to the north. On some telescopes there is a reversible slow-motion control for fine-tuning the telescope's position to compensate for any misalignment on the celestial pole.

The fork mount has all the same movements as the German mount, except that the telescope is mounted in a fork fixed on both sides at the declination axis (Figure 4.3). A fork mount is the best for ease of viewing overhead stars. The German mount runs the telescope into the pier near the zenith.

Setting up the polar alignment

Polaris, the north pole star, is currently about 9/10th degree from the North Celestial Pole (NCP). It is on this pole that the polar axis of a telescope in the northern hemisphere must be aligned to ensure accurate tracking with minimum declination error. A finely aligned telescope is a tremendous aid in astrophotography, and the following methods have been developed to make this adjustment easy.

1. Level the base of the telescope.
2. Adjust the telescope in declination so that its tube is parallel to the polar axis, that is, to a declination reading of 90 degrees on the setting circle.
3. Set the altitude of the telescope to the angle equal to the exact latitude at your observing site. You can find this in any good atlas at your local library.
4. Roughly align the finder telescope with the main telescope. It is not necessary to have it exactly aligned, as it is the finder scope that is to be used for alignment of the polar axis. A finder eyepiece that has cross hairs is an asset, as explained earlier.
5. Centre *Polaris* in the field of the finder by adjusting the altitude and azimuth only.
6. Unlock the right ascension axis and place the telescope tube on the west side of the mount. Observe *Polaris* through the finder as you swing the telescope back to the east side of the mount.

If *Polaris* appears to move from side to side, it is an indication that the finder is not parallel with the polar axis; you will need to adjust the finder half the distance the star moved away from the centre. Recentre *Polaris* the rest of the way, adjusting the horizontal or azimuth position. Repeat this until the side-to-side motion of *Polaris* is eliminated when the telescope is swung from east to west or rotated about the polar axis.

Polaris will, most probably, also move up or down depending on which side you swing the telescope. Move the altitude half the distance that *Polaris* is off centre, and adjust the other half by moving the declination axis to recentre *Polaris*. Repeat the procedure if necessary.

You should now be able to rotate the telescope from the east side to the west side without there being any motion of *Polaris* in the finder. In fact, *Polaris* will appear to pivot dead centre of the cross hairs during the swing from west to east or vice versa.

Polaris is not, however, the north celestial pole, the NCP being about 9/10 degree away. A line connecting *Thuban, Delta Draconis* and *Polaris* intersects the pole itself. Instead of *Thuban* you could use *50 Cassiopeia* if it is more available. Thus a sweep in declination will join *Polaris*, the NCP, *Delta Draconis*, and either *Thuban* or *50 Cassiopeia*.

Lock the RA axis, and relocate *Polaris* using the declination axis only. If you have fine-setting circles the rest is easy. With *Polaris* centred and the RA axis locked, move the declination axis 9/10 degree in the direction opposite to that of the celestial pole. Then lock the declination axis.

Now recentre *Polaris*, using only the latitude and azimuth axes of the telescope. This translates the polar axis to the spot in the sky occupied by the NCP.

If you do not have setting circles, use an eyepiece of a known field and shift *Polaris* slightly less than two full Moon widths across the telescope eyepiece in the direction opposite to that of the NCP. Recentre *Polaris* using only the azimuth and altitude of the mount. Of course, you can use the finder instead of the main telescope, but it is slightly more difficult.

The next time you have the telescope out and the Moon is up, see how much space the full Moon takes up in your finder scope. Its width equals half a degree on the sky.

Alternatively, use the fact that the NCP is located near the centre of a right-angle triangle made up by two 7th-magnitude stars and *Polaris* (Figure 4.4). The stars can easily be seen in binoculars or the finder scope. After completing steps 1 through 6 above to get your telescope parallel with the polar axis, just point it at the centre of the triangle using only your azimuth and altitude axes. This may be a little more convenient for you.

There is yet another method for achieving the accurate polar alignment that is so important for good astrophotographs.

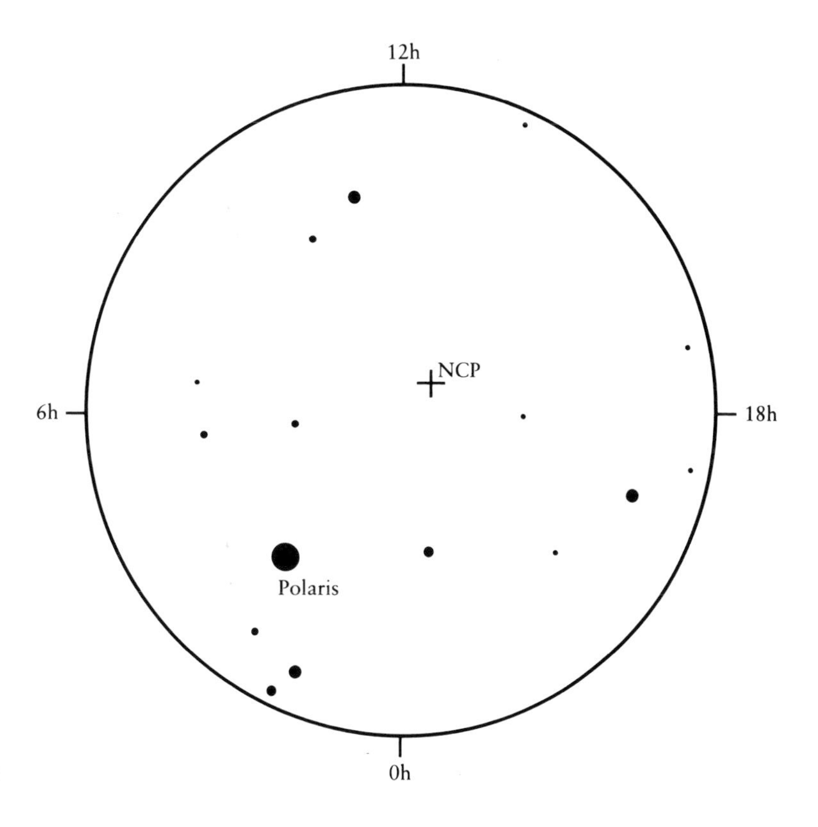

Figure 4.4. The north celestial pole in a 2 degree finder-field.

Plate 96. Polar-aligned telescope serves as guide for piggybacked camera (see following chapter).

1. Point the polar axis as close to the NCP as possible.
2. Centre a star that is near the celestial equator and also near the meridian with your most powerful eyepiece–Barlow combination. Turn on your clock drive. Now note in which direction the star drifts in your eyepiece. If the drift is northward, the polar axis is pointed west of the pole; if southward, it is too far to the east. Adjust the telescope using only the azimuth adjustment, and repeat the procedure until there is no drift at all.
3. Next, centre a star that is near the celestial equator and near the eastern horizon. If the star drifts to the north, the polar axis is above the pole, and if it drifts southward your polar axis is below the pole. Make the correction by adjusting only the altitude axis of the telescope.

This method is quite accurate, but it takes me much longer than the first two methods.

It is not necessary to have the main telescope tube parallel with the polar axis. I used to shim the telescope cradle of the main telescope and use its optics for aligning. Then I discovered that the finder scope works just as well and is much less difficult to handle. Once the alignment is completed, the finder can be realigned with the main telescope and the set-up is complete.

The first time you do this, it may take more than an hour. However, after a little practice it can be done in about ten minutes. Several manufacturers supply precision polar axis finders that employ cross hairs in a reticle offset by the amount that the NCP is offset from *Polaris*. I can set up a telescope on the NCP in three or four minutes using this special finder.

For general viewing, the equatorial mount needs only rough alignment with the NCP, but for astrophotography it is most desirable to have an accurate alignment.

4.4 Easiest guided photos: the 'piggyback' system

The piggyback system has definite advantages over the simple camera and tripod method for astrophotography.

You attach the camera to the body of the telescope with a 6 mm ($\frac{1}{4}$ in.) stove bolt, available at most hardware stores. Drill a slightly larger hole in the top or upper side of the dew cap. The standard tripod-mount threading takes the stove bolt. Adapter rings that do the same job can also be obtained.

Your telescope should be focused on a bright star in the same field as that viewed through the camera's finder, and moved along

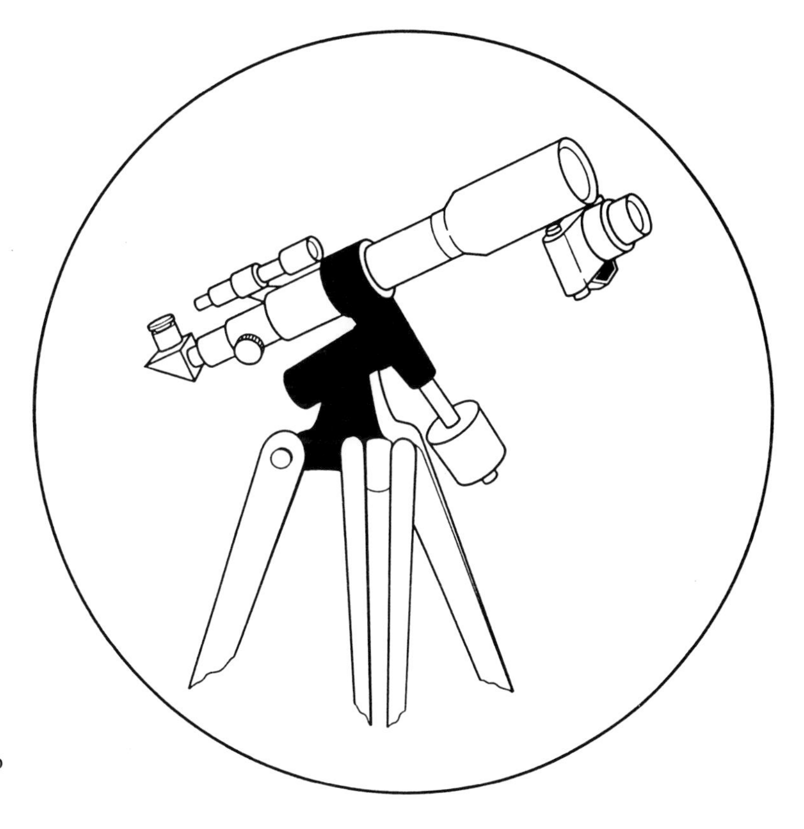

Figure 4.5. Mounting the camera piggyback. The simplest approach to guided photography.

Plate 97. Effects of lens focal length: views with 55 mm, 200 mm, 450 mm and 2000 mm telephoto lenses.

55 mm

200 mm

450 mm

2000 mm

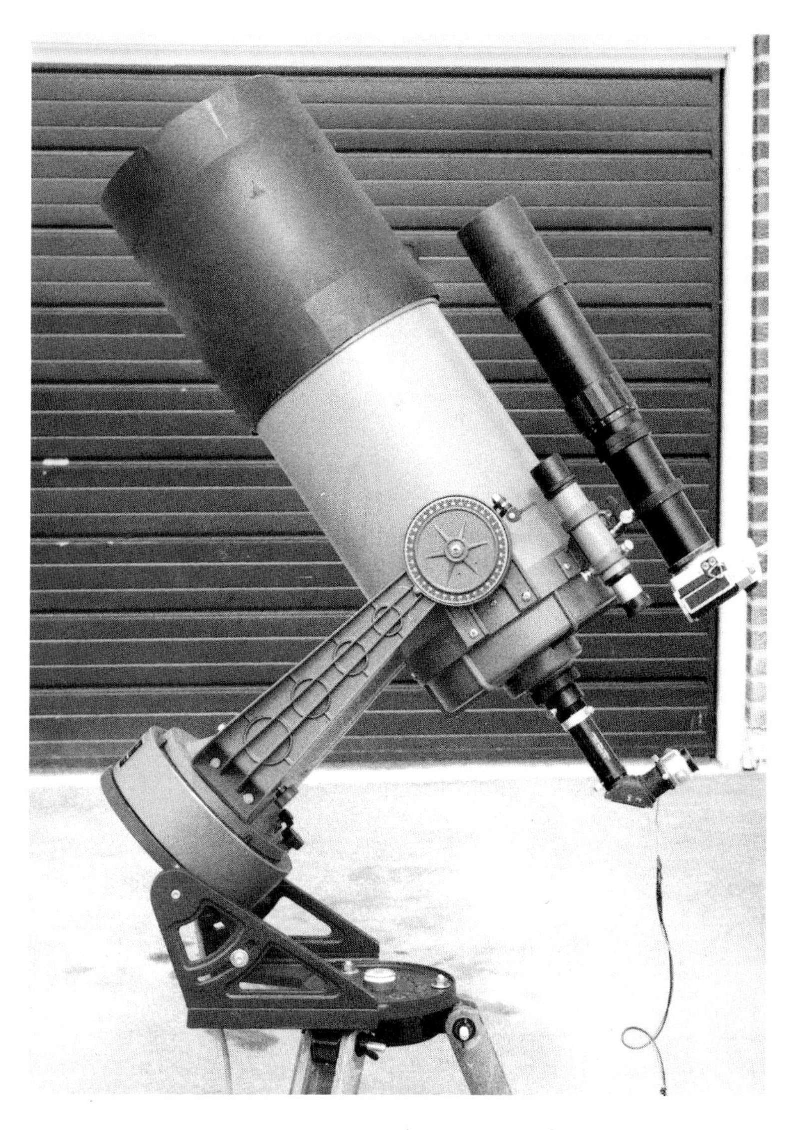

Plate 98. Piggybacked telephoto lens.

to compensate for the Earth's rotation. The altazimuth-mounted telescope restricts the camera to the shorter focal lengths, such as 135 mm or the regular 50 mm camera lens. Short focal length means little magnification, and thus less sensitivity to a bit of movement.

For your first attempt at guiding your camera, centre a bright star with your most powerful eyepiece, and rack the star out of focus to fill the eyepiece. This lets you see when the star drifts out of the field, and allows you to make an instant correction. The length of exposure will depend on the sky conditions, film type and, most of all, your patience and ability to keep the star centred. Using an eyepiece with cross hairs in the reticle makes the tracking job much easier. It is not difficult to place cross hairs in a 6 mm or 12 mm eyepiece. Most standard eyepieces (Kellner or orthoscopic types) have their image plane beyond the last lens element. This means cross hairs can easily be placed at this point. In fact, many eyepieces have a light-stop that screws down the barrel; this is a threaded ring tapered to a sharp edge. Here your cross hairs can be glued to the light-stop and screwed down the barrel until they are in perfect focus.

Plate 99. The Moon and the Pleiades: 5 s exposure on Tri-X film.

Another way to find the image plane is simply to move a pin or needle up the barrel toward the last lens element. But do be careful not to scratch the glass of the lens! Where the pin comes into sharp focus is where the cross hairs should be placed. However, it is possible to buy a variety of eyepieces with cross hairs already in place, some with a light source to illuminate the reticle, which is essential for ease of use.

A 12 mm eyepiece is the longest focal length that should be considered. With some practice you can produce very good astrophotographs by this method. Try a five minute colour shot at f4 of the Orion Nebula. You will be amazed and proud of the outcome.

Photography of the celestial objects mentioned in Chapter 4.2 can only be improved by using the piggyback system and compensating for the Earth's rotation. Our planetary daily motion is so fast that stars near the celestial equator are moving about 15 degrees an hour; that is, about 30 Moon-diameters every hour.

When adjusting the height of your telescope with the tripod, consider the object to be photographed in relation to where you are sitting or standing, so that guiding will not be awkward. A comfortable guiding position makes the job much easier and helps to avoid straining the neck or back – most of the time! A lawn chair or painter's stool is recommended for sitting during long exposures.

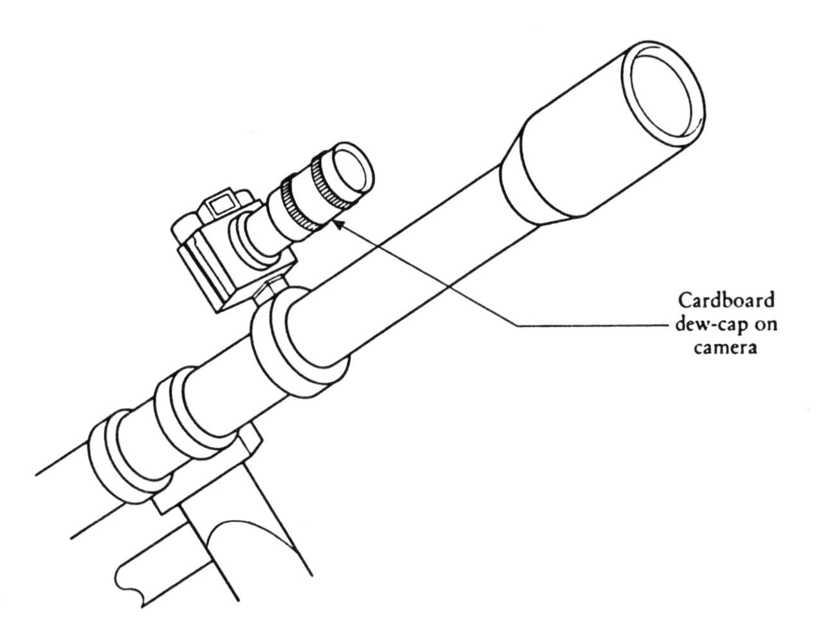

Figure 4.6. Simple cardboard dew cap prevents fogging of camera lens.

The weight of the attached camera will throw the telescope out of balance, but almost anything can be used as a counterweight to remedy the problem. For example, a book could be taped to the telescope tube. A large roll of masking tape is always handy to have with you for such uses. To prevent moisture from forming on the cold glass of the camera lens as it adjusts to the outdoor temperature, a dew cap can easily be fashioned from a piece of black cardboard about 10 cm long. Roll it tightly, so that it fits snugly around the camera lens, as shown in Figure 4.6. This cap will also prevent unwanted light from entering the camera, such as from a car driving up with headlights on.

If your observing site has an electrical outlet nearby, it is a good idea to use a portable hair dryer to take the moisture off the camera and telescope. And of course, don't forget to take along your warm clothes and some insect repellent.

4.5 Through the telescope: guided photography

With your equatorial mount carefully and accurately aligned on the celestial pole, guided astrophotography through the telescope becomes a possibility. As essential accessories you will want to have a 12 mm eyepiece with illuminated cross hairs and a Barlow lens*.

Your first experiments with guided photography may best be attempted, not through the telescope itself, but through the piggy-backed camera, equipped with a 135 mm to 600 mm telephoto lens.

The one motion of the equatorial mount can be guided by the main telescope using a suitable eyepiece and guide star. The guiding eyepiece should be rotated so that the guide star drifts along one of the cross hairs. This direction will be the right ascension axis of the telescope. The other cross hair becomes your declination axis. These are now your tools for guiding the telescope. The cross-hair illumination can be adjusted to suit the brightness of your chosen guide star.

If the telescope has an electric clock drive, it is a good idea to add a variable frequency control. This piece of electronic wizardry permits you to vary the speed of the drive motor, thus compensating for local power discrepancies, and any drive-gear abnormalities or misalignment of the NCP.

A suitable guide star should be located, close to the central object in your camera's field of view. The camera usually has very little latitude for adjustment. In other words, if your guide star is out of the field of your camera, you may not be able to redirect the camera to the same area of sky that you wish to photograph. With the guide star located and centred, direct the camera to the desired area of the sky. Recentre your guide star, and allow your drive to catch up.

Quite often as much as half a minute is required for your motor to take hold and start moving the telescope; this is generally due to sloppy gears in the drive train. Look to make sure that you have full control of your drive before starting the exposure. Loading up

* The conventional wisdom calls for a guiding magnification of about five times the telescope's focal length in inches. The 'magic' combination of 12 mm eyepiece and standard Barlow lens yields approximately this magnification when used with a telescope *of any focal length*.

Plate 100. 300 mm telephoto shot of NGC 1499, The California Nebula.

the drive train a little in the direction opposite to the sky's movement helps make the drive work. This means that you should have slightly more weight on the eastern side of your RA axis, so that it takes out and compensates for slip or variation in the drive train.

If the equatorial mount is not aligned accurately on the pole, and your guide star is too far away from your subject, the stars appear to revolve around your guide star.

For astrophotography through the telescope the process for guiding the primary instrument is very similar to that for guiding the piggyback system. The only change is that your main telescope is now the telephoto lens. Of course, now you will need a second telescope to guide the main instrument.

Figure 4.7. T-adaptor with threaded connecting ring links camera to focusing mount.

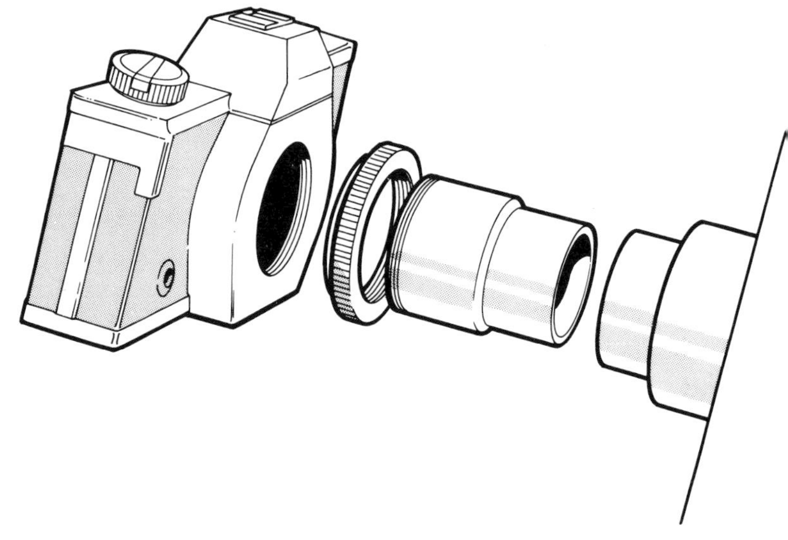

The camera must now be attached to the telescope. A standard T-adapter will do the job (Figure 4.7). The T-adapter is a short tube, 31.75 mm (1.25 in.) in outer diameter, that attaches to the camera where the camera lens goes. An adapter ring for each camera, whether threaded or bayonet mount, is available to couple the T-mount to the camera. For this type of photography, the 50 mm (2 in.) focuser is preferred. A 50 mm opening is sufficient to illuminate the full area of the film in a 35 mm camera. This system is, however, only good for prime-focus photography or negative projection, which are discussed later.

The guide telescope should have as large an aperture as possible; a 60 mm (2.5 in.) refractor with 1000 mm focal length might be a minimum. Remember that the guiding magnification should be twice the focal length in centimetres (or five times the focal length in inches) of the main telescope or telephoto lens used for photo-

Plate 101. The Andromeda Galaxy: 20-min. exposure with a 30 cm telescope, on Tri-X film.

graphy. For example, let us assume a 150 mm (6 in.) f8 reflecting telescope with a focal length of 1200 mm (48 in.). Therefore 120 × 2 (48 × 5) = 240, say 250-power, is required for good guiding. So our 1000 mm guide scope with a 3 × Barlow lens and 12 mm eyepiece equals 1000/12 or 88.33 × 3, or 249.99-power. Therefore, 250-power would be ideal. This is why a 12 mm illuminated reticle eyepiece is most desirable to reach the higher powers required.

My deep-sky photographs taken with the 450 mm lens were all guided with a 75 mm Tasco, 1200 mm focal length, and a 12 mm eyepiece with a 3 × Barlow. This is about equal to 300-power, which I used for guiding.

The guide telescope should be firmly attached to the telescope tube using large rings. If the main telescope tube is fibreglass or cardboard sona-type tube, large support plates should be used to prevent flexure between the guide scope and the telescope. The large three-screw adjustment rings should permit the guide telescope to move about four degrees maximum on the sky. This will ensure that at least a minimum of a fifth magnitude guide star is available for guiding.

Cameras are discussed in greater detail in Chapter 4.6. A single-lens-reflex (SLR) camera, however, is best suited for this job. The SLR camera with the lens removed is used as a film holder for the main telescope. Some cameras have a self-timer triggered by a cable release, allowing the camera's mirror to fall into place ten seconds prior to the shutter's opening. This way, most vibration will stop before the film is exposed. This also allows the observer time to move from the camera at one end of the reflecting telescope to the guide scope at the other. Thus you can have the guide star properly recentred before the camera shutter is opened. The best system for working between the guide scope and the camera for photographing deep-sky objects follows.

Deep-sky objects: targets beyond the solar system

Nebulae and galaxies are generally too dim to be focused on the ground glass of the camera. This means that your telescope-camera must be centred and prefocused on the object prior to the exposure. If there are two objects in the field, such as M84 and M85, centre between the two so that both will be on the picture. Use your regular low-power eyepiece to locate the object. Then point the guide scope to a medium-bright star and check back to make certain that the object is still centred in the eyepiece of the main telescope. Lock your guide scope and check again. Take the eyepiece out and insert your camera on the main telescope. Rebalance the main telescope if necessary. Your camera will not necessarily be focused, and in fact probably is not. Check your guide star and memorize its position.

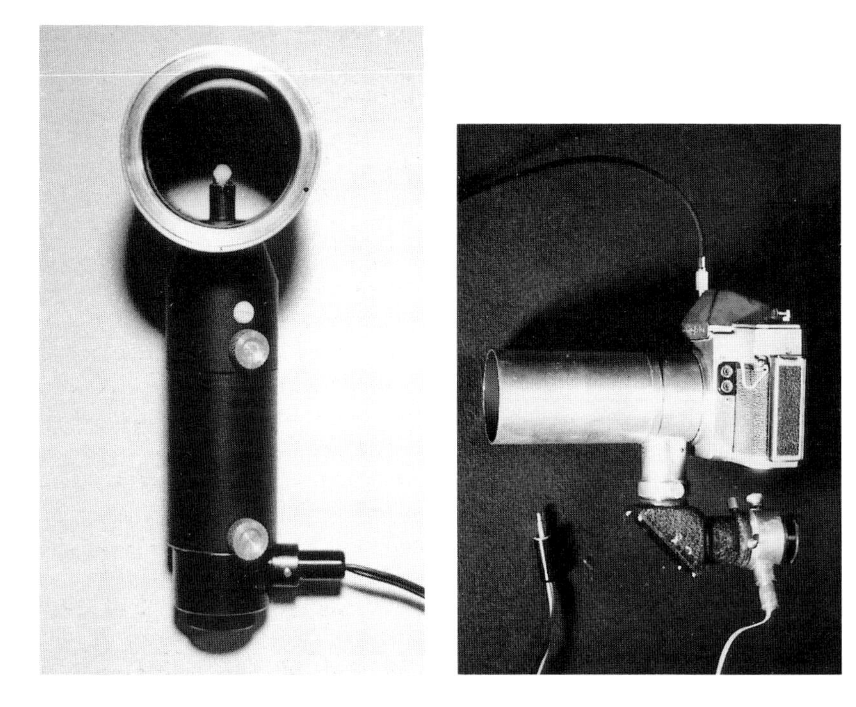

Plate 102. A guiding head.

Plate 103. Off-axis guider.

Now move the telescope to a bright star at approximately the same altitude as the object, and carefully focus the camera on the bright star. Having memorized the guide star's position earlier, relocate and centre the guide star in the guide scope. The deep-sky object should now be centred on the film and in perfect focus. Adjust the brightness of your cross hairs so that the guide star and the cross hair can be seen clearly, although you should use only the minimum amount of light necessary on the cross hair for comfortable viewing. Rotate your guiding eyepiece so that the star drifts in right ascension along the cross hair. This can be accomplished by rocking the telescope gently in RA. Adjust the speed of the clock drive and recentre the guide star. The system is now set for photography.

Because the magnification in the guide telescope is so great, the field of view is very small. This often makes it difficult to locate a good guide star. First, use a 30 mm to 40 mm eyepiece (without the Barlow) or other suitable low power, and a wide field of view, to locate your star. Then step up the power for guiding just prior to starting the photograph.

The guide telescope system for photography is probably the most commonly used, but it has its drawbacks. The guide scope must not change position or flex during the exposure. The atmosphere plays tricks with large aperture differences between the main and guide telescopes.

Most guide telescopes are restricted to the bright stars because of their small apertures. The off-axis system eliminates many of these problems. This system uses a mechanical device to locate a guide star near the edge of the field, permitting use of the main telescope as both the camera and guide scope. The device is basically a tube that has a small prism mounted near the edge of the field and that directs a small portion of the light to an eyepiece alongside the camera. Celestron and Lumicon sell units designed for Celestron's 200 mm (C8) and 355 mm (C14) telescopes.

The one I use has a 3× Barlow lens built into the guiding system. This off-axis guiding head has a 12 mm eyepiece–Barlow combination mounted on the telescope focusing platform, with a series of adjustable prisms. The prisms direct light from the telescope mirror through the eyepiece of the guiding system. By rotating the focuser and adjusting the prisms, I can choose a guide star almost anywhere within the field.

The deep-sky object I wish to photograph is located and the guiding head is rotated until a guide star near the edge of the field is found. My prefocused camera is inserted, and the horizontal cross hair is aligned on the RA axis. The exposure is then made.

The main disadvantage of this method is that most commercially built units do not have the Barlow lens built in, and therefore do not give sufficient magnification for good guiding. Also, the guide stars are seldom available near the edge of the field. This results in the object you have photographed often being badly off-centre in your picture.

When I started using my new 500 mm f5 reflector, I found its finer figure and resolution showed a diffraction disk surrounding the brighter stars that had not been apparent with a smaller telescope. Several causes were suggested, by my experiments indicated that the disk was caused by the optical window within the cold camera; an optically flat window with antireflective coating was required.

I was also experimenting with Kodak VR 1000 film at the time, and this proved a very fast film in the cold camera but so sensitive to light pollution that it could be exposed for a maximum of only eight minutes at my location with the large telescope. I tried my old light-pollution filter and it gave a dramatic improvement on a picture of the Veil nebula. I contacted Lumicon and they recommended their Deep-Sky filter; it had an antireflective coating to reduce internal reflections and was expected to eliminate the disk around bright stars.

Still experimenting, I replaced the glass window of the cold camera with this Deep-Sky filter, but was concerned about whether it would affect the colour balance of the VR 1000 film or limit light transmission from distant nebulae. Fortunately, the high sensitivity of the film to the blue, coupled with the new filter's design, made an almost perfect combination. Stunning results are given by exposures of 30 min., with no ghosting of star images. The magnitudes recorded on VR 1000 are now comparable with those reached with Ektachrome 400 in shots of 20 min., but the grain of the VR 1000 is slightly coarser. With these promising early results, my experiments continue.

4.6 The camera

The single-lens-reflex camera is the most suitable for photography through the telescope. The major advantage of the SLR is that you can focus the camera through the telescope lens, making it very easy to focus on stars or planets.

Most SLR cameras now have behind-the-lens light metering systems for average or spot readings. This feature enables you to focus on the Moon and then take a light-meter reading to set the correct camera speed. The more expensive SLRs have mirror lock-up and interchangeable ground glass for focusing, some designed especially for astrophotography.

Camera weight can be a factor, depending on the size and mass of your telescope. Nowadays there are some very light SLR 35 mm cameras, the Olympus being one.

The Polaroid camera also lends itself to astrophotography. The SX-70 model can be adapted for instant pictures. Fast films, such as AS 3000 are ideally suited for this work. The only disadvantage of the Polaroid is that you have no negative and the film must be at room temperature for processing. A metal plate is supplied with some models to utilize body heat for this purpose.

The Schmidt model of camera is the best of the fast telephoto lenses. Celestron sells two models: 140 mm (5.5 in.) f6.5, and 200 mm (8 in.) f5. This fast-lens system can capture the North America Nebula on an SA 200 film in fifteen minutes, and the nebulosity in the Pleiades after only two minutes.

This system has to be loaded one frame at a time. The small convex film holder is inserted at the focus for each photograph. By placing the film inside the lens, the normal diffraction caused by the secondary mirror's spider support is eliminated. A notable feature of the Schmidt is its very flat field with good round star images right to the edge of the photograph.

In very cold weather, extra care should be taken with film and camera equipment. A camera should be checked out for cold operation by placing it in a plastic bag and putting it in the home freezer for four hours. Make sure the shutter operates both open and closed under freezer conditions. Nothing is worse than taking several photos in very cold weather and, on returning indoors where the camera warms up, hearing a loud click as the shutter finally closes to end the exposure.

Film gets very brittle in the cold, and can scratch or even break inside the camera. Static electricity can expose the film by its flashes inside the film cassette. Both of these problems can be greatly reduced by advancing the film very slowly.

Cameras and telescopes should be placed in plastic bags before being brought into the warm house. This allows the moisture to condense on the outside of the cold plastic instead of on the equipment.

Larger-format cameras are sometimes used for wide-field shots of the stars and Milky Way. Aerial photography cameras used for mapping are also excellent for this purpose. The camera must be mounted on an equatorial mount and guided. Because of the large film holders and film format, darkroom facilities are almost essential.

Footnote on choice of camera

When selecting a camera for astrophotography, consider whether the shutter is electrically held open on the 'B' setting, or whether it is mechanical. The electronic shutter can drain an expensive battery in one night's work.

If you do have such a camera, it will pay to attach a larger battery of the six-volt size, on cameras that use that voltage. Long fine wires from the large battery can be held in place by the camera's own battery. This system will last for months of astrophotography.

The cold camera

Most film emulsions suffer reciprocity failure when exposed for more than a few seconds, resulting in less increase in image density than would be expected for the increased exposure. The emulsions simply cannot cope with gathering the light over several minutes; the process deteriorates in effectiveness beyond a certain point. But chilling the film with dry ice at −78°C enhances the storage capability of the emulsion crystals, almost eliminating reciprocity failure. Colour film is especially susceptible to this problem, so a cold camera is particularly advisable for colour astrophotography.

A cold camera, with dry ice applied to the back of the film, keeps the three emulsion layers of colour film in balance during the exposure, resulting in truer colour and better definition of stars and nebulosity in your picture. I have used two types of cold camera, which I designed with George Ball.

The first was well documented in *Astronomy* magazine in February 1981 (Figure 4.8), and was built to handle 35 mm film in a standard cassette. An optical plug was cut from 50 mm plastic or

Plate 104. Cold camera and telephoto.

plexiglass sheet, and sandwiched the film against the copper end-plate of the cold-chamber. The camera body was fabricated from a black 5 mm plastic sheet and contained the film and the cold-chamber, a space to hold the dry ice. It fitted over a shutter assembly so that the camera could be removed without altering the focus.

The cold-chamber was constructed of ABS pipe with a threaded cap. Through the cap was a spring-loaded plunger to keep the dry ice against the copper plate located at the end of the chamber. This plate is to help remove the heat more evenly from across the film. A shutter release was made from ten-speed bicycle brake cable. The whole camera was constructed for less that \$20.

The second cold camera is a dry-nitrogen variety. It consists of a shutter assembly containing a heated window and dry-gas chamber. The chamber is open at one end, and pins the film against the cold-chamber. The camera body, made from black plastic, is open at the front with a removable cap, and is also designed to accommodate 35 mm film cassettes. On the back of the camera is the cold-chamber. It utilizes a screw cap with a spring-loaded plunger that feeds the

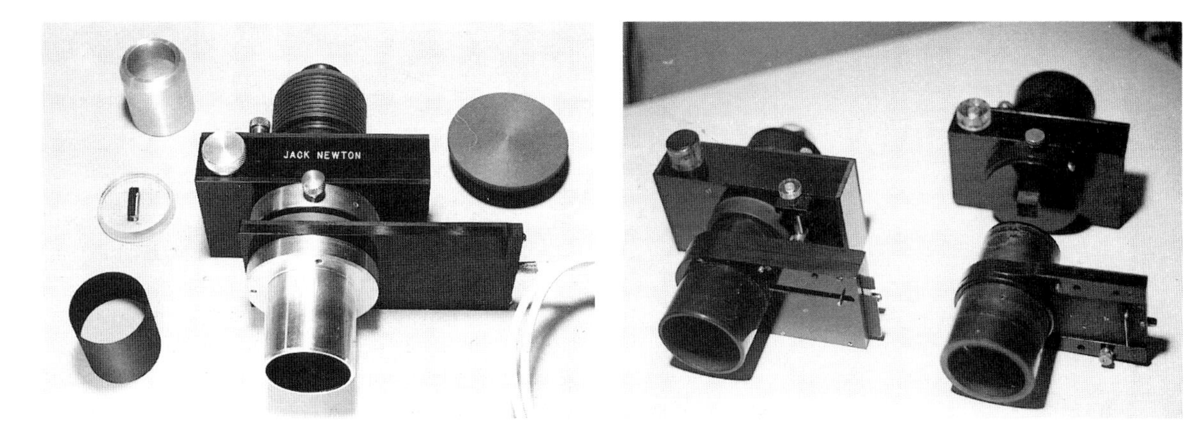

Plate 105. Cold cameras.

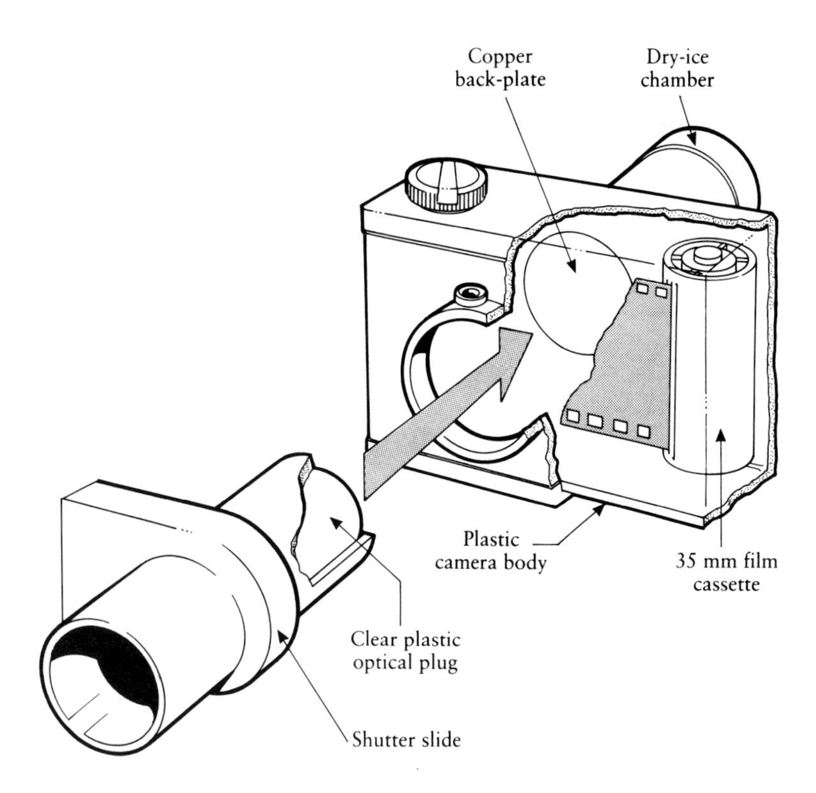

Figure 4.8. Optical plug cold camera.

dry nitrogen to a copper plate in the end of the chamber. The dry nitrogen keeps frost from forming on the film during the long exposure.

The camera from the original optical-plug model has gone through constant evolution to what it is today. Later on, we will cover the question of film for cold cameras.

4.7 Moon and planets

When photographing either the Moon or the planets, higher magnification is usually desirable. This means that a negative lens, Barlow lens, or eyepiece protection is preferable to the small image scale of prime focus. The most simple adapter for eyepiece projection is a tube that fits into the telescope and holds both the camera and the eyepiece. Extension rings are then placed between the eyepiece and the camera, giving a great deal of latitude in the amount of eyepiece projection that you can achieve.

Plate 106. The Moon is a fine astrophotographic target.

It is possible to buy chrome drain extensions that join the sink to the trap in the bathroom that measure 31.75 mm (1.25 in.) outside diameter at one end and 31.75 mm (1.25 in.) inside diameter (for the eyepiece) at the other. An eyepiece can be fitted in the top end, and the other end attached to the telescope. The standard adapter ring, or one section from a set of extension rings, can be used with the drain extension, and a little machining, to make a camera adapter for eyepiece projection. A plug must be machined to fit inside the adapter ring and hold the drain extension. A set of extension rings allows adjustment for the desired amount of projection.

Not all eyepieces are suitable for projection, and some give a much-flatter field than others. A Kellner or orthoscopic eyepiece seems to work best. Test several short-focal-length eyepieces to determine which one best suits your needs.

The negative lens system can be used with a straight camera adapter. The Barlow or negative lens is inserted in the telescope, and the camera and adapter merely inserted in the Barlow. This

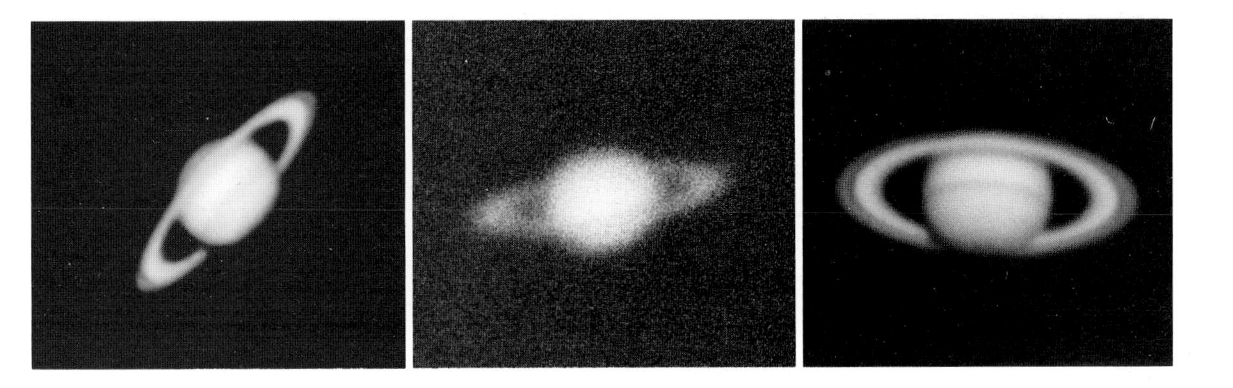

Plate 107. Amateur photographs of Saturn. Image at far right by Donald C. Parker.

provides a rather fixed magnification; the only adjustment is the travel of the Barlow lens in the tube itself. The negative lens system projects a flat field and is particularly well-suited to lunar photography.

A clock drive is mandatory for good lunar and planetary photographs. The Moon can be successfully photographed without a clock drive, but only at very low magnification with very fast film, such as hypered (hydrogen gas sensitized) TP 2415, Tri-X pushed to 1200 ISO or VR 1000 colour. 'Pushed' means that forced pro-

Plate 108. Jupiter: good planetary photography requires large focal length and careful focusing.

Plate 109. Venus: $\frac{1}{8}$ s through a 30-cm telescope.

cessing is given by a photo lab to achieve a greater effective film speed.

The clock drive should be allowed to run for a few minutes before the exposure is started, to allow time for the slack to be taken up by the drive motor. The camera must be fitted with a cable release, as without it the vibration from your touch would be disastrous. I suggest that you hold a hat in front of the end of the telescope to block off the light, but be careful not to let the hat actually touch the telescope. The cable release should be depressed and locked to keep the camera shutter open. When the vibration stops, after letting go of the cable release, carefully flip the hat from in front of the telescope and expose the film. Take care not to bump the telescope. Then replace the hat in front of the telescope and close the shutter with the cable release.

Some cameras have a mirror lock-up that eliminates the vibration caused by the impact of the mirror in the SLR when it hits the top of the camera. The Ricoh and Canon cameras have a self-timer that releases the mirror prior to exposing the film. The air release is best for high-power photography because it virtually eliminates vibration caused by the hand on the camera. This is essential if the tip-the-hat method is not being used.

With eyepiece projection of an object such as the Moon, the SLR camera with behind-the-lens metering will give a fairly accurate reading. However, the camera may have been set for an average reading rather than a spot reading, and so may overexpose the Moon. Always bracket the exposure, by taking a shot one or two speeds slower and another at one or two speeds faster than the meter reading suggests.

Planetary photographs vary greatly depending on the magnification, telescope aperture, etc. With the larger-aperture telescopes, you can afford to use the slower finer-grain films and still expose them at a relatively fast shutter speed.

The moons of Jupiter and Saturn require much longer exposures than the planets themselves, as you might imagine. You will need about 12 s with a 75 mm refractor, 12 mm eyepiece, and 50 mm projection, using a film such as high-speed Ektachrome.

The Kodak SO-410 and SO-115 emulsions are especially well suited for planetary photography with fair speed and a fine-grain structure.

Venus can be successfully photographed in broad daylight. The best time, however, seems to be just after sunset or just before sunrise. Venus will then be at its highest point in the sky with good contrast. This is the brightest planet, and for that reason a slower fine-grain film can be used; Kodachrome 50 or Plus-X will give good results. Venus also has the largest change in size of all the planets, ranging from 9.6 arc-seconds to 64. For comparison, the full Moon is 1800 arc-seconds.

You will find Mercury a very difficult planet to photograph, because its low altitude makes atmospheric distortion almost

Table 4.2. For your guidance, here are the maximum and minimum apparent sizes of the planets. These sizes will serve as an indication of the relative difficulty that will be encountered in the attempt to project and photograph the various bodies

Sun	31.4'	32.5'
Moon	29.5'	32.9'
Mercury	4.6"	12.2"
Venus	9.6"	65.6"
Mars	3.5"	23.4"
Jupiter	30.4"	49.6"
Saturn (rings)	35.1"	48.4"
(planet)	15.1"	20.8"
Uranus	3.1"	3.8"
Neptune	2.2"	2.4"
Pluto	0.2"	0.3"

'indicates arc-minutes, "indicates arc-seconds.

impossible to deal with. Its small size, from 5 to 13 arc-seconds makes this planet a real challenge. With phases like Venus and the Moon, it will appear slightly different every day or two. Several photographs taken over a one-week period will record these changes in phase.

4.8 Photographing the Sun

The Sun is one of the easiest objects to photograph as it is the greatest source of our natural light. **But** this great source of light can be **extremely dangerous.**

A filter **must** be used. The safest type is a solar filter, designed to cover the whole objective lens or mirror, or most of it (see Figure 4.9). Some solar filters are mounted in a cap that covers the end of the telescope. These solar filters can be purchased in the USA from advertisers in *Astronomy, Sky & Telescope* and other popular astronomical journals.

The safest way to observe and photograph the Sun is by eyepiece projection. The Sun's image is simply projected onto a white card. Many telescopes come with a solar projection screen. The Sun's image size can be changed by increasing the distance of the card or screen from the eyepiece to give a larger image, or by using an eyepiece of shorter focal length. For instance, using a 6 mm eyepiece instead of one of 12 mm would give a larger projected image. A light-meter reading can be taken directly from the Sun's image on the card. A very slow fine-grain film should be used, such as high-contrast copy film. Again, you should bracket each exposure and record all particulars.

The Sun photographed through a full-aperture filter presents a new difficulty. Most full-aperture filters are designed for direct viewing. This means that the amount of light permitted to pass through the telescope is sufficient for the eye, but is much too dim for photography. If this type of filter is used for photography, VR 1000 film will do nicely.

Some manufacturers sell a controllable filter that appears to have solved the light-limitation problem. You just dial the amount of light required for each magnification and expose accordingly.

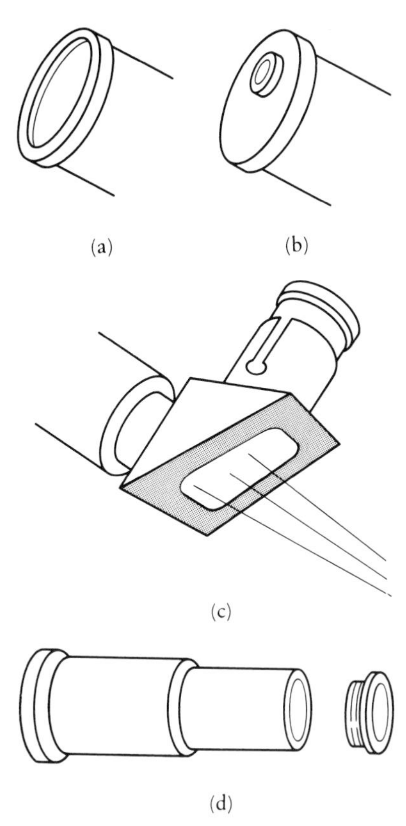

Figure 4.9. Solar filters: *(a)* full aperture reflective filter; *(b)* partial aperture reflective filter (less expensive); *(c)* the Herschel wedge 'spills' energy through open backed, non-aluminized prism (use in conjunction with solar eyepiece-filter); *(d)* a bad idea! Eyepiece filters for solar observation, used directly in the telescope can be *dangerous*.

Plate 110. Solar equipment: the Herschel wedge.

The SLR camera with a behind-the-lens metering system is ideal for solar photography. Regardless of the filter combination, the light-meter in the camera will indicate the approximate shutter speed for the best exposure. Always take a few exposures on either side of the recommended one.

The filter–film combination should allow sufficient light for use of a shutter speed not less than $\frac{1}{125}$ s. This is fast enough to prevent blurring due to slight wind vibration.

Welding-mask glass makes an excellent sun filter. Number 5 density glass is normal for welding, but is not dark enough for viewing the Sun. I would recommend not less than No. 14 for naked-eye viewing. Two pieces of No. 7 glass are equal to No. 14 if you place one tight against the other. Welding glass can be tilted to slightly vary the density while visually viewing the Sun. Filters can be cut and shaped with a pair of scissors; the trick is to keep the scissors, filter and hands completely under water and just chip away a small piece at a time until the desired shape is made. This method is quite safe, although you may prefer to use a glass-cutter to etch the circle and then, under water, break away the excess glass with a pair of pliers.

Each magnification used on the Sun requires a different filter density for the exposure time to remain above $\frac{1}{125}$ s. A filter sandwich is constructed by stacking together different densities of filter glass, and merely shuffling them to get the correct density combination for each magnification. The sandwich should be placed beneath the eyepiece on a refractor of less than 75 mm aperture. A larger instrument must be stopped down to this limit, or the filter sandwich will instantly crack in the intense heat.

Sun filters that are supplied with small refractor telescopes are usually designed to fit over the top or underneath the eyepiece (Figure 4.10*d*). **This filter system is very unsafe**. The concentrated sunlight can easily crack the filter, and the resulting burst of light can severely burn the eye. This type of sun filter is only designed for refractors of maximum 60 mm aperture. Larger aperture telescopes will crack them instantly.

There is always the danger of a filter cracking: therefore, you must **never use this method for observing. Instant blindness can result**. If the filter should crack, the heat may warp or melt the shutter inside the camera.

Another popular system for reducing the Sun's light is the Herschel wedge. It is mounted like a star diagonal with the back left open (see Figure 4.9*c*). This reflects only a small portion of the sunlight to the eyepiece, with the rest of it passing through the back of the wedge, where a hole in the holder permits the light to escape. The Herschel wedge will greatly reduce the chance of breaking filters, but it still is not infallible. On one occasion, I was observing the Sun for only a few seconds using a Herschel wedge, when my tie burst into flames.

Welding glass should be used also to cap the finder telescope. Visitors have a tendency to occupy themselves by looking through the finder when you are busily engaged with the main telescope. Again, this could mean instant sight impairment for the unsuspecting observer.

4.9 Solar eclipse photography

Telephoto lenses are ideal for photographing an eclipse of the Sun. Welding glass can be taped to a home-made cardboard cap, which is placed over the objective of the telephoto lens. A series of test photographs should be taken a few weeks before the event to confirm the exact exposure and clarity of the pictures.

The camera and lens can be mounted on a tripod and should be equipped with a cable release. A camera platform can be made for an equatorial mount. Two or more cameras can be mounted on the platform. A battery and DC converter are used to run the clock drive from remote areas. Using a clock drive permits longer time exposures for the outer corona, stars and planets. Photographs are recorded with a wide-angle or regular camera lens at f1.4 and exposed for one to eight seconds. The partial eclipse is photographed by capping the 450 mm telephoto lens with a No. 10 welding filter; the lens is set at infinity at f32. You should set the camera for ASA 200 with a shutter speed of $\frac{1}{500}$ s.

The totally eclipsed Sun is photographed through the telephoto lens with the sun filter removed and the lens opened up all the way, at f8 in this case.

The 'diamond ring' is the last burst of light before and the first burst after totality, and should be photographed at $\frac{1}{125}$ s. 'Baily's beads' are the last bit of sunlight flickering through the lunar mountains and valleys. This beautiful effect should be photographed at $\frac{1}{60}$ s at f8 or $\frac{1}{125}$ s at f4 using a film of 200 ISO. Remember that both the diamond ring and Baily's beads only last for a few seconds.

The solar prominences are jets of gas very close to the limb of the Sun. An exposure of $\frac{1}{250}$–$\frac{1}{125}$ s is best. Longer exposures show more of the corona, but will overexpose the prominences.

The corona should show a range of exposure times, starting at $\frac{1}{125}$ s and moving through each shutter speed on the camera, continuing up to a maximum of 4 s. Each photograph will show more of the corona than the shot before, and all of them will be most interesting. The giant outer coronal streamers will show up on exposures of 0.25 s.

Plate 111. Solar eclipse shadow bands: $\frac{1}{125}$ s on Tri-X film with 50 mm f2.8 lens.

One telescope can be used to photograph the prominences and corona. A 200 mm Celestron or Dynascope at prime focus f10, using high-speed Ektachrome ISO 160, can record the prominences at $\frac{1}{60}$ s and the corona at about 6 s.

Shadow bands are waves of very dim shadows a few centimetres apart that move over the ground just before and just after totality. They appear to move at 10–11 km/h in the same direction as the Moon's shadow. My only successful photograph of shadow bands was on the wing of a DC-3 aircraft just after totality at Baker Lake, Northwest Territories. The photograph was taken through the aircraft window above the wing. The sideways drift of the plane slowed the strong shadow bands and permitted them to be photographed. A device on rails or wire that would allow a camera

Plate 112. Solar eclipse projected onto sidewalk, through trees.

to travel with the shadow bands could, in effect, stop their motion and allow for a longer exposure. A simple alternative is to pan the ground, as one might pan to follow a race car.

A series of multiple exposures can be taken to show the whole eclipse from start to finish on only one picture. Using a wide-angle or regular 50 mm camera lens and tripod, align the Sun so that the first exposure will be at one edge of the field. The Sun's predicted path through the sky should be centred in the camera. A No. 12 welding glass at f15 for $\frac{1}{50}$ s on Ektachrome ISO 400 film should be used. A cap should be fashioned to cover the camera and the welding glass. With the cap in place, lock open the shutter with the cable release. Just after first contact, carefully remove the cap and expose the film for 0.25 s. Every ten minutes repeat the exposure until totality. At that point, remove the cap and the filter and expose the Sun for one second at f1.8. Replace the filter and cap after totality, then repeat the 0.25 s exposures every ten minutes. With last contact make sure that you have the same number of exposures on each side of totality. The complete eclipse is now recorded on one photograph.

Trees will act as pinhole cameras and project hundreds of tiny eclipses on the ground.

The oncoming shadow of the eclipse can also be photographed with interesting results. A tape recorder will enhance the memory by recording the reactions of people and of any wildlife in the area.

Make sure that you are not set up near a street light, for it may be automatically activated during totality and ruin your observing. And keep a flashlight handy to read your camera when it is dark. A variety of telephoto lenses, ranging from the standard 135 mm to the giant 600 mm, are useful for recording the event.

Tele-extenders that amplify the power of the lens can also be incorporated. When using a tele-extender, the f-ratio of the lens changes drastically and should be compensated for. The safest method for checking the filter system beforehand is to photograph the Sun at the same elevation as it will be during the eclipse with a range of exposures and focuses. You can then adjust pictures of the corona, etc, by these factors; this, as mentioned earlier, should guarantee good results on the day of the eclipse.

The best exposure guide for the stars and planets near the totally eclipsed Sun is the same exposure used to record stars in the vicinity of a full Moon. The light of the full Moon is as close to duplicating the total-eclipse conditions as we can get in the night sky.

Welding glass (No. 14) can be used to cap each side of a pair of binoculars for viewing the partial eclipse. The No. 14 welding glass is handy for viewing the Sun with the naked eye; several layers of black and white film, fully exposed, is also satisfactory. *Do not use colour film for this purpose.* Remember your eyes are your most precious possessions, so whatever method you use be extremely careful.

Plate 113. Eclipse of the Sun.

4.10 Photographic films and filters

Most amateur astronomers start off with a refracting telescope that has an altazimuth mount and a focal ratio of about f15. The usual first targets are the Sun, with appropriate filter, the Moon, and the brighter planets. A telescope without a clock drive must rely on film and shutter-speed combinations to overcome the effect of the Earth's rotation.

Films will vary depending on the type of telescope, the focal ratio and the object being photographed. The ISO or DIN rating of a film is the standard way of indicating a film's sensitivity to light. A slow film would have a low number and a fast film a higher one. This rating provides a basic standard to set the camera by. When we talk about pushing a film, we mean that we will be developing it longer or in a different way, so as to increase the normal rating. Tri-X developed in Accufine will push the ISO from 400 to 1200. Ektachrome 400 can be pushed to ISO 800 or 1000 by your local photographic laboratory. The focal ratio or 'f' number of your telescope essentially indicates how bright an object will appear. The lower the number, the brighter the object.

Slower film with finer grain could be used, but at the cost of taking longer exposures. Good films to try are Kodak Plus-X, ISO 125, and Ektachrome 400 colour. A better investment would be a 150 mm (6 in) homebuilt reflector at f8, or a 20 cm catadioptric telescope. This now opens the door not only to short exposures for lunar and planetary photographs, but also to the infinite expanse of deep-sky photography. The 2000 mm Celestron 8 is one of the most popular commercially-built telescopes sold in North America. This compact telescope has a focal length of 2000 mm at f10 and has accessories such as camera adapters and off-axis guiding for long time exposures. This is an excellent tool in the hands of the amateur astronomer. Tri-X or 103aF spectroscopic film and a 30 min. exposure will show surprising detail in distant galaxies or emission nebulae. The Moon or Venus can be successfully photographed on Kodak fine-grain positive black and white film. This is very inexpensive film, each cassette working out very cheaply. You can buy the empty cassettes from any mail-order house or perhaps locally – wherever you purchase your bulk film. Fine-grain positive is the highest contrast film that I have used. It is monochromatic

Plate 114. Gas-hypered exposure of NGC 7000, The North America Nebula.

and can be used with a standard safelight; development in Dektol developer takes eight minutes. Fine-grain positive is excellent for making black and white slides from negatives.

A telephoto lens piggybacked on a guide telescope requires medium to fast film, such as Fujichrome 200 or 400 or Ektachrome 200 or 400 colour, or Plus-X for fine-grain black and white, or Tri-X for higher speed black and white. However, a regular camera lens at f4 can take Kodachrome 64 colour.

The main problem in deep-sky photography is the reciprocity failure of the film. Cold cameras, described in Chapter 4.6, solve most of this problem. But also film can be soaked in nitrogen or hydrogen for a time, thus removing the oxygen and water vapour

Plate 115. NGC 2237, The Rosette Nebula, on hypered film.

which causes the reciprocity failure. This process is gas-hyperization, and is carried out in a pressurized tank which can be made at home.

Take a 150 mm length of plastic plumbing pipe and cement a threaded plastic cap to each end. Through one of the threaded caps, mount a bicycle valve retrieved from an old inner tube. This can be accomplished by merely drilling the appropriate sized hole through the cap; you may require a rubber washer, cut from the inner tube, to seal the valve. From a welding supply store, a pressure gauge is obtained and cemented into the cap at the other end of the pipe. You now have an airtight tank which you can access from either end. The hyperization process requires that the film be 'baked'. A styrofoam picnic cooler, equipped with a thermostat and a 60-watt light bulb, will hold the hyperizing tank for this purpose.

The film cassettes are placed in the tank and flushed three times

Plate 116. Hypered film aids in the capture of this faint object (NGC 1449, The California Nebula).

with the forming gas mixture; 92 per cent nitrogen and 8 per cent hydrogen is used because it is non-combustible. Then the tank is pressurized to 1 kg/cm^2 (15 lb/in.^{2}), and placed in the baking chamber for 6 h at 50°C (120°F). The tank should be flushed twice during this time with the gas mixture. When the film cassettes are removed, seal them into plastic containers and store them in your freezer until they are needed. Storage time should not exceed 30 days, so estimate how much film you might use in that period before hyperizing it. Kodak black and white 2415 can equal Tri-X in speed and is very fine grained. Fujichrome 400 colour also hypers very well.

Use of spectroscopic film and filters

The spectroscopic films from Kodak are probably the best compromise to avoid either cold camera or hyperizing. The following are all designed to reduce substantially the reciprocity failure rate: 103aO, blue-sensitive, 103aE, red-sensitive; and 103aF, all-purpose. These films are available from advertisers in *Sky & Telescope* and *Astronomy* magazines.

All spectroscopic emulsions may be developed in Kodak D19 developer. These films have a carbon backing that will require a gentle rub with alcohol after processing. Spectroscopic emulsions are particularly well suited for filters to reduce sky fog and unwanted light. Most light-pollution or deep-sky filters are specifically designed to block mercury- and sodium-vapour lights and are excellent for astrophotography.

A Wratten No. 25 filter with 103aF film can be used to photograph red emission nebulae from city locations. To look at the deep-red filter it is difficult to imagine how any light gets through, but the red hydrogen-alpha emission spectrum comes streaming through as through there was no filter at all. Blue mercury-vapour emission stops at 580 nm, and the Wratten No. 25 red filter starts transmission at 590 nm. Thus the blue mercury-vapour light does not penetrate to the film. If sodium-vapour lights are used in your area, an RG 645 or Wratten No. 92 red filter will solve the problem. Sodium-vapour light emission ends at 640 nm and the RG 645 filter starts transmitting at about 650 nm.

To photograph with filters, some special care must be taken. The camera should be focused with the filter in place, as there is a slight change in focus caused by the filter. A very bright star should be used to focus the filter and camera. Also, because the filtered camera will probably not yield a visual view of the faint object being photographed, a bright guide star should be used to ensure that the telescope and camera are directed to the intended field. Be sure you relocate the correct guide star. I have many interesting pictures of star fields that unfortunately do not resemble what I had in mind!

Blue objects are very difficult to shoot from the city because of the light interference, but luckily a large number of the most interesting objects are at the red end of the spectrum.

Tri-X falls short at the red end of the spectrum and does not perform well with deep red filters. Kodak 2415 black and white, when hyperized, is excellent with deep red filters. The new Konica SRV 3200 colour print film is very good for astrophotography, and it works well with light pollution filters.

The use of spectroscopic film and filters may enable many backyard observatories, long closed to photography because of increased light pollution, to be reactivated. I suggest you try setting up experiments using filters and decide for yourself what combinations best suit your neighbourhood.

The Lumicon Deep-Sky filter now enables the astrophotographer to pierce light pollution and record both blue and red from city locations. These new light-pollution filters can be purchased for under $100, and they outperform the inexpensive Wratten 25 and 92 filters. The new oxygen-3 filter (for visual use) is also suited to black and white photography; it reveals views of the Veil nebula hitherto unavailable to most observers.

Both Ektachrome 400 and Fujichrome 400 can be improved by using a *Sky-Light* filter that will shift the colour balance of the film from a blue-green to a more neutral blue-black background. The actual colour of the night sky is greenish; this is probably caused by mercury vapour colouration, or perhaps by ionized oxygen.

4.11 Darkroom procedures

After taking successful astrophotographs, the next step for most amateurs is to set up their own darkroom. A second-hand enlarger, trays, etc., can be purchased through advertisements in most newspapers.

The darkroom can be any space from a light-tight closet to the family bathroom. Just block off the window completely with black poster paper and push a rug up against the door to seal out all light. The space required is very small and can be set up in minutes. Very little preparatory reading is required before starting the developing and printing of your own astrophotographs; libraries are full of good books on how to get started.

The developing of black and white film is very straightforward. The developing tank has to be loaded in complete darkness. The trick is, when you rewind the film in a 35 mm camera, do not rewind it all the way back into the film cassette. Listen for it to release from the winding spool, then when you open the camera the film tongue will be sticking out of the cassette and you just snip off the excess to make the end square. In normal room light, start the film end into the winder of the developing tank. When the end is started, turn out the lights, wind the film right into the holder, and place it in the developing tank. This simple process eliminates all of the fumbling around with the film and cassette in the dark.

A pail of water at room temperature (20 °C or 68 °F) should be set aside for rinsing. D19 developer is used for high-contrast negatives, and needs four minutes for developing Tri-X or 103aF spectroscopic film. The developer is drained from the tank, and water from the pail used to rinse out the tank. The stop bath is then poured into the tank and left about three minutes for Tri-X. Pour out the stop bath, and again rinse the film with the room temperature water from the pail. The fixer is then added for about five minutes, and then poured out.

After fixing, the developing tank can be opened, and the film placed for washing in a 500 ml plastic yogurt tub or similar with a small hole drilled or cut in the bottom with a sharp knife. Then add water from the pail for about fifteen minutes of washing. A hypo eliminator can be used after the fix to reduce the washing time to about five minutes.

Because astrophotographs are so seriously affected by scratches, it is wise to use Photo-Flo (or a similar anti-spotting solution) after the washing to avoid having to wipe the film. Then hang it to dry from a clothes peg; attach a second clothes peg to the bottom to hold the film straight and prevent it from curling.

Tri-X film developed in Accufine will have an effective ISO of 1200, and is ideal for lunar and planetary photographs through small refractors. D76 developer will give a fine-grain structure and lower contrast for both Plus-X and Tri-X films. There always seems to be a trade-off among slow films, fine grain, and longer exposures. A small telescope with a good drive will afford longer exposures on slower film. But for larger telescopes with a poor mount, you need to use fast films and short exposures to compensate for vibration and tracking errors.

Printing the negatives is the most rewarding part of the darkroom process. Here at last you get to see the results of all your time and effort at the telescope.

Deep-sky photographs should be printed on paper of a contrast No. 5 or No. 6 – Ilford No. 5 or the highest of the variable contrast papers. Lunar or planetary photographs usually look good on the medium, No. 3 or No. 4, and the higher contrast papers also.

Dodging

Clear dust-free negatives are essential for good prints. A glass negative carrier should be redusted for every change in negative.

Many photographs can be greatly enhanced by a process known as dodging. This process is used to lighten or darken parts of the print by varying the amount of light falling on the paper during printing. Your darkroom should have a variety of dodging tools, which are dodging masks or sheets of cardboard with rough holes torn in them. By varying the size from a small 10 mm to a large 150 mm, these masks are used to burn in dark areas and bring out hidden detail. In the Orion nebula photograph (Plate 118), for instance, the mask was used after the initial exposure to the paper.

Then a beam of light from the enlarger is allowed to fall on the darker areas to lighten these burnt in or dark areas on the negative.

The size and location must be constantly changed during dodging to prevent any distinct marks from the dodging tool forming on the final print. This is usually prevented by constantly moving the mask up and down and around.

To remove the lighter edges due to aberration or lens defects, an irregular circle of cardboard taped to the end of a coat hanger is used to block off the light to the centre of the print. This darkens the edges. After a little practice, dodging becomes a normal part of darkroom procedure.

Dodging the Moon is one of the most difficult jobs. The challenge is to keep the light intensity about the same along the terminator as for the rest of the Moon. A coat hanger with a 15 mm strip of

Plate 117. M42, The Orion Nebula, 'plain'.

Plate 118. M42, 'dodged'.

cardboard, about 300 mm long, taped to it will assist in this procedure. This improvised dodging tool is used to block out the light along the terminator after sufficient light has exposed that region. The tool can be bent into a slight curve to accommodate all phases of the Moon. Try to design it so that the cardboard covers only the Moon, and the wire the dark sky. This will reduce the chance of leaving a white mark on the dark sky directly above and below the terminator.

Plate 119. M42, 'unsharp masked'.

Unsharp masking

An alternative way of bringing out faint detail on a print is to create an unsharp mask, which is basically an unfocused positive copy of the original negative. It has the effect of dodging your whole print at the same time. Professional astronomers use spectroscopic emulsion on glass plates 1.6 mm ($\frac{61}{1000}$ in.) thick. That thickness was suitable to produce the mask. The amateur can use a 1.6 mm piece of window glass instead of the photographic plate.

The unsharp mask is made by copying the original negative through the 1.6 mm glass, using a slow-speed copy film such as

Plate 120. Black Eye Nebula, 'dodged'.

Panatomic-X or Kodak 2415. The emulsion sides of the negative and the copy film should face the same way, and a light source should expose through the original negative onto the copy; bracket your exposures by time and by the distance of the light source. The developed mask is then secured to one side of the glass and aligned with the original negative on the other. A positive can now be produced by copying the original, with its emulsion in contact with the emulsion of the copy. This positive can be printed or recopied to produce a final negative.

A more simple method of producing an unsharp print can be used. Mount the original negative in a slide mount and cut off the top of the mount along the top of the negative. This edge will enable you to align the copy film in the dark. Attach the leader of the copy film into an empty cassette so that you can wind exposed film into this light-tight cassette while subsequent exposures are taken.

Use your enlarger as a light source, and place the slide with the original negative on top of the copy film, emulsion side up. Then in total darkness, use the take-up cassette to wind on the film. Align the trimmed edge of the slide with the edge of the copy film and expose the film. Bracket the exposure by varying the height of the enlarger head or the exposure times, or both. You should now have one or two exposures of suitable density to use as an unsharp mask. The thickness of the slide mount provides the right distance to unfocus the mask.

The new mask should be aligned with the original negative and both taped to a piece of glass to make the job easier. Now simply make a print from the combined original negative and mask. M42 was my first attempt at unsharp masking (Plate 119).

Integration printing

Integration printing is a system of using several negatives one after the other to produce one print. The advantage is that the location of the grain on each negative is different. This means they tend to cancel each other out and produce a print with a very fine-grain texture, significantly increasing the faint detail. The negatives should be about the same exposure density to ensure that the background sky fog is the same.

Cut out a white cardboard mask the exact size of the area to be printed, and tape it down on one side, so that it can wing out of the way when you are printing. Place the unexposed paper beneath the mask, and then project the first negative onto the mask. Mark the position of several stars on each corner.

For the approximate exposure when using three negatives, divide the normal exposure time by 3. Add 0.4 of this time to the first negative and 0.2 to each of the others. For example, if the normal exposure is 60 s, the first negative would $\frac{60}{3} + (\frac{60}{3} \times 0.4)$, or 28 s. The second negative would be $\frac{60}{3} + (\frac{60}{3} \times 0.2)$, or 24 s, and the third negative would be the same. This basic formula also works for any number of negatives. If you have five negatives, for a normal exposure of 60 s, your first exposure will be 16.8 s (rounded to 17 s) and the other four will be 14.8 s (rounded to 15 s).

Next, flip the mask out of the way and make the first 28 s exposure. Replace the mask, line up the second negative on the mask using the marker stars, and then expose the second negative for 24 s, and repeat that process for the third negative. This process takes a bit of patience, but the results are well worth the effort. When integrating planetary photographs, the easiest method is to take two or three negatives and sandwich them together. Then merely print through this sandwich the same as you would a single negative. The results are usually better than using the separate negative method because it is very difficult to line up the planets on the mask as there will be no marker stars.

The handling of negatives and cleaning and mounting of prints is well covered in most popular books on the home darkroom. However, a fine-pointed felt-tipped pen is a handy tool to cover unwanted dust marks that always seem to appear on the dark skies of astrophotographs. Always stipple these with many small dots – never attempt to paint them out.

Enhancement by copying

Astrophotographs can be greatly improved by copying, using a slide copier or an enlarger. Commercial slide copiers are available for about $70 (£35).

Take a slide that you wish to copy, and rephotograph it using slow, fine-grained film such as Kodak's VGR 100 (for turning slides into negatives, for printing) or Ektachrome 100 (for slide copies). Negatives can also be copied using Kodak Vericolor film and an appropriate filter combination. This will produce a high-contrast slide. I use a regular flash unit on top of a flat surface, and take a variety of copies, with the full range of flash-positions, to achieve the best exposure for the copy. This is used for both VGR 100 and Ektachrome. Vericolor film requires a filter pack and floodlight. An example shown on p. 235 is NGC 7000, the North America Nebula.

Most slide copiers allow you to enlarge the image and frame the object. This also enhances the photographs.

The darkroom enlarger can be used also for copying by removing the camera lens and projecting the object directly into the camera, using a right-angle viewer.

PART 5 Amateur astronomy in the electronic age

5.1 The CCD: a quantum leap

In a book on the nature of scientific discovery* science historian Gerald Hawkins notes that recent centuries of human endeavour show a highly significant pattern. Our perception of the Universe tends to evolve, not in a steady upward curve, but on a generally level line that periodically takes a sudden step upward.

Quantum leaps in knowledge, Hawkins points out, invariably follow the introduction of radically new observational tools. Thus, a revelatory new view of deep space was the immediate result of the invention of astronomical photography in the mid-nineteenth century. Photographic emulsions were able to accumulate an image during a long exposure and some emulsions were sensitive to wavelengths never seen by the human eye. A new Cosmos of nebulae and intricately structured galaxies was suddenly accessible.

In recent years a similar leap forward has occurred, in the introduction of the CCD camera, an electronic innovation that has expanded the horizons of first the professional astronomer and then, almost immediately, the amateur as well.

The CCD camera

The *charge coupled device* (CCD) is an electronic camera designed to capture light and, in conjunction with a computer, to produce an image. The heart of the CCD is a silicon chip with thousands of light-sensitive sites called pixels. Each pixel converts incoming photons of light into electrons, which are stored in the pixel. At the end of the exposure these electrons are converted into digital form, and sent to the computer. The number for each pixel represents a shade of grey. The computer reads the numbers and displays them together, forming an image on the computer monitor.

* *Mindsteps to the Cosmos*, Harper & Row, 1983

Plate 121. CCD image of galaxy M81.

The greater the number of shades of grey produced by the bit configuration of a chip, the smoother an image appears. Table 5.1 shows some typical values for CCD cameras with which I am familiar. All of these use various Texas Instruments chips.

Table 5.1. *Some available CCDs*

Company	Model	Bits	Pixels	Shades of grey
Santa Barbara	ST-4	8	192×164	250
Instrument	ST-4×	14	192×164	16,384
Group (SBIG)	ST-5	14	320×240	16,384
	ST-6	16	375×242	65,536
Lynxx		12	192×164	4,096
Meade	Pictor 208	8	320×240	250
	Pictor 216	16	320×240	65,536
	Pictor 916	16	768×512	65,536
	Pictor 1616	16	1536×1024	65,536

The beauty of having images composed of numbers is that you can make unlimited numbers of identical copies, with no integrity loss to the original. You can use the computer to change the numbers and manipulate the image by adding a constant value to each pixel. The computer software provided with a CCD camera is like having a darkroom built into the computer!

The CCD camera has a built-in electronic cooler intended to reduce electronic noise, or 'dark current'. Cooling of the chip cuts electronic noise in half with every 8 °C drop in temperature.

Plate 122. CCD image of M81. Comparison with Plate 121 shows the supernova 1993J among the triangle of stars above the galaxy.

The CCD image has tremendous dynamic range. Objects such as The Orion Nebula, containing both a very bright central region and very faint outer filaments, can be enhanced using image processing.

Experiments with the CCD

Three years ago I received my first CCD camera (the 'ST-4') from the Santa Barbara Instrument Group in Southern California. I purchased it for use as an auto-tracker for my 25-inch (64-cm) Newtonian telescope. However, when it arrived I was intrigued to see what results might be achieved using the camera for imaging. I chose M57, The Ring Nebula, as my first subject and couldn't believe my eyes as the image appeared on my computer monitor. Since that moment I have been hooked!

As I am a keen colour astrophotographer, I wanted to see if I could use this new CCD camera to take colour images. To do this I recorded three black and white images using red, blue and green filters. At that time there were no programs available to combine the black and white images into colour. I had been working with Richard Berry, former editor of *Astronomy* magazine, to provide him with CCD images to use in developing his Image Pro software for the ST-4 and Lynxx CCD cameras. With the tri-colour images as a special incentive, Richard wrote the ST-4 colour program to combine the images.

In September of 1992 I received the prototype engineering model of SBIG's 'ST-6', a CCD camera specifically designed for imaging, but capable of tracking as well. I was able to reach 21st magnitude stars (or fainter!) taking 15-minute exposures. I used tri-colour filters to image M57 and M27, hoping again to entice Richard Berry to write me a special program to combine the images into colour. Fortunately, Richard was spared this task, since we received a timely introduction to 'PhotoStyler', a desktop publishing program provided with the purchase of scanning equipment. In October Richard and I electronically processed my 'first light' ST-6 colour images, and my image of the Dumbbell Nebula appeared on the cover of *Astronomy* three months later.

I was sent the pre-production model of the ST-6 in time for my travels to the Winter Star Party in Florida in January. I imaged M65, M42 and NGC 2903. Then I received the first full production model and have spent every available night since experimenting with colour balancing the camera, using different filters. To date, I have achieved the most accurate renditions using the SBIG dichloric filters, or the standard No. 25 red, No. 58 green and No. 47 blue filters with a Corion NR-400F near-infrared (blocking) filter.

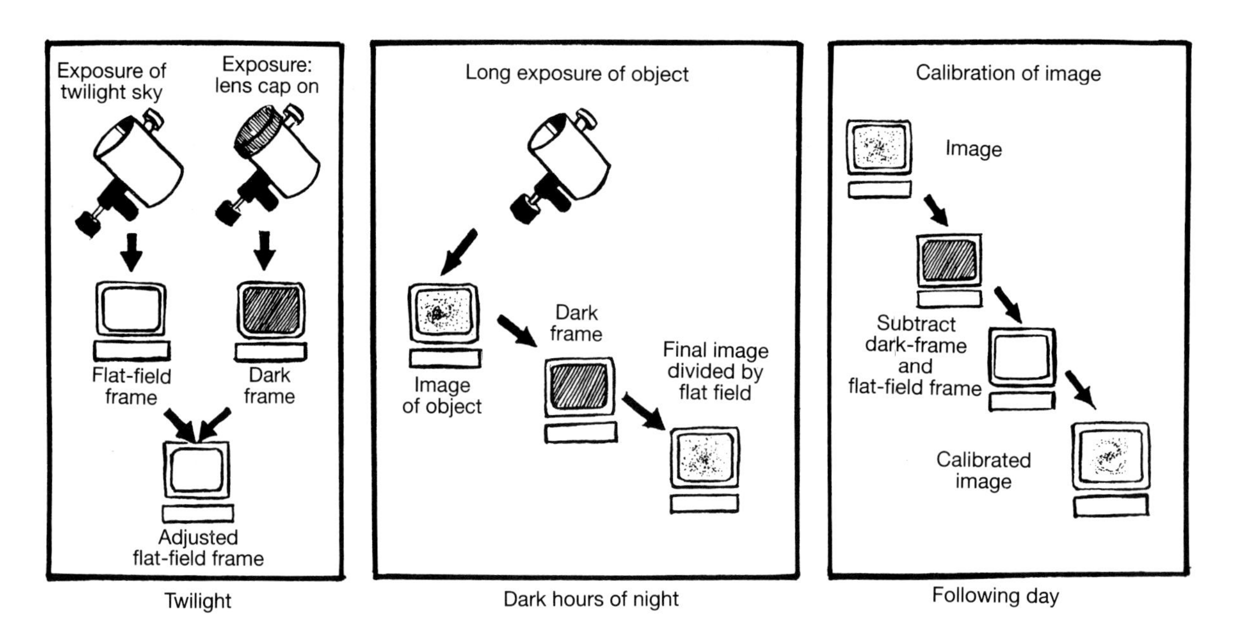

Figure 5.1. Acquiring & processing a CCD image.

Both the green and blue filters open up in the near infrared, causing the famous 'red leak'. The 58 green opens up to red at a wavelength of 700 nm and the blue at 740 nm. The ST-6 uses a Texas Instruments TC241 computer chip whose efficiency reaches 50% at 635 nm in the red, and is still 20% efficient at 850 nm in the near infrared. I place the Corion NR filter over the end of the ST-6 and image the three colours with this filter in place. To colour-balance the ST-6 terrestrially I used a colour photographic print of a woman holding a grey card (Plate 124). I imaged the photograph

Plate 123. Grey scale test model. Imaged with the ST-6 CCD camera.

through my Celestron-8 telescope using the tri-colour and Corion filters. Each colour exposure is adjusted until the pixel density on the grey card is identical. This means that each colour filter is passing an equal amount of light. The results were near-perfect colour balance in daylight. With the CCD camera colour balanced, I could turn my attention to the night sky.

Tri-colour images are moved into Richard Berry's Image Pro program, where I remove electronic noise and unsharp-mask them.*

* See Chapter 4.11 for details of unsharp masking.

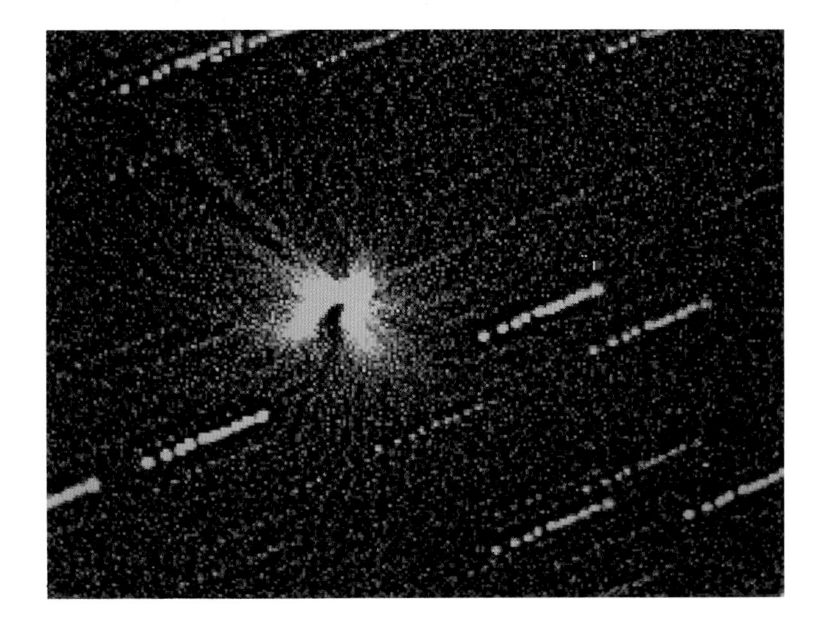

Plate 124. CCD image of Comet Swift–Tuttle, 1992.

Plate 125. Jupiter, a CCD image by Donald C. Parker.

Plate 126. Saturn, a CCD image by Donald C. Parker.

I can also manipulate images with the ST-6 software to remove individual pixels that have registered too bright or too dark ('warm' or 'cold' pixels), and to sharpen or adjust the background. I save the images as square TIF files (a graphics storage software) and move them into PhotoStyler to alter lightness/darkness and to adjust the 'register' or congruence of the three colour images so that the final tricolour product will be sharp. The red, blue and green colours are then combined into the final colour image. Range, colour density or hue can now be adjusted.

Results and potential

The CCD camera is very good at recording extremely faint objects. The ISO speed of the CCD compared with film speed is only about 80 to 100. When a regular camera lens is placed on the CCD and an image taken in daylight indoors, the lens setting and shutter speed are equal to Kodak 100 film. When the CCD camera is attached to a telescope and long time exposures made, the results are dramatically different than with film. A ten-minute exposure with the ST-6 and my large 25-inch (64-cm) f5 telescope produces results equal to those from a 90-minute exposure on hypered Kodak 2415 film! The CCD chip is linear, and records twice as much image in twice the exposure time. Film suffers from reciprocity failure and does not record in proportion to the exposure time, because it chemically 'peaks out' and fails to record accurate detail beyond a certain point.

Pixel size in the ST-6 is 27 microns by 23 microns in rows of 375 pixels by 242. That equals 8.6 by 6.5 millimetres. In my 25-inch telescope the field is 10 arc-minutes.

Plate 127. Spiral galaxy M100, a maximum entropy image processed by Ajai Sehgal's *Hidden Image* software (CCD image).

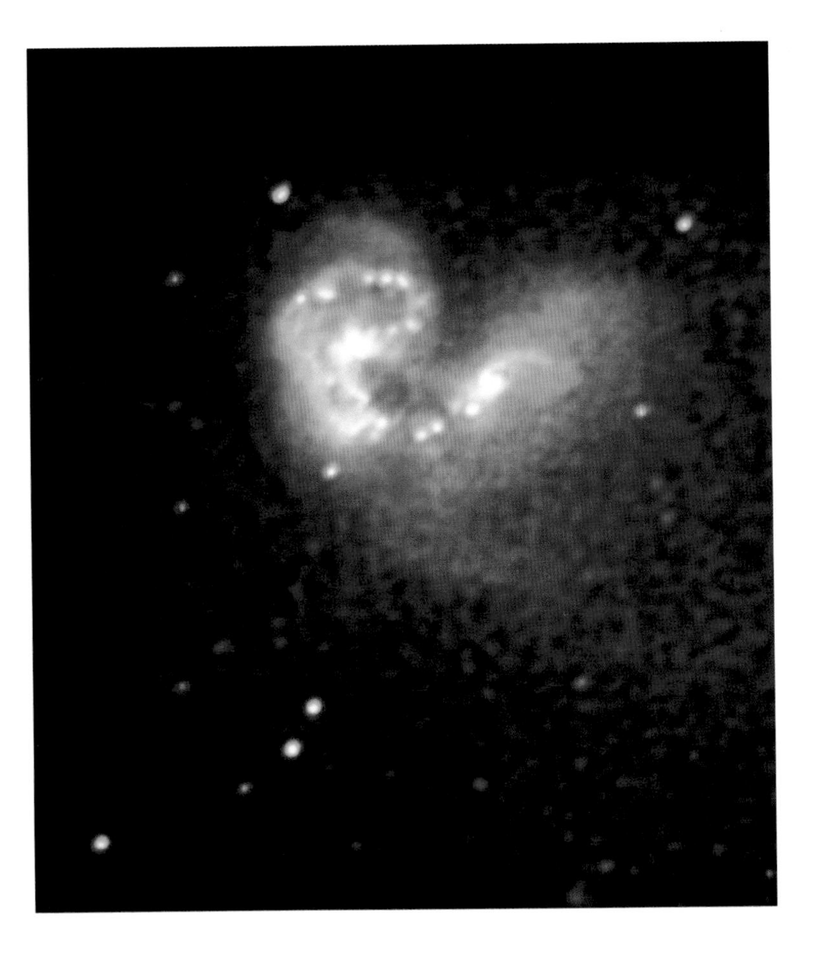

Plate 128. 'Colliding' galaxies NGC 4038/39, a maximum entropy image processed with Ajai Sehgal's *Hidden Image* software.

Plate 129. CCD raw image.

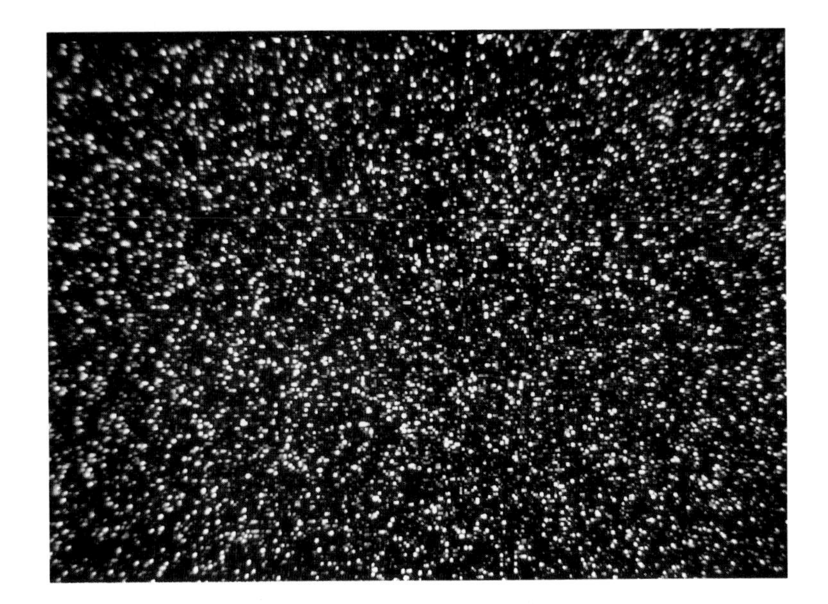

Plate 130. CCD dark frame.

I find that the CCD camera's greatest advantages are offered in resolution. For lunar or planetary imaging, eyepiece projection works very well. (Details of eyepiece projection appear in Chapter 4.7). The ST-4 and Lynxx models have pixels measuring 13 × 15 microns. The brightest part of an Airy disc (the visible 'dot' of a stellar image) is an excellent telescope at f40 is about 50 microns, which covers two pixels. Thus, you can get near-diffraction-limited images with a CCD camera. This method is best used on nights of extremely good seeing. Small bright planetary nebulae can be greatly enhanced using this eyepiece projection system of imaging.

S-BIG's new manual colour filter wheel is designed to be used in conjunction with the ST-6 CCD camera. The filters are custom

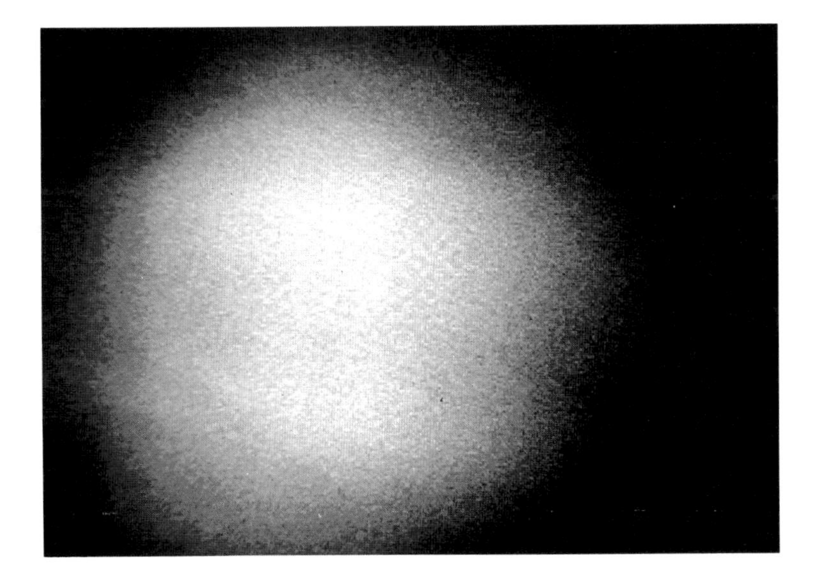

Plate 131. CCD flat-field frame.

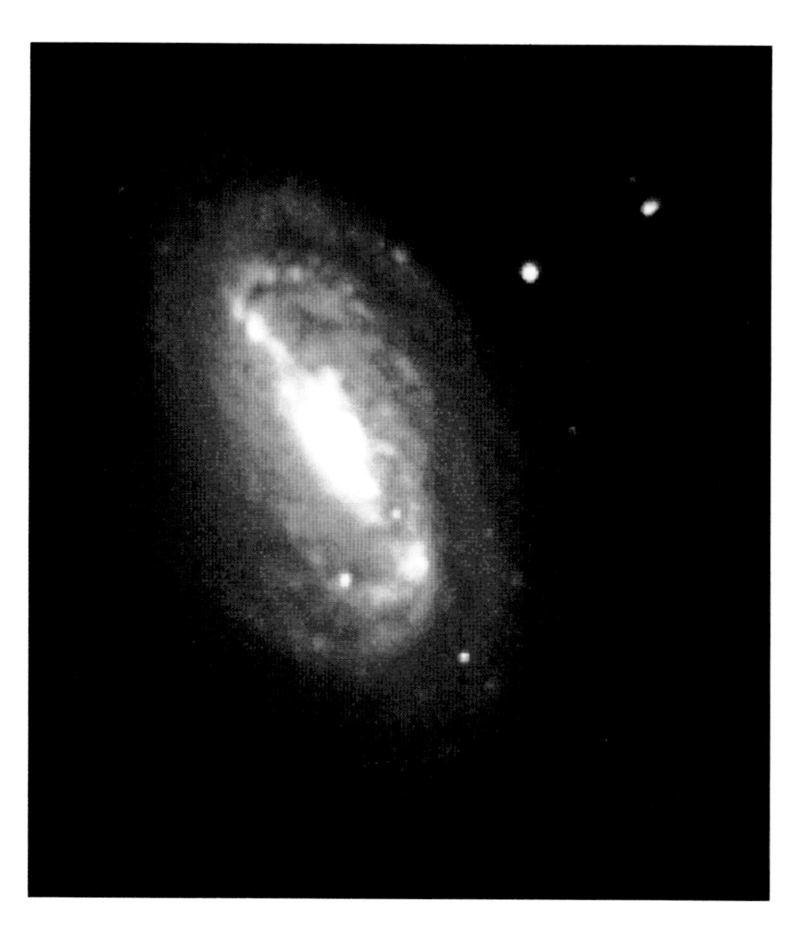

Plate 132. NGC 2903, the finished CCD image.

at 500.7 nm, which is missed between the green and blue layers of conventional colour films. The filters cover the wavelengths of:

Red: 585–680 nm
Green: 496–585 nm
Blue: 380–502 nm.

The colour wheel has four openings for filters plus a star diagonal for viewing. An infrared blocking filter is mounted in the light path of all the filters. The colour wheel software will automatically align the three images to produce a final colour image.

Terminology

Raw frame — Unprocessed image taken straight from the telescope.

Dark frame — An exposure equal in length to a raw frame taken with the lens cap on so that no light will fall on the chip. This dark frame records the buildup of thermal noise in the CCD, and is subtracted from the raw image.

Flat-field frame This frame is a one- or two-second exposure of the twilight sky produced densely enough not to saturate the pixels. It is divided into the image to flatten any uneven pixels, and remove vignetting or dust-shadows.

A note about CCD image processing software:

All CCD images except as noted were processed with Richard Berry's *ST6 PIX* program. For information, contact Willmann-Bell Inc., PO Box 35025 Richmond, VA. 23235, USA.

The other principal image processing software used was *Hidden Image*; for information, contact Ajai Sehgal, 1776 Windflower Way, Gloucester KIC 5Y9, Canada.

5.2 Astronomy and the software revolution

In 1984 Canadian broadcaster Lister Sinclair was asked to comment on the accuracy of George Orwell's vision of the future, a 'future' in which we now find ourselves living. Sinclair pointed out that Orwell, for all his visionary perspicacity, had totally failed to predict the most revolutionary development of the era, the pervasive effect of computers in our lives.

Writing in 1949 Orwell could not possibly have foreseen the sudden evolutionary change that computers – especially the personal computer – would have on the way we store, manipulate and communicate information. As Sinclair remarked in his 1984 commentary, some of the applications of this electronic manipulation of data are as sinister as anything about which Orwell warned; other applications have turned out to be a gift of the most beneficial kind.

Among the most intriguing possibilities are the breakthroughs that have their application in astronomy and space technology. Without the most sophisticated computers the exploration of our solar system's remotest corners by Voyager would have been a mere pipedream in the realm of science fiction or fantasy. It is only

Plate 133. Astronomical software is a colourful and instructive aid to the amateur.

through highly evolved software that the faint, noise-scrambled signals from Voyager and other probes have been processed into the revelatory images that everyone has seen and enjoyed. Because complex problems of orbital calculation and positional astronomy are child's play to the computer, we now make routine use of occultation predictions, asteroid ephemerides and cometary data that were never before so readily available.

For the amateur astronomer a somewhat different computer application is currently in vogue as the focus of interest. This popular activity involves the avalanche of commercially marketed astro-software designed especially for the astronomy enthusiast who owns a personal computer.

The Cosmos on your PC

If you are the average stargazer, the infinite depth and subtlety of the night sky is your primary interest. What can the computer offer that will be of sufficient interest to lure you indoors and away from the eyepiece of your telescope?

For a quick overview of the exciting range of possibilities aimed at by scores of software publishers, you need only scan the colourful. advertisements in the astronomy periodicals. Vast sky-catalogues for the observer, stunning graphics that explain the workings of the Universe, animated displays that show tonight's positions of planets and deep-sky objects – all of these functions utilize the special powers of electronic data storage.

In our own sampling of typical astronomical software packages on the market during the year of writing, we have found that most fall within three broad categories of type or purpose. One major group is the *tutorials*, programs that give you a dramatically graphic perspective on the nature of the planets, galaxies and other aspects of the Universe. Another type of software offers what might be described as the *planetarium* function, allowing you to display the planetary positions and astronomical events of past or future dates. A third category combines the functions of *star atlas and catalogue*, providing identification and observational data on thousands of stars and deep-sky objects.

Most are available as packages that include one or more diskettes accompanied by a detailed instruction manual. A smaller number are in CD ROM (compact disc) format, requiring a CD-drive as an accessory to the home computer.

Do these astronomical programs for the personal computer offer a kind of information access that you cannot find easily in conventional print sources such as handbooks, atlases, etc.? Some of the packages that we have sampled (especially the most basic tutorials) seem to provide little that might not as easily be pursued in a good pictorial book. In others you will find exciting access to a realm of data manipulation that reveals the computer's true potential for the amateur astronomer.

In our own use of a variety of astro-programs we have come to a perhaps rather surprising conclusion: those that will initially stun you with their heavy emphasis on elaborate pictorial content may prove to be of smaller long-term practical use than others whose content is largely tabular data.

A software sampler

We have used many of the astro-programs currently on the market. During the preparation of this book we undertook a simultaneous exploration of a variety of these offerings. We are grateful to Mike Scott of Computer Operations at the Greater Victoria Public Library. He provided an array of work stations and assisted us in our comparison of all these products. It was great fun, and we believe that you will find the products described below especially useful.

Bear in mind, however, that the software industry is extremely volatile; changes occur rapidly and continuously. To learn about the latest products, peruse the software reviews in current magazines or write to the software producers themselves.

Table 5.2. *Software sampler*

Software title	Comments	Supplier
Dance of the Planets	A versatile solar system planetarium, with tabular information on deep-sky objects as a bonus. Fine, high-precision atlas of the night sky with animation to show date-specific planetary positions. Individual planet displays feature Voyager-style pictorials, and animated positional data on satellites. Beautiful and useful.	Arc Science Simulations PO Box 1955 Loveland, CO 80539 USA.
The Earth Centered Universe	This is a star atlas that interfaces with the amateur's telescope, for automated finding of objects in the actual night sky.	Nova Astronomics PO Box 31013 Halifax NS, Canada B3K 5T9
EZ Cosmos	Combined atlas and planetarium functions. The basic environment is a full-sky chart for current date and your location. A zoom feature permits large-scale close-up of selected sky region. A box-cursor can be positioned over any star/object to display identification and description of object, plus photos of some of the objects. An especially useful feature is the solar system animation that allows prediction of planetary positions for many centuries.	Future Software Trends 1508 Osprey Drive, Suite 103 De Soto, TX 75115 USA.
HyperSKY II	Our favourite: a superdetailed atlas–catalogue for the telescopic observer. Tens of thousands of stars and objects (including all NGC) can be targeted for identification and descriptive data. On command, very full documentation on each object is supplied – a veritable library of recent material. An example of the minimal graphics/high-content style of practical working software.	Willman-Bell Inc. PO Box 35025 Richmond, VA 23235 USA.
Lodestar Plus	Sky display that permits identification of deep-sky objects and location of current planetary positions. An interesting feature is capability of adding your own selection of objects. Many manipulations possible, but not available on the very limited demonstration disk that we have viewed.	Zephyr Services 1900 Murrary Ave., Dept. A Pittsburgh, PA 15217 USA.
Night Sky	A combined atlas and planetarium package that can be set up for your own geographic position and for any date between 1 AD and 4000 AD. Looks useful, but we have seen only a limited demonstration disk.	Zephyr Services 1900 Murray Ave., Dept. A Pittsburgh, PA 15217 USA.
Orbits	A highly graphic solar system tutorial. Shows comparative sizes and structures of the planets, provides animated orbital diagrams that permit conjunction predictions, etc. Even includes demonstration of the celestial mechanics of interplanetary space flight. Although colourfully pictorial, this instructional software also includes a large textual content. Excellent introduction for the young amateur.	Software Marketing Corp. 9831 South 51st St. C-113 Phoenix, AZ 85044 USA.
The Sky	Extremely handy and user-friendly atlas–catalogue. Can be set up for your observing location simply by selecting your city's name from menu (we found our small home city on the list!). Displays superfine star maps, with zoom feature to focus on segment that you choose. Pointer selects individual objects for identification, observational data and very good photographic pictorials. Shows current-date planetary positions, and also permits fast animation to future dates.	Software Bisque 912 Twelfth St., Suite A Golden, Colorado 80401 USA.

Further comments

The above products have an attractive feature in common: they all present a clear invitation to practical exploration in the night sky.

The best of the astro-software will greatly expand your knowledge of the sky. We have found that to move efficiently among the charts of the atlas programs you must quickly sharpen your familiarity with the celestial co-ordinate system. In most of these programs, rapid movement from one constellation to another is achieved by specifying the RA and Dec. upon which you wish to centre the chart. Thus, to focus quickly on the sky region that includes the Pleiades you must know that the constellation of Taurus (which contains the Pleiades) straddles roughly RA 4h and Dec. +20°. The process of gaining this facility is an enormously entertaining and instructive game.

Yet this colourful, information-packed material appears to have one rather alarming aspect. We see around us an increasing number of amateurs whose fascination with the home computer has become an end in itself, replacing actual observation of the night sky.

As one participant in a recent astronomical society software display was overheard to remark, 'These images of galaxies are far more beautiful than the real objects, viewed through the telescope!' It is important to beware of the computer's seductive power, which has turned some addicts from active telescope observing, in favour of sitting mesmerized before a glowing, beeping screen.

Enjoy and benefit from your home computer, by all means. But if its shimmering displays of planets, stars and nebulae begin to hold you in thrall, remember that they are only a surrogate. The real Universe is still overhead, outdoors!

5.3 Telescopes that think

The age of the 'smart' telescope has arrived.

In the opening chapters of the present book we described a fundamental skill (and a principal delight) of amateur astronomy: the technique of star-hopping to locate faint objects in the labyrinth of the night sky. This procedure is instructive fun, but it can also be difficult, especially when the objects being sought are faint or subtle. Recent years have seen the evolution of an alternative to the basic method, in the form of telescopes that 'think'.

More precisely, the new technology is a mounting that 'knows' the night sky, and can accurately direct the telescope toward any one of a pre-programmed catalogue of nebulae, galaxies or individual stars.

Details of operation

In the early 1980s the Meade Corporation introduced its 'CAT' – 'Computer Assisted Telescope'. This unit was a computer that communicated with the telescope through optical encoders attached to the right ascension and declination shafts of the telescope's mounting. Pulses from the encoders were translated by the computer into the position of the telescope in RA and Dec.

The CAT was programmed with all 7000 NGC objects, the full Messier catalogue and 250 of the brightest stars. Although Meade and other suppliers have since introduced somewhat more refined versions of this basic idea, the CAT continues in widespread use among amateur astronomers. I have found it to be an indispensable aid to locating faint and elusive objects.

Before the CAT can be used, the telescope must be set up in accurate polar alignment. The CAT itself assists in this procedure. You must first enter your date, Universal Time, longitude and latitude; the computer then instructs you to move each axis (right ascension and declination) until you detect an audible beep. This signatures the optical encoders to the CAT. The computer then indicates the location of Polaris. By adjusting the azimuth and altitude of the telescope until Polaris is centred in the field of view you can automatically align the mount on the celestial pole.

As a final preparatory step the computer selects a star on which it will instruct you to centre the telescope. When this star has been located press 'enter', and the CAT is ready to assist you in your observing session.

To locate a target object you first enter its Messier or NGC number. This activates an illuminated cross on the face of the computer and as the telescope is moved toward the required object the CAT responds by lighting up more of each axis of the cross. When the intersection of the two axes becomes illuminated, the target-object will be found in the eyepiece.

You may find that the object is not centred within the eyepiece field, or even that it lies slightly outside the field. If this is the case, you merely centre the object and then press a 'sync' button, which recalibrates the encoders to the object's position. In this way the alignment of the telescope is refined, to permit more accurate pin-pointing of subsequent targets.

When movement from a present target-object to your next selection is a relatively short jump, the accuracy of the CAT (or any similar system) will be high. However, if you are moving between objects separated by a very large area of sky you will probably need to make a slight final adjustment to bring the object into view. Having shifted the telescope in this way, you will want to use the 'sync' command again in order to correct the encoders' calibration.

My own use of the CAT has been gratifying. This version of the 'smart telescope' has saved me hundreds of hours in searching out faint objects as targets for CCD imaging.

Many such targets present a special challenge because of the CCD's ability to record objects too faint to be easily detected visually with the telescope. The fact that some of these 'challenge' objects are not found in the CAT's preprogrammed catalogue is not a deterrent to their location by means of the system. An example is *Cass. A*, a faint radio source in Cassiopeia. In order to find this, the CAT was synced on an NGC object near the target, and then moved to the position of *Cass. A* by using that object's co-ordinates. A 60-second image of the area on the computer screen showed part of *Cass. A*. The computer CCD chip provides coverage of only 11 arc-minutes; the CAT computer was shown in this instance to have pointed the telescope with an accuracy of about 5 arc-minutes or better.

Some other products

In the years since the CAT's welcome appearance on the market 'smart telescope' technology has not remained static. A newer generation of products has evolved in the direction of even greater sophistication combined with more compact format. The most recent version of this move toward compactness is the built-in controller, a miniaturized computer that is an integral part of the telescope mounting itself.

One recent variant is Lumicon's NGC Sky Vector, a control-computer for the amateur astronomer that includes an impressive catalogue of 12 000 objects in its memory. As is typical in the newer products, this system is easier to set up than the CAT computer.

Both the Meade and the Celestron companies now market smart telescope drives that add an extra dimension of automation to the process of locating objects in the night sky (Table 5.3 shows addresses of these and other sources). These drives, when set up, not only guide your finding efforts, but actually move the telescope to the chosen sky location. You merely select an object by entering its NGC number into your computer, and the telescope will move automatically to the desired object.

Perhaps the most remarkable development of this new computer technology is its application to telescopes that are not equatorially mounted. Some units now available can be used with the popular Dobsonian telescopes; these require only the date and time and three orientation stars located in separate parts of the sky to allow pinpointing. With these 'very smart' control systems, polar alignment is no longer essential.

Table 5.3. *Sources: a selection*

Supplier	Product
Celestron International 2835 Columbia St. Torrance, CA 90503, USA.	*Compustar*: an 8000-object integrated telescope–computer system.
JMI Inc. 810 Quail St., Unit E Lakewood, CO 80215, USA.	*NGC-MAX* guiding computer that aids setting telescope on NGC objects, double stars, etc.
Lumicon 2111 Research Drive, 5 Livermore, CA 94550, USA.	*NGC Sky Vector*: a guide system with 12 000-object memory, adaptable to fork-style equatorial mountings and also to Dobsonian telescopes.
Meade Instruments 16542 Millikan Ave. Irvine, CA 92714, USA.	*LX 200 'Smart Drive'* system permits automated slewing of telescope to positions of 747 preprogrammed objects (for use with Meade telescopes).
SkyComm Engineering 4691 N University Dr., Ste 329 Coral Springs, FL 33067, USA.	*Sky Commander*: Pocket-sized locator for use with either equatorial or azimuth mountings.

I am often asked: doesn't my use of a telescope that thinks take some of the adventure out of astronomy? My answer is that I have paid my dues; I have enjoyed (and learned from) the experience of star-hopping for thirty years. Today I would not trade my computer for anything!

5.4 The backyard photometer

Here is a project that will interest the home astronomer who enjoys using simple electronic instrumentation. When the professional scientist wishes to detect and measure variability in stars or other objects, he or she does not depend on rough visual estimates. Instead, a photoelectric photometer is coupled to the telescope to yield super-precise values for the light fluctuations in stars, rotating asteroids, etc.

The photoelectric photometer is a reasonably simple and physically compact electronic detector that fits onto the telescope's eyepiece holder, in place of the visual eyepiece. Its function is to translate the incoming stream of photons from an astronomical object into a stream of electrons, and to amplify this electron flow in order to activate some kind of a readout device – a chart recorder or perhaps a digital meter.

Difficulties – and delights

Let us begin by mentioning some of the limitations of amateur photometry.

The typical small photometer is a far less efficient detector of low light-levels than the human eye coupled to the telescope. Starlight will be measurable down to a limit that is about four or five magnitudes brighter than can be detected visually. Let us say that a variable star observer's 20 cm telescope permits him or her to make visual brightness-estimates of stars as faint as magnitude 14; his or her photoelectric photometer, although vastly more accurate than his or her eye, will cease to be useful on objects fainter than perhaps magnitude 9.5.

A second constraint involves the mechanical design of the telescope itself. Unlike the eye, which can tolerate considerable vagaries in the telescope's mounting and drive, the photometer demands a good level of field-steadiness and smoothness of tracking. This is because of the high degree of accuracy with which the incoming beam of starlight must remain centred on the tiny aperture of the photometer's entrance-diaphragm. If you wish to experiment seriously with photoelectric photometry, you will need a telescope that is sturdy enough to support, and to drive smoothly, a fairly heavy package of instrumentation. A small photometer will be at least four times the weight of a camera.

And what are the joys of amateur photometry?

The overwhelming delight is the professional level of accuracy that the amateur can achieve in certain kinds of observational projects. As optical engineer E. Pfannenschmidt of the Dominion Astrophysical Observatory recently asserted in a conversation with the author, 'an amateur who still does visual photometry is like a sailing man who still uses oars for auxiliary power!'

The best variable star observers can detect brightness differences of about 0.1 magnitude. By comparison, a single photoelectric photometer will yield measurements accurate to 0.01 magnitude, or even better. This degree of precision opens the door to a number of fascinating projects. For example, the variable star observer can monitor very small, short-term fluctuations that characterize the behaviour of some irregular, long-term variables. By maintaining a patient surveillance he or she may even discover slight variability in stars not previously known to be variable.

An amateur equipped with a photoelectric photometer can find, in certain of the asteroids, rotational light curves too subtle to be detected or measured visually.

Events that require precise timing (occultations, binary star eclipses, mutual interactions of planetary satellites) can be charted and measured in a manner that is far less subjective than visual monitoring.

Particularly when the amateur has access to a telescope of aperture large enough to overcome the magnitude-limitations of the photometer his or her work in the above-mentioned fields can make a genuine contribution of scientific data.

The photometer itself

The photoelectric photometer is not an unduly complex electronic device. At its most fundamental it involves an off-the-shelf photomultiplier tube, in which incident light induces electric current, plus a simple amplifier that raises the induced current to a level sufficient for practical use. The practical use may be as basic as activating an ordinary galvanometer, or it may be something more sophisticated, such as a small chart recorder.

In practice, the photometer will contain six essential parts, illustrated in our simplified sketch (Figure 5.2), as follows.

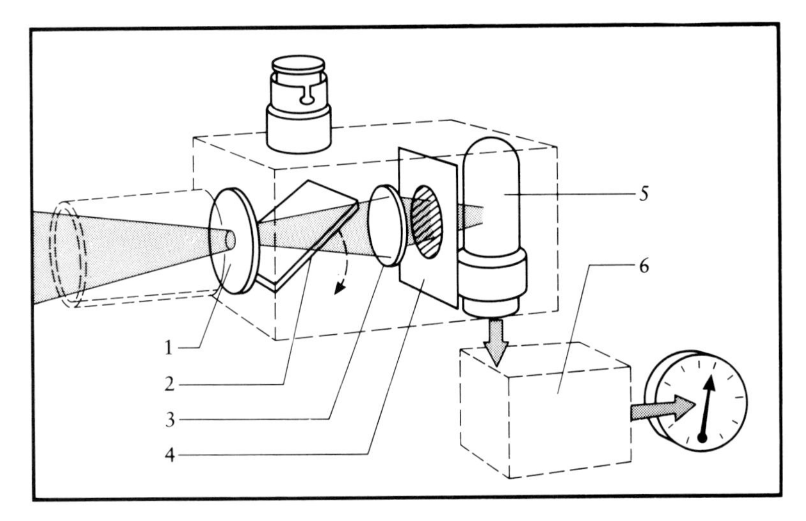

Figure 5.2. Elements of the photoelectric photometer. 1–6 defined in text.

1. *Diaphragm* with tiny (<0.5 mm) aperture to mask out sky background light
2. *Off-axis eyepiece* (with removable diagonal mirror) for visual centring of target
3. *Filter*, to admit selected wavelengths of light
4. *Fabry lens*, which distributes light in such a way as to minimize effects of the beam's movement on the photocell
5. *Photomultiplier tube*
6. *Amplifier*

The electronics hobbyist can build a photometer, but more commonly nowadays the amateur acquires one of the relatively inexpensive ready made models available from a number of commercial suppliers. As a very rough guide to cost: one popular photometer for amateurs is priced at about half the current retail value of a standard 22 cm catadioptric telescope. At the time of writing, mail-order sources of photoelectric photometers include Optec Inc. of Lowell (MI), USA and Cecil Instruments Ltd of Cambridge, in the UK. Other suppliers advertise occasionally in the pages of the astronomical journals.

Amateur photometric technique

When applying the photometer to a specific astronomical project, the amateur will most likely adopt a method that is essentially the same as that used by the visual observer of variable stars. In the technique called *differential photometry* the astronomer establishes the brightness of one object by measuring its relationship with a second object of known magnitude – the comparison star of the ordinary variable star procedure.

Suppose that we are attempting to measure the brightness of a variable star at some specific moment, using the photoelectric photometer. We choose at least one non-variable star of precisely known magnitude (from detailed field-charts) to serve as the comparison. For convenience, and to ensure that both the comparison star and the variable are subject to closely similar conditions of atmospheric extinction, we choose a star very near in the sky to the variable under study.

Moving back and forth from one to the other, we measure each several times, recording the photometer reading at each attempt. The final step is a measurement of the photometer reading with just the sky background centred upon the aperture of the diaphragm (no star in view). The latter value must be subtracted from each of the individual stars' measured values, to eliminate the component in each measurement that represents background 'noise' – that is, ambient sky light.

Reduction of our findings will take the form of a calculation of the *magnitude-difference* between the known and the variable star. The standard equation for this reduction is

$$\mathrm{MAG}_{\mathrm{var}} - \mathrm{MAG}_{\mathrm{comp}} = -2.5 \log (d_{\mathrm{var}}/d_{\mathrm{comp}})$$

In this expression, d_{var} and d_{comp} are the galvanometer readings, or readings of a digital pulse-counter, recorded for the variable and the comparison star, respectively. As in the usual method for visual estimation of a variable's brightness, we will have arrived at a value that represents the fraction of a magnitude by which the variable differs from the comparison star's known brightness.

If the measurements have been made through a narrow-band-pass filter (which blocks all light except a selected wavelength), the value that we have found will show the two stars' magnitude-difference in only one wavelength. The measurements can be repeated with a range of different filters, to compare the stars in a full range of wavelengths. If only a single comparison is to be attempted, however, we will use a yellow 'V' filter, which approximates the effect noted by the visual observer.

The above description of photometric technique is a simplified overview. If you are attracted to serious photometry, you will want to familiarize yourself with detailed procedure and refinements as presented in, for instance, Henden and Kaitchuck's excellent *Astronomical Photometry* (Van Nostrand Reinhold, 1982; reprinted, with corrections, Willmann-Bell, 1990).

An amateur's project

Typical of work attempted by some amateur photometrists was a recent project undertaken by Royal Astronomical Society of Canada member David Kopriva. It is interesting as an example of the power of photoelectric photometry applied to special astronomical events.

In 1985, when the orbital planes of Jupiter's satellites lay aligned with our line of sight, numerous mutual satellite encounters (occultations, eclipses) occurred. Kopriva's project involved photoelectric monitoring of these events in an effort to achieve very precise timings that might yield refinements of the satellites' orbital elements. The telescope used was the 50 cm Cassegrain reflector of the University of Victoria, in British Columbia. The unusually sophisticated monitoring system employed by Kopriva linked a computer with the photometer, the computer's clock having been synchronized with the shortwave time signal. The computer was programmed to read and record the photometer's counter at precise five-second intervals.

In at least one of the events, an occultation of one satellite by another, the computer-recorded photometric data identified the minimal point of the occulted satellite's light curve to within an accuracy of plus or minus five seconds. This precision of timing allowed the body's orbital position to be fixed to within a matter of *a few tens of kilometres* – a creditable result for an informal amateur exercise!

PART 6 The build-it-yourself astronomer

6.1 The build-it-yourself astronomer

To several generations of amateur astronomers the name 'Stellafane' has had an almost mystical significance: it is a magic word that has conjured up the creative spirit among builders and users of astronomical telescopes.

Stellafane, on Breezy Hill in Vermont, was the pioneer gathering place of amateur telescope makers, the first of the clubs to which people from hundreds of miles around gathered annually to display and use homebuilt telescopes and to learn from each other the secrets of simple, handcrafted optics that were capable of penetrating the remote depths of space. For almost three-quarters of a century astronomically minded visitors have flocked each year to the August rallies at Stellafane, inspired by the example of almost legendary pioneers such as Albert Ingalls and Russell Porter.

Traditionally, the mystique of amateur astronomy – a strong element of its essential spirit – has been the pride and joy of hand-building the equipment with which we make our excursions into deep space.

In the generation that saw the first Stellafane gathering, every amateur built his or her own telescope, because commercial equipment was virtually unavailable. More recently, during the present authors' own novice years in astronomy, most amateurs built at least one telescope before eventually acquiring a factory-made instrument.

For many of us the rich satisfaction of crisply defined lunar craters or the rings of Saturn observed through simple optics created by our own hands has never been surpassed. The views obtained through more sophisticated commercial telescopes are usually superb; yet they never supply the thrill of the (perhaps rather crude) equipment that we built in order to enjoy our first glimpse of the Cosmos.

Plate 134. The homebuilt telescope may be as simple or as elaborate as your taste requires.

Nowadays, that thrill is largely unknown, for the home construction of telescopes has passed almost completely out of vogue.

There are understandable reasons for the demise of amateur optics. One of them is the increasingly widespread availability of mass-produced telescopes and other equipment at prices that do not greatly exceed the cost of building in the home workshop. Another is the current waning of do-it-yourself enthusiasm in general. We seem to have moved into an age that favours look-alike mass products of all kinds, in preference to the fascinating individuality of things designed by the owners to suit their own tastes.

Before you decide to buy, rather than build, your own telescope, consider this: when we bypass the excitement of discovering the night sky with equipment constructed by our own hands we are cutting ourselves off from a tradition that began with Galileo. His crudely hand-figured lenses in their narrow pasteboard tube comprised a pretty rudimentary telescope. Yet it was his own – and it was sufficiently powerful to stir up a storm of delightful trouble!

6.2 A telescope of your own: grinding and figuring the mirror

Many of today's amateurs buy ready-made telescopes, or at least assemble their telescopes from commercially produced optics. It seems to me that there are two groups of amateurs: those who build telescopes, and those (often a quite separate class) who look through them. Some of the finest telescopes are built by people who derive all their pleasure from the actual construction itself.

I have ground and polished mirrors from 20 cm to 40 cm in size, and have constructed a variety of telescopes and the observatory buildings to house them. Nevertheless, I get far more pleasure looking and photographing through these instruments than I did from building them. I build telescopes as a means to an end.

My approach is pragmatic. I often say that if something doesn't fit, I'll get a bigger hammer. I am not noted for building telescopes that look beautiful; my only concern is that they be functional. It is this pragmatic outlook that is the basis of the instructions that follow.

How to begin

What is involved in producing a high-quality primary mirror at home, and by hand?

There will be, for one thing, a considerable amount of work. This work, however, is of a not unreasonably demanding sort; the manual skills required for the shaping of a telescope's most important optical element are surprisingly small. Patience is more essential than technical expertise. A second thing that will undoubtedly be involved in the process is satisfaction; there is a special thrill to be experienced when you observe the first starlight through a telescope whose mirror you have shaped with your own hands.

Plate 135. Telescope maker Frank Shinn, intent on polishing.

The first step, if you choose to grind a mirror, is to acquire suitable materials in the form of a mirror-kit, which will contain a pyrex mirror blank, a matching disc of glass for use as the grinding tool, abrasives for grinding, and the pitch and cerium oxide used in final polishing. The abrasives will form a large part of your kit. They will include carborundum grits in grades Nos. 80, 120, 220, 320, 420, and 600 and ALOX (3 grades) for rough-shaping and step-by-step smoothing of the mirror's concave surface. Some mirror-kits may also contain a completed diagonal mirror, ready for use in your telescope, when it is complete.

Purchase of these basic supplies will almost invariably be a mail order transaction. The popular astronomical journals advertise numerous suppliers. My own source has often been the Edmund Scientific Company (Barrington, NJ 08007, USA, and 3500 Bathurst St., Toronto, Canada M6A 2C6). This firm also supplies an outstanding pictorial instruction book, Sam Brown's *Homebuilt Reflector Telescopes*, one of the simplest guides available for the beginner.

The actual work of creating your telescope mirror will comprise the following steps:

1. Rough grinding, to turn the flat surface of the pyrex blank into a concave curve of the desired radius.
2. Fine grinding (in several stages) to produce a smooth glass surface.
3. The making of a 'pitch lap', which will be used as a polishing tool.
4. Polishing, to achieve the final mirror-surface.

5. 'Figuring' the surface. This polishing stage corrects the mirror to a precise optical form (spherical or parabolic).
6. Aluminizing the surface. This step is usually done by a commercial mirror-coating firm.

Choosing the configuration

Before ordering your mirror-kit, you will have decided on your telescope's aperture. Pyrex blanks come in a number of standard diameters: 10 cm, 15 cm, 20 cm, 25 cm, etc.

The second decision concerns your mirror's focal length. Amateurs typically select focal ratios ranging from as small as f5 (focal length equal to five times mirror-diameter) to as large as f10, or f12. My choice would be a 20 cm mirror of f9 – a focal length of 180 cm. There is a major advantage in undertaking a mirror of this length of focus, rather than shorter. Unlike mirrors of f8 or less, which need a precise paraboloidal figure if they are to perform adequately, an f9 will be excellent if it is left with a simple spherical figure. This is an attractive option, because the natural polishing strokes used by a hand craftsman tend automatically to produce a surface that is spheroidal to an accuracy of $\frac{1}{4}$ of a light-wavelength.

One wavelength (λ) equals only about half a micron, or a few ten-thousandths of a millimetre (0.00005 mm) and a quarter λ is equivalent to about 0.0000125 mm. Although an accuracy of ten-thousandth of a millimetre may seem an improbable goal for hand-figuring techniques, it turns out that such results arise fairly automatically from the standard patterns of grinding and polishing strokes used in the normal amateur method. If you can arrive at a mirror surface that does not depart by more than $\frac{1}{4}$-wavelength from a smooth spheroidal figure (not an especially demanding task) your telescope will produce first-class star images.

Grinding the rough shape (Figure 6.1)

Grinding requires a workbench that you can walk around. It is essential that the bench or pedestal be extremely stable. I have used a 45 gallon (180 l) drum partly filled with water, but one might choose to make the working pedestal from heavy pipe on a base weighted down by sandbags.

Grinding is done with the glass 'tool' fixed to the pedestal-top, and the mirror worked face-down on top of the tool. Use three small blocks of wood to chock the tool into position, taking care not to hit the glass tool with the hammer while securing the blocks. (You may find your project held up for many weeks if you have to wait for a replacement to arrive!)

Have a pail of water, a supply of paper towels and a squirt-bottle on hand.

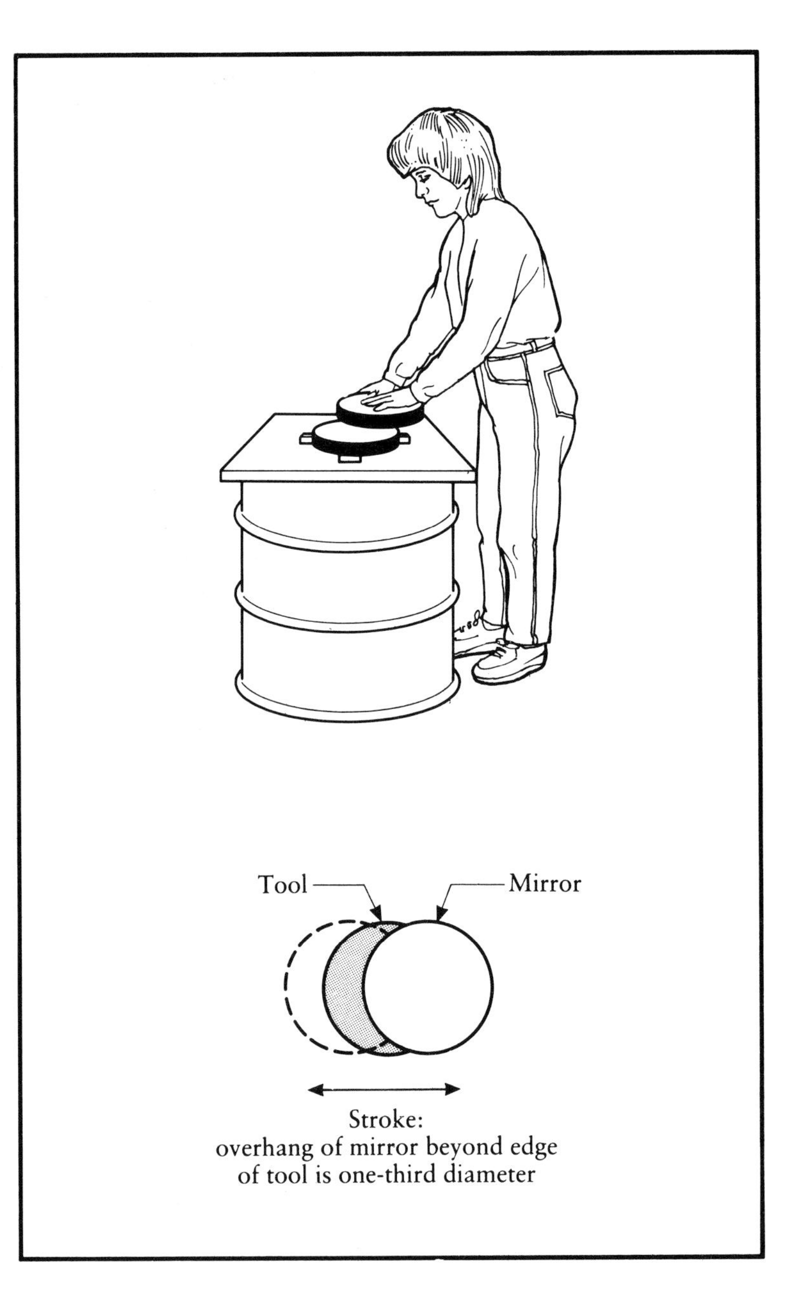

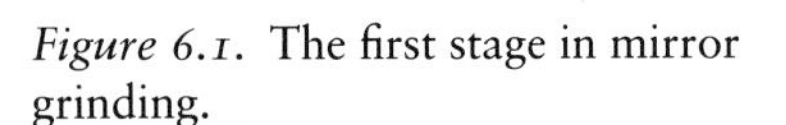
Figure 6.1. The first stage in mirror grinding.

To begin grinding, squirt a small amount of water onto the tool, and sprinkle a light scatter of the coarsest grade (No. 80) of carborundum into the water. Place the mirror face-down on the tool, and slide the mirror back and forth, centre over centre, in a stroke whose length causes the mirror to overhang the tool-edge by about one-third of its diameter.

After a few strokes in one direction, rotate the mirror about 10 degrees in one direction, step in the opposite direction about 20 degrees, and do a few more strokes. The mirror will ring with the sound of rough grinding! After each few strokes, the mirror is rotated and your working position is altered so that, in effect, you are circling the workstand as you progress with the grinding. A few circuits of the pedestal will use up the charge of carborundum, and this brief session of work is called a 'wet'.

After a 'wet' of a minute or two, the carborundum will have broken down to a smooth mud, at which time the tool and mirror must be washed in your pail of water, and a new 'wet' must begin. This cycle will be repeated until the desired depth of curvature has been achieved, a goal that I have sometimes reached in only about three hours.

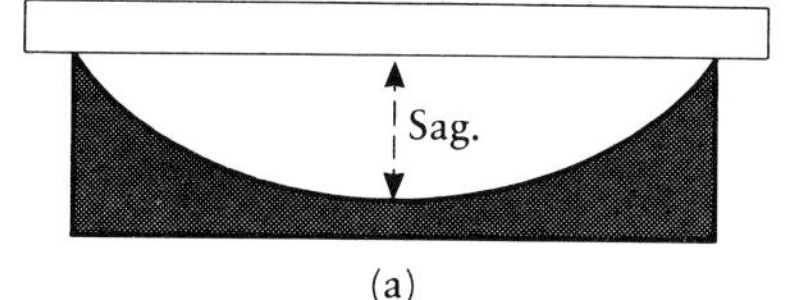

(a)

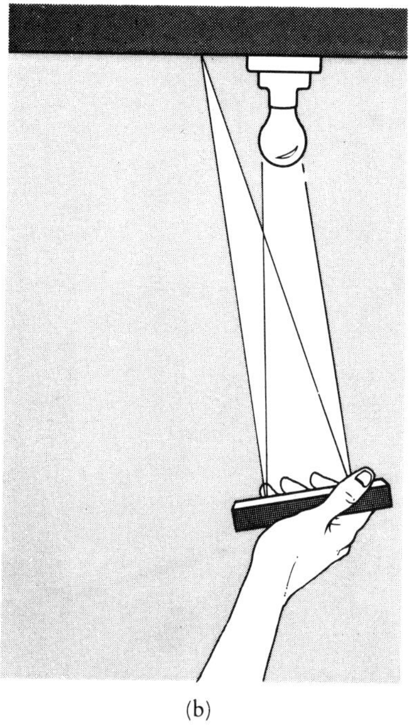

(b)

Figure 6.2. *(a)*. Depth from straight edge to centre of mirror is called the 'sagitta' of the curve. *(b)* Finding the approximate radius of curvature of a partially completed mirror.

At this stage there is a small precaution that should be taken. As the curvature of the mirror deepens, the edge of the blank will become quite sharp and fragile. To prevent accidental chipping, you should chamfer the edge, using an oil-stone and water.

Also at this stage, you will want to determine the radius of curvature that your rough grinding has achieved. The correct radius of curvature for a given mirror will be *twice* its proposed focal length (our 20 cm, f9 mirror, with a focal length of 180 cm, will need a spheroidal curve of 360 cm radius). What depth of curve across the mirror-blank's 20 cm face will correspond with the desired radius? There is a simple equation that lets you know how deep a concavity you must grind:

$$\mathrm{SAG} = \frac{r^2}{2R}$$

where Sag. (*sagitta*) is the central depth of the mirror (Figure 6.2 (*a*)), r is half its diameter, and R is the radius of curvature. In the 20 cm telescope described above, the desired sagitta will be

$$\frac{10\,\mathrm{cm}^2}{2(360\,\mathrm{cm})} = 1.39\,\mathrm{mm}$$

The various standard thicknesses on a wire spark plug gauge include one of 1.37 mm. If you place a straight edge across the top of the mirror, and if the 1.37 mm gauge just slips under it at the centre, you are very close to the desired curvature.

When you are close to the correct depth, shorten your grinding stroke in order to smooth the spherical curve. You can check your progress in the following manner: with a waterproof marking pen, draw a large X over the whole mirror-surface. If this X disappears evenly as you continue grinding, you are achieving a reasonable spheroid.

First optical test

You may now wish to do a simple optical test that will quite precisely confirm the radius of curvature (and the focal length) that you have achieved in your mirror.

The first step involves a sheet of honeycomb base, a material that can be obtained from beekeepers' supply stores. Press this onto the tool, and with some cerium oxide or optician's rouge, polish the mirror for about ten minutes. This will put enough shine on the roughly ground mirror to reflect the image of a lightbulb. Using a bare ceiling light, reflect the image of its bulb from your mirror back up to the ceiling (Figure 6.2*b*). Measure the distance from the mirror surface to the image of the bulb on the ceiling, to determine the radius of curvature. The focal length is half that distance.

Outdoors in daylight hours, you may vary the procedure. Tilt your mirror skyward, to project the Sun's image onto the underside of a house's eaves. In this case, the distance from the mirror to projected image is equal to the focal length itself (not the radius of curvature).

Smoothing the surface

The next steps involve successive stages of finer grinding. You use the standard grinding stroke as above, but with the finer grades of carborundum, one at a time. Each time you change to a finer grade of abrasive, everything must be washed and cleaned up with a degree of thoroughness that you might expect in an operating room. If just a single grain of the previous (larger) grain of abrasive remains to contaminate your mirror, it will leave a *huge scratch*! The only remedy for this disaster is to revert again to the larger grain of carborundum, rough grinding until the scratch has disappeared.

As you work with successive stages of finer abrasives, you will want frequently to check the surface for evenness (that is, to ascertain that you are not departing from the perfect spheroid achieved from the beginning). Examine the surface by reflecting the image of a frosted lightbulb at a very low angle (Figure 6.3*a*). As you move the image of the bulb across the surface, the light may apparently dim; if so, you have hit a 'valley' or depression in the mirror's figure.

Figure 6.3. (a) Examining mirror in low-angle light. (i) Smoothly lit surface is probably a good spheroid. (ii) Shadows and highlights indicate problems. *(b)* Short W-stroke used in final smooth grinding stages.

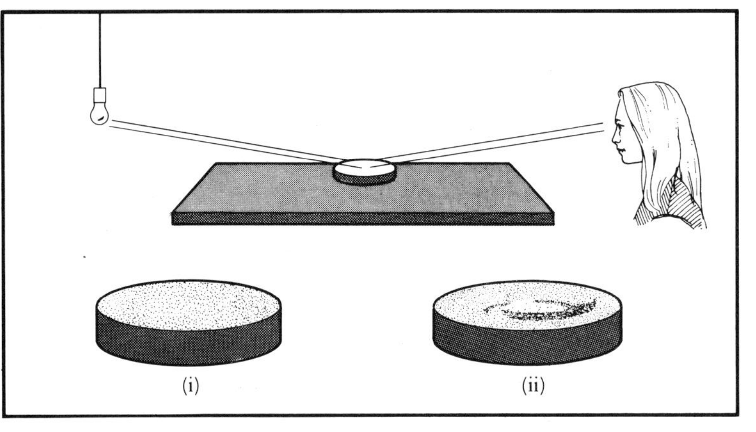

(a)

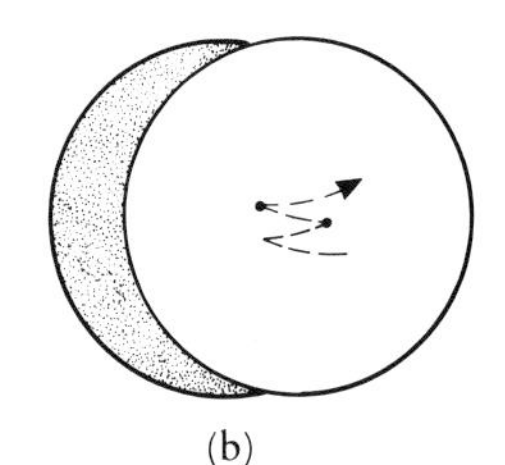

(b)

A local brightening of the light suggests a hill or ridge in the glass. If the light remains uniform as it moves over the whole surface, the mirror is adequately spherical.

To maintain the desired focal length, rather than continuing to shorten it as fine grinding progresses, alternate the positions of tool and mirror. In other words, work for equal periods of time with the tool on the bottom, and on top.

During the final stages of grinding – those that use the very fine ALOX – a slight W-stroke will help smooth the surface, with no more than a 12 mm overhang on the edges (Figure 6.3*b*). The finest grinding stage will leave the glass surface satin-smooth, and nicely spherical in figure. At this point you will have done perhaps ten hours of work on a 15 cm mirror, or twelve to sixteen hours on the 20 cm glass. (Determination helps. I once hogged out a 40 cm mirror to f5 in one weekend.)

Pitch lap and polishing

The mirror's reflective finish and its ultimate, high-precision figure will be arrived at by polishing with cerium oxide or optician's rouge. The tool that will apply this polishing agent will no longer be the bare glass disc that was used with the grinding abrasives, but a softer 'pitch lap', which is poured onto the glass tool's surface.

The pitch should be heated slowly in a coffee can to which a wooden handle is attached, or on which a handle is formed by cutting the tin and bending it to form a handle (see Figure 6.4).

Figure 6.4. Pouring a pitch lap. A coffee can is cut and bent to form a handled pouring vessel. Melted pitch is poured onto the glass tool, which has been encircled by a paper collar to contain the pitch as it cools and sets.

The hard pitch should be broken in a bag with a hammer, and placed into the melting tin in small fragments. Once melted (by heating on top of a hotplate), the pitch must be further softened by adding one teaspoon of turpentine for each pound of pitch. After melting, the pitch should be taken off the heat and allowed to cool until it has the consistency of cool melted honey or warm molasses.

When its consistency is right, it is poured onto the glass tool. The mirror itself will then be pressed onto the pitch surface to give it a convex form that matches the concave figure of the mirror. This will be the surface that will be used for polishing.

To do this the mirror and tool should be first warmed in hot water, and dried. A coat of water-and-rouge paste applied to the mirror will prevent sticking. (One part rouge to four parts water will suffice.) A wax paper collar must be placed around the perimeter of the tool and held by a rubber band; this 'dam' will contain the soft pitch on the surface of the tool when it is poured. Pour the pitch directly onto the warm tool, working from the outside in, and adding slightly greater depth at the centre to accommodate the mirror's concavity.

Now quickly press the mirror, with its covering of rouge paste, onto the soft pitch (Figure 6.5). In about five minutes slide the mirror off the lap. The soft pitch now has to be grooved, using a

Figure 6.5. Making a pitch lap. Mirror (rouged to prevent sticking) is pressed face down onto lap while the pitch is still soft. Lap is completed by addition of impressed channels, producing a regular surface pattern of 20 mm squares.

straight edge such as a wooden ruler. A pattern of channels on the pitch-lap's surface will enable the build-up of spent rouge, during polishing, to be carried away from the mirror. The first grooves will be pressed in at 20 mm intervals, a second set of these channels being placed at right angles to the first. The resulting square facets will be offset (that is, arranged so that no intersection of grooves falls precisely at the lap's centre), so that rings will not be generated on the mirror during polishing.

Some kits provide a rubber mat that can be laid over the pitch and pressed into its surface. The mat must be coated with rouge paste as a release agent, to prevent its adhering to the soft pitch. I have had success with an alternative method of using such a mat: I cut the mat in a circle to fit the tool inside its collar of waxed paper, in such a way as to place its square facets off-centre. The pitch is then poured over the whole assembly. If this approach is used, one must use only enough pitch just to cover the mat. Finally, the mat is removed, leaving behind a clean, deep pattern of grooves.

When the lap is finished, any pitch must be trimmed from the edge using a large knife or a chisel. It is wise to wash the lap and the mirror, and to coat the mirror again with the rouge-and-water paste and then to press the mirror onto the lap overnight with a moderate additional weight (a large dictionary or two will be about right).

To begin polishing, apply a loose mixture of water and rouge or water and cerium oxide with a small paint brush. The polishing stroke is similar to the final grinding stroke, in that it should have a slight W- or zigzag pattern that prevents the mirror and tool from crossing exactly centre over centre every time. As in the grinding procedure, polishing is done while slowly walking around the lap, and slowly turning the mirror in the hand. Be sure to avoid overhanging the mirror more than 30 per cent of its diameter, or you may turn down its edge, thus spoiling its figure.

A 20 cm mirror will acquire a full, overall polish after about ten to sixteen hours of work. If your work-sessions are about one hour, press the mirror on the lap between the sessions in order to keep the lap in perfect match with the mirror. Over a period of a day or so the channels in the lap will compress and gradually begin to disappear. A knife can be used to chip and reshape the grooves.

Testing the polished mirror

Periodically during polishing the mirror should be tested for accuracy.

The equipment required is a light source housed inside a container such as a small juice can which is given a fine slit to permit a very narrow, directed beam to escape. A wooden holder should be made to support the mirror on edge, facing the source of light. A razor blade should be mounted on a free-sliding wood block, to be moved

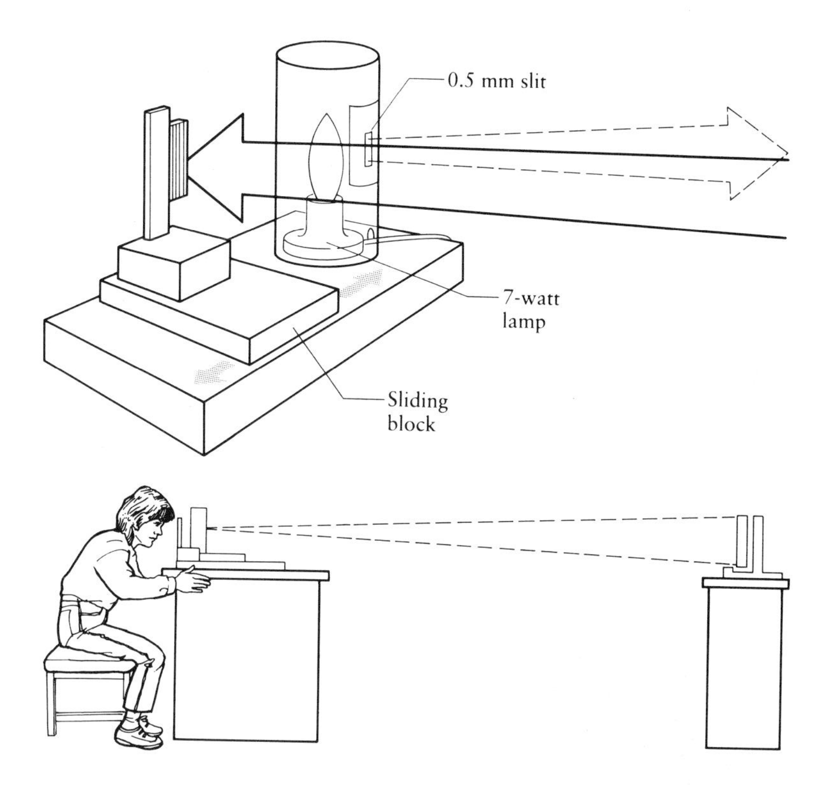

Figure 6.6. Razor-edge tester.

Plate 136. Testing the mirror's figure for a paraboloid mirror.

across the path of the reflected light in such a manner as to cut the beam. See Figure 6.6 for details. A word of caution: the light and its surrounding can may become very hot!

The procedure is to shine a narrow beam of light at the mirror and to bring the reflected beam back from the mirror to a point at which the razor blade can be slid into its path. With your eye very close to the blade, locate the light on the mirror; the whole surface of the mirror will be fully lit. As you bring the razor-edge into the light beam, a shadow will appear to cross the mirror. If you have placed the razor blade at a distance that is outside the focal point of the mirror, the shadow will cross from one side to the other. If the blade is inside the focus, the shadow's direction of movement will be reversed. When you manoeuvre the blade to precisely the focal distance, then if the mirror is spherical, the light on its surface will blink out all at once, evenly. (See Figure 6.7.) If this occurs, you have ascertained two facts: the mirror's precise focal length and the fact that its figure is adequately spherical.

DON'T FORGET . . .
The TEST LAMP is HOT!

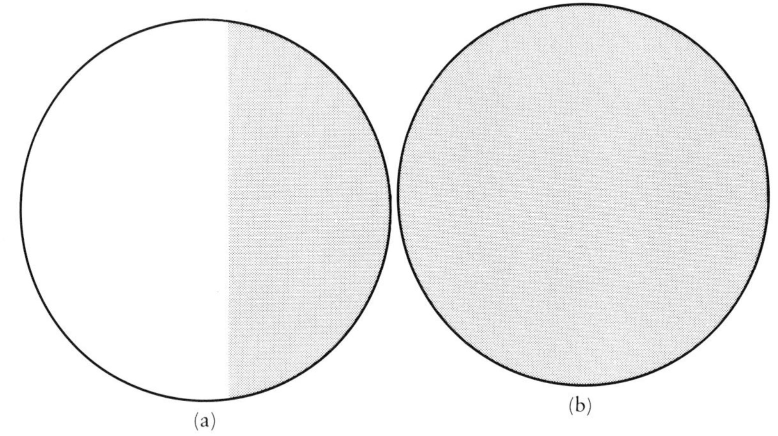

Figure 6.7. (a) Straight-edged partial darkening as blade cuts beam outside (or inside) focal distance. *(b)* Instantaneous uniform darkening of spherical mirror, when blade intercepts beam at focus.

There is another way to perform a quick test on your mirror. In place of the razor blade, use a fine petrol filter-screen with a mesh of 20 lines to the centimetre, or alternatively a 40-line optical grating purchased from one of the telescope supply outlets such as Edmund Scientific. When the filter or grating cuts the returning light beam, the mirror will appear to be crossed with lines, the number of visible lines being greater as you move the grating further from the focal distance, and fewer as you more nearly approach the exact focal point. Position the grating so that you find about three to five lines across the full mirror diameter, just outside of focus. If the lines have a straight edge, you have a perfect sphere. If the lines are straight except at the edge of the mirror, and if they curve abruptly in that region, you have unfortunately turned down the edge of the mirror. Perhaps the easiest solution to this problem is simply to mask the outer edge of the mirror, or to chamfer off the offending edge.

Remember that a mirror of focal ratio f9 or longer should be left spherical. When your tests indicate that you have a good spheroid, you can regard the mirror as finished; it will produce a superb telescope.

Even when spherical, a 15 cm f10 mirror will be better than $\frac{1}{4}$-wave (the accuracy of figure normally deemed acceptable to the discerning eye).

Parabolic mirror

If you have opted for a telescope of shorter focal length (f8 or less), you have a moderately tricky additional step to complete, before your mirror is usable. In 'parabolizing', the shape of the mirror must be changed to a curve (the paraboloid) that very subtly varies from the spherical figure. You will have to use a special polishing stroke that will deepen the centre of the mirror.

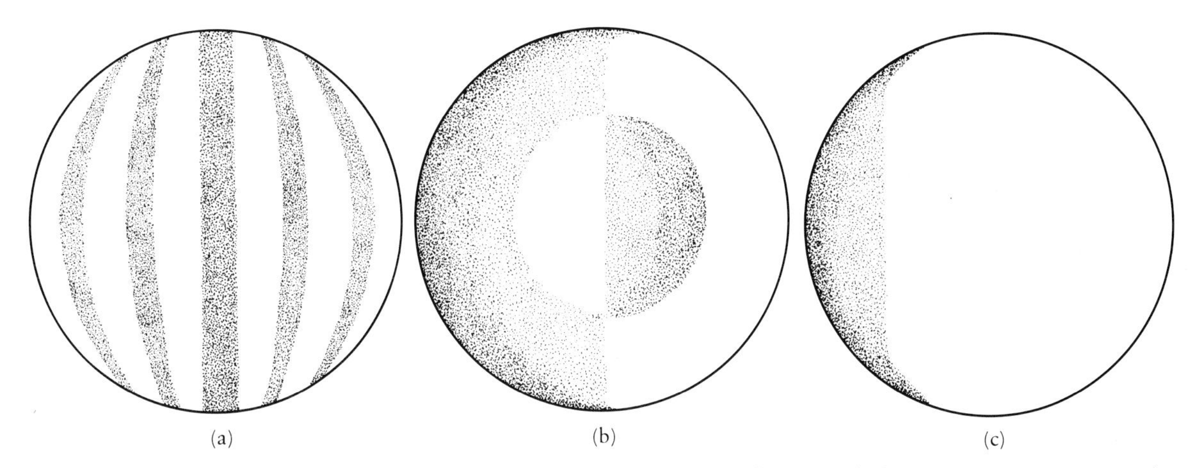

Figure 6.8. *(a)* Grating test of paraboloid, showing smooth gradation of line curvatures. The extreme line curvature here is typical of a very-short-focus paraboloid. A long-focus mirror will show a more subtle effect. *(b)* Knife-edge test of paraboloid. Shadow pattern here is that of blade-edge at focal plane. *(c)* Knife-edge shadow on mirror with turned down edge. This is the shadow pattern (on both spherical and parabolic mirrors) of blade-edge entering beam.

The following discussion of the parabolizing process is necessarily brief. Entire books have been devoted to the techniques of polishing a parabolic figure, and testing for accurate results. If your mirror is one that will require a parabola, you would be would advised to visit your local library for a copy of one of the amateur guides to this work. A classic and favourite example is Jean Texereau's *How to Make a Telescope* (2nd edn, Willmann-Bell, 1984).

The parabolizing stroke, in the final stages of polishing, is again a **W**-pattern, but this time with an elongated stroke that also includes a broader zigzag, or **W**. After only about five minutes of this work, you will want to test the mirror's progress.

In the filter or grating test that we described above, a parabolic surface will show a characteristic pattern of lines. Instead of being straight, as in the case of the spheroid (or straight at centre, with curving edge-lines in the case of a spheroid with turned edges), the paraboloid will show an array of lines that begin curving outward from the centre. The central lines will be slightly curved, and the degree of curvature will smoothly and gradually increase as the lines move outward towards the mirror's edge.

If a razor blade is being used for testing, its intrusion into the returning light beam will produce a pattern of light and shadow that appears doughnut-shaped. In more detail: at one distance from the mirror, the razor blade will produce a circular shadow at the centre of the mirror; as you move the block away from the mirror, this shadow will roll over the surface until it curves around the mirror's edge. In effect, you are seeing the result of a slight difference of focal-distance in the central area of the mirror and in its outer rim.

In a correct paraboloid, the difference in focal distance between the centre and the edge will be a precise ratio that is determined by the mirror's diameter and its focal ratio. The formula is r^2/R, where r is the radius of the mirror and R is the mirror's radius of curvature. Thus, for example: if the mirror is a 15 cm, f5, its total correction (centre-to-edge difference of focal distance) will be

$$\frac{r^2}{R} = \frac{56\text{ cm}}{150\text{ cm}} = 0.37\text{ cm}$$

In practice, tests of the focal lengths of three zones are checked – the centre, a region of the mirror 70 per cent of the radius outward from the centre, and the edge. When a figure has been achieved that gives a smooth curve of focal distance from centre to edge, the parabola is a good one, and the mirror will make an excellent telescope.

6.3 The 'people's telescope': a Dobsonian

Imagine an astronomical telescope so powerful that it can resolve a globular cluster or show the dustlanes in a galaxy, and yet be so simple that a novice without experience and with only a few basic tools can easily construct it.

To the traditional telescope maker, such an instrument might seem like a pipedream. Large, powerful optics are usually found suspended in mechanically complex 'flotation cells', within a precision-formed metal alloy tube. In its classic form, this kind of telescope has normally been fitted atop an equatorial mounting whose moving parts are machined from steel, to exacting tolerances.

Plate 137. Binoculars, giant size! (Paired 17-in. telescopes).

It is at this stage of construction that many amateur telescope projects fail, even after the builder has successfully completed his or her own astronomical mirror.

In the past decade, however, a revolution has swept the amateur telescope-making field. It began when a resourceful member of the San Francisco Sidewalk Astronomers, in search of a cheap and easy way to build large reflecting telescopes, decided to challenge some of the optical and mechanical dogmas.

John Dobson found that there are inexpensive alternatives to the usual pyrex mirror-blanks, whose standard thickness the experts had always quoted to be at least 1–6 cm (1 cm for each 6 cm of diameter). Dobson made thinner, cheaper mirrors from surplus plate glass port-hole-covers, supporting these (perhaps) slightly flexible optics on a cushion of carpet, and in a sling strap. His mirrors held their figures very well. Unable to afford an elaborate machine stop, he constructed telescope tubes and mountings of plywood, inventing ways to make their mechanical movements surprisingly smooth and precise.

In the 1970s Dobson began to display unprecedentedly huge reflecting telescopes at star parties. Looking through these plywood giants, many people enjoyed their first taste of really large-aperture deep-sky observing. When they learned that some of these fine instruments had cost their builder virtually nothing, the revolution began. Soon hundreds of 'Dobsonians' were being built by amateurs around the world. Telescopes of this kind ranging from 20 cm to 75 cm aperture are seen nowadays at star parties everywhere.*

* My own equatorially mounted 50 cm reflector shows some of John Dobson's influence: it is a relatively thin-mirrored telescope, with a glass only 6 cm thick. (JN)

Plate 138. Jeannie Clark of Sarasota, Florida, and her 36-in. (91-cm) 'yard scope'.

Dobsonian design

This style is the simplest and most economical for the amateur who wishes to build a large-aperture telescope. Mechanically, it is built entirely of plywood and optically it will yield adequate performance with a mirror of a thickness-to-aperture ratio of only 1:16.

The mounting is a simple box, open at one side and on top. It achieves its smooth *altazimuth* movements through two bearings. The azimuth motion is obtained through rotation of the box on top of the base-plate, to which it is attached by a 12 mm through-bolt. To ensure easy rotation, the bottom of the box is covered with Arborite, and the base-plate is topped with three equidistant 5 cm^2 patches of Teflon. The altitude motion is obtained by rotation of the tube (on its large-diameter pinions) in the semicircular cutout yokes on the box's top edges. The pinions are 15 cm circles cut from heavy plastic pipe. They rest on Teflon pads within the semicircular yokes.

Can anything so simple actually work? Try it, and you will be quite amazed at the smoothness of motion and precision of control that so basic a mounting can provide, even when the telescope is being used at high magnification. An added bonus is portability. It is amusing, when everybody is preparing to depart at the end of an amateur gathering, to watch some owners laboriously unbolting and dismantling their complex equatorial mountings, while the Dobsonian enthusiasts simply lift the two segments of their instruments apart, for quick storage in a car.

One limitation, however, must be mentioned: being an altazimuth style of instrument, the Dobsonian is a visual telescope only (not suited to astrophotography). Yet for visual use, the large-aperture Dobsonians deliver enormous satisfaction at minimal cost.

Your personal Dobsonian

Let us suppose that you have made (or purchased) a mirror of 20 cm aperture and focal ratio f8 – that is, 160 cm focal length. The design of your telescope will begin with these primary mirror specifications.

To accommodate your mirror's diameter and focal length, you will need a tube at least 160 cm long and about 25 cm square. The primary mirror itself will be mounted on a piece of plywood (thickness: 2.5 cm), which is in turn attached to the tube's 2.5 cm

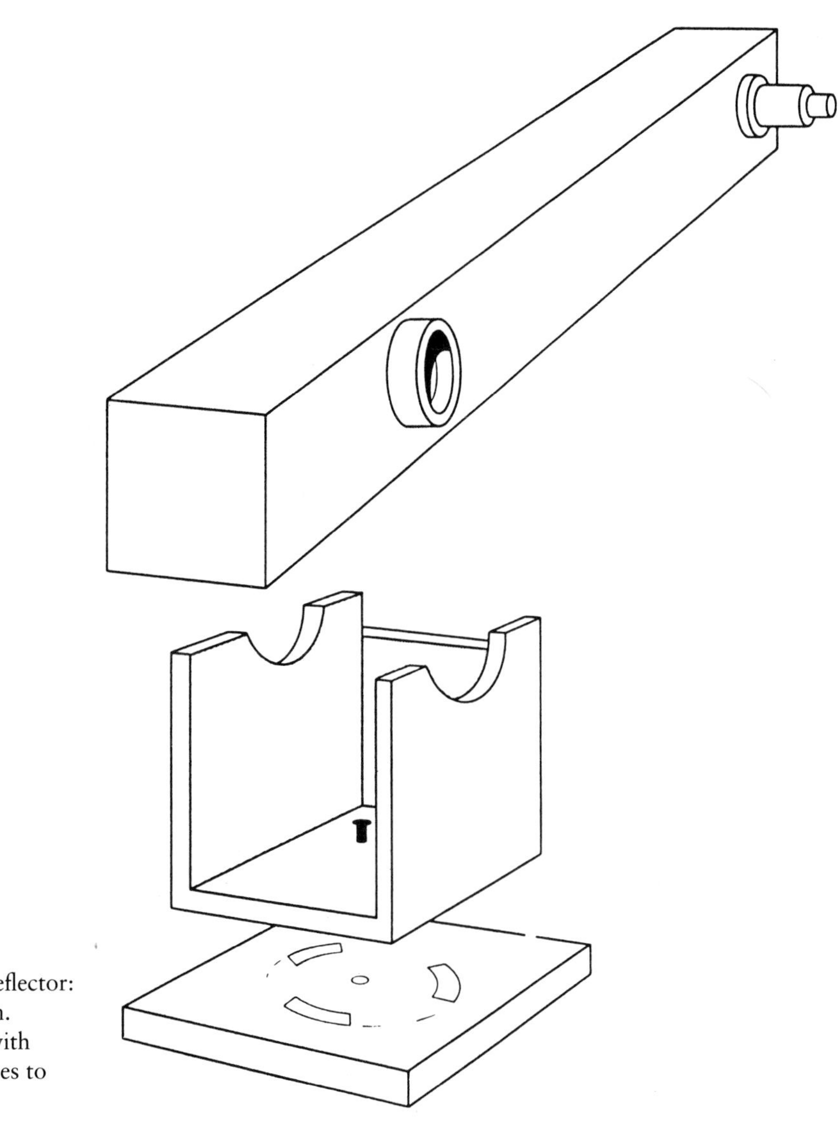

Figure 6.9. The Dobsonian reflector: principal features of its design. Construction is of plywood with Teflon pads on bearing surfaces to ensure smooth motion.

plywood endboard. This attachment is by means of three spring-loaded 12 mm bolts, by which the mirror can be collimated, or 'squared on'.

The distance of the diagonal mirror from the front (reflective) surface of the primary should be equal to the primary's focal length *minus* distance '*d*' in Figure 6.10, which must be sufficient to allow the focus to fall about 7 cm outside of the telescope's main tube, or whatever distance may be required so that the focal point lies at about the midpoint of mechanical travel in the rack-and-pinion eyepiece mount that you will be using. Focusing mount and diagonal mirror support can be purchased from commercial suppliers, or they can be constructed in the home workshop. (Some builders improvise focusing devices from smoothly threaded pipe fittings.)

Since all measurements within your Dobsonian telescope's optical tube will depend on the position of the primary mirror, it may be wise to design (and perhaps actually assemble) the mounting-board for that mirror before doing anything else. The finished dimensions of that structure will determine the precise location of the mirror inside the plywood tube.

Figure 6.10 illustrates all essential details of the primary-mirror mounting. The mirror rests against its plywood backboard with a

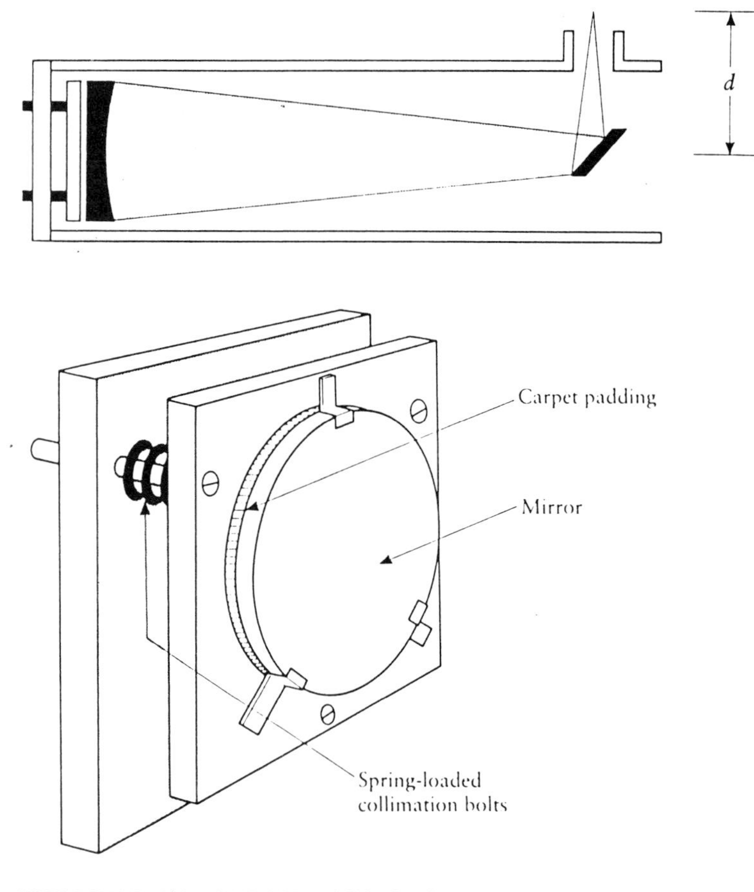

Figure 6.10. Essential details of primary-mirror mounting.

circle of carpet (or foam rubber) lying between the glass and the wood. If clips are used to hold the mirror in place, *they must be loose*. The slightest pressure from these retaining clips will surely distort the mirror. (If someone were to shake my own 50 cm telescope, the primary mirror would rattle!)

An alternative method of holding the mirror in place is the use of a soft leather sling that encircles the bottom half of the mirror-edge, and attaches to the backboard at the upper ends. This is adequate if the telescope-tube is never tilted to lower than the horizontal position. The sling method for distortion-free mirror support was used by John Dobson in his original telescopes of this kind.

The outer plywood backboard, to which the mirror-board is attached by three spring-loaded adjustment bolts, should be hinged to the end of the plywood tube. This permits easy access to the mirror, and allows the primary mirror to be quickly lifted out for transport. Alternatively, some designs have the mirror in a separate box that detaches from the tube-end, for storage. This mirror box can be covered with a detachable lid, and carried like a suitcase. Such an arrangement, although very common on the largest Dobsonians, is probably unnecessary in the 20 cm version that we are envisioning here.

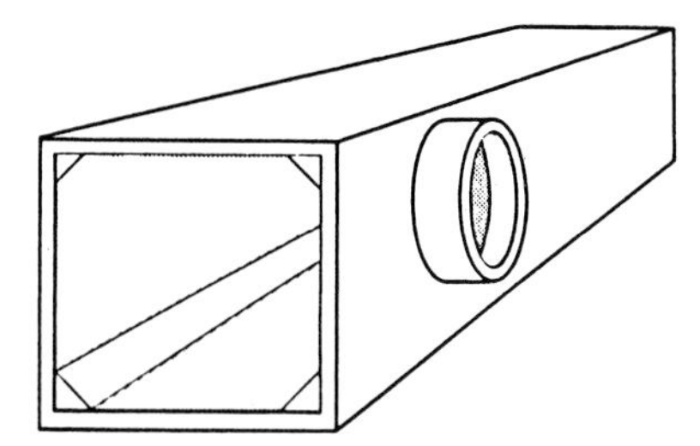

Figure 6.11. Mounting box for Dobsonian.

The main tube itself must be rigid, but it should also be kept light, for easy portability. For a small- to medium-sized telescope such as our 20 cm model, 9 mm plywood should suffice, framed at its corners by 2.5 cm × 2.5 cm glued strips as shown in Figure 6.11.

The mounting-box will be made of heavier plywood, of 18 mm thickness. The height of the mounting is a critical dimension. Before you can establish this measurement, you must discover the point along its length at which the optical tube assembly (with mirror in place) will balance. Place a wooden dowel on a flat surface, and find the point at which the telescope will balance on this dowel. Measure the distance from the bottom (mirror end) of the tube to the dowel. This distance, with about 5 cm added for clearance, is the inside height of the base-box.

The closed end of the base-box must be cut low enough to allow the telescope to reach the horizontal position. When the box has been fitted onto a heavy plywood baseboard by means of a 12 cm centre bolt, the Arborite-sheathed box should turn smoothly on the Teflon-padded board. Set the optical tube's plastic pinions into the Teflon-padded 'yokes' of the mounting box, and your Dobsonian telescope is ready for first light.

6.4 Constructing a cold camera

As mentioned in Chapter 4.6, a cold camera permits longer astronomical exposures without the reciprocity failure that normally limits the usefulness of films when the shutter is open for long periods. Deep chilling of the film (to below −50°C) is the key to performance in these circumstances, but a camera must be designed that will prevent the film surface from frosting up when exposed to air at such low temperatures.

The camera illustrated earlier in Plate 105 and Figure 4.8 uses an internal optical plug made of 50 mm-thick plastic to pin the film against the chilled back-plate of the camera, and to fill the space that would otherwise contain moist air. Thus, freezing moisture is kept from condensing onto the film emulsion. About 25 mm of the plug's exterior will frost over, but it will remain clear at both ends.

When an exposure has been completed, the camera body is detached from the shutter and plug; it is capped and brought to ambient temperature with a hair dryer before the next frame is advanced, for another exposure.

Simple cold camera design

A simple cold camera can be constructed from a sheet of three or four millimetre black plastic and some plastic pipe, available from sign shops, plastic plumbing supply outlets or various other sources.

The camera consists of two major parts: the body with dry-ice compartment at its back, and the section that comprises shutter and optical plug. The latter segment attaches to the telescope, while the camera body can be removed. The body should be just large enough to hold two 35 mm film cassettes spaced so that the film is transferred from one cassette to the other across the 50 mm opening occupied by the plug, without any other intervening space. A major advantage is thus gained over a conventional 35 mm camera, in that if the camera is accidentally opened the only frame that is affected is that which lies between the two cassettes. This allows easy removal of

Plate 139. The cold camera.

the film at any time if, for instance, one wishes to change over from colour to black-and-white film.

Begin construction with the body, cutting out its four sides from plastic sheeting, and cementing them together. A flat table with waxed paper on top of it will help keep the camera body square during assembly. The back of the camera will be made up of two layers of plastic, one cut to cover the back, and the other to fit snugly inside; this arrangement will ensure that the back cover is light-tight.

The cold-chamber is made from a plastic plumbing fitting called a drain inspection pipe; it is for 50 mm pipe, and is fitted with a threaded cap. One end of this short piece of inspection pipe will pass through the back of the camera, which will need a snugly fitting circular hole for the purpose (the portion of the pipe that passes through this will have an outside diameter of about 60 mm). At its outer end, the chamber will be covered by the threaded cap. Its inner end, against which the film will lie, will be closed by a disc of 2 mm brass plate, cemented into place with silicon bath-sealing compound. The whole assembly is glued into the camera back.

A spring plunger can be made to fit inside the pipe, and mounted through the threaded cap, to keep the sublimating dry ice pressed firmly against the brass plate.

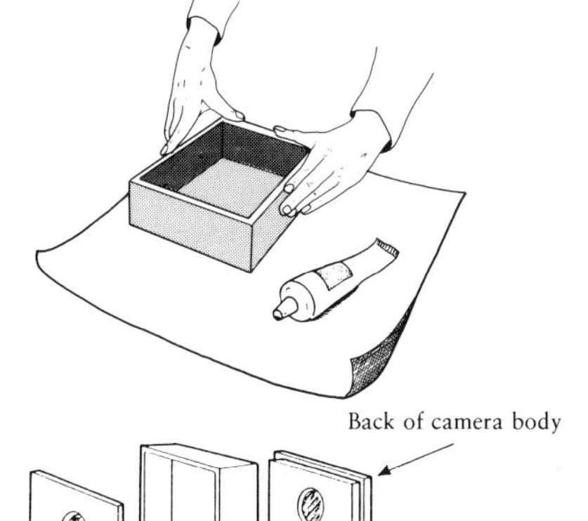

Figure 6.12. Constructing a simple cold camera.

Inside the camera, plastic film guides and cassette chambers fashioned from sheeting can now be cemented into position. Now mount bolts measuring 6 mm into plastic blocks inside the camera body, and the film-advancer can be made from stacked or laminated plastic, machined round.

The camera's front plate is now cut and fitted, from a single thickness of 4 mm plastic sheet. A circular opening at the centre of this plate will accommodate the 40-mm-long plastic guide tube, which will slide into the shutter assembly, to keep that assembly

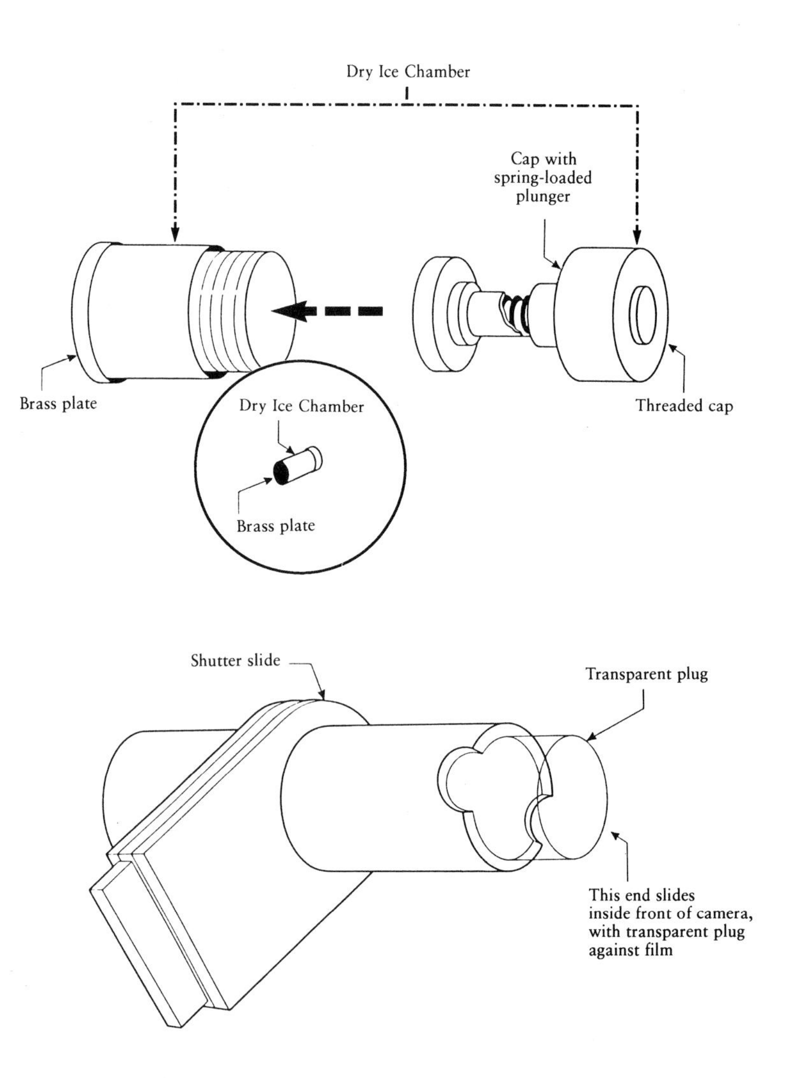

Figure 6.13. *(a)* At back of camera: the dry-ice chamber; *(b)* at front of camera: the shutter slide.

and the film in alignment. Over this guide tube, and cemented to the front of the camera, is an 80 mm laminated supporting ring. Drill and thread a 6 mm hole through the edge of this ring for a bolt that will be used to clamp the camera body to the shutter assembly, when the two are mated together. This will also be used to secure a cap over the camera body when it is detached from the shutter segment. Make the cap from a slice of pipe closed by a disc of plastic sheet.

Shutter assembly

This also is made up from plastic sheet and pipe, with a shutter that will itself be constructed of 4 mm aluminium plate. The first step is to select a pipe that will fit into your telescope's eyepiece holder – preferably a 50 mm eyepiece holder, for astrophotographic purposes. This tube will be 4 cm in length.

The shutter itself will be constructed of three layers of plastic sheet, the central one being cut to provide a channel or passage in which the aluminium shutter-plate will slide, from its open to its closed position.

A second tube will be fitted to this shutter assembly, to slide into the camera body, stopping one millimetre short of the brass plate at the film-plane. Both tubes – that which fits into the camera and that which fits into the telescope's eyepiece holder – are cemented into place on either side of the plastic shutter assembly. Inside the tube that fits the camera, there must be a plastic stop-ring of about 48 mm outside diameter glued into place near the tube's shutter-end, to provide a stop for the optical plug. Two half-moon cuts at the top of the tube will allow your fingers easily to extract the optical plug, a solid cylindrical prism of clear plastic that fills the tube in order to exclude moist air that would deposit condensation on the film surface. The shutter-slide itself can be fitted with a cable release made from bicycle brake-cable.

Finally, the optical plug is made, to fit inside the camera-end tube of the shutter assembly. Optical plugs should be cut from 5 cm thick plexiglass sheet, leaving the protective paper on the plastic while cutting (remember, the surfaces of the plastic will become optical surfaces, in your finished camera). I cut my plugs with a bandsaw, and then machined them on a lathe to a finished diameter to fit the shutter's tubing. As a last step, I threaded and flat-painted (black) the plugs' edges, as a means of stopping light from flaring around these edges during photography.

Finder-eyepiece for the camera

I made an eyepiece holder that would fit over the shutter tube, to facilitate location of deep-sky objects that are to be photographed. A 25 mm eyepiece is mounted in a plastic tube, to mate with the shutter assembly.

Finding procedure with the eyepiece is as follows. When the shutter is first placed on the telescope, a medium-bright star is located with the eyepiece, which is then removed. The optical plug is inserted, and a razor block is used to null-focus* the star, by adjusting the position of the shutter assembly. The required target object is now located, and centred in the field for astrophotography.

The photographic procedure itself has been described in Chapter 4.5.

* A razor-edge passed across the beam of starlight exiting from the telescope will appear to darken the image gradually from one side to the other when the blade-edge is not precisely at the point of focus. When the blade is precisely at the focal position, all parts of the lighted image will darken or 'null' simultaneously, thus indicating the correct position of the camera for a perfectly sharp image.

6.5 The home observatory

An observatory dome of your own is not essential to the enjoyment of astronomy. Many advanced amateurs do a lifetime of scientific observation with telescopes set up under the open sky, unprotected. Yet if the instrument is larger than, say, an 8 cm refractor or a portably mounted 20 cm reflector, it will probably be permanently set up, and will require a measure of shelter from the elements. The need for a permanent mounting is especially a factor if the telescope is used for astrophotography.

The present chapter will look at a few of the amateur's building options, ranging from the crudest lift-away cover for the telescope to an elaborate dome that is a miniature of those found at the mountain-top sites of the world's major observatories. First, however, there is the important initial consideration of which site to chose.

Choosing an observatory site

Most amateurs have little choice in the matter of an observing site; it will almost inevitably be the back garden at home.

If a special location is to be selected, however, it should offer at least a few basic advantages in terms of astronomical 'seeing' conditions. The amateur observatory-builder will probably not go so far as the professional, who tests prospective sites by means of long-term photometric studies of sky-transparency, etc. Nevertheless, darkness, transparency and relative atmospheric steadiness are goals to be sought.

To find reasonably dark skies one must place the observatory, if possible, beyond the city and away from intensely lighted areas such as shopping centres and major highways. At a minimally decent site, objects such as the Double Cluster in Perseus and the Beehive (M44) in Cancer should be easily located with the naked eye. One should be able to trace the outline of a constellation as dim as Equuleus (the little rhomboid of 4th- and 5th-magnitude stars southeast of Delphinus) without difficulty. These are not stringent tests. The night skies above some suburban locations may be acceptable, by these standards.

The other major 'seeing' requirement is more difficult to quantify. Atmospheric steadiness is the quality that will permit your telescope, on the best nights, to detect fine planetary details or to split double stars near the theoretical limit of resolution. If stellar images are typically so large and wobbly as to prevent your 20 cm instrument from ever clearly dividing the $2\frac{1}{2}$-arc-second pairs in Epsilon Lyrae, the site is probably too turbulent for either visual astronomy or good astrophotography. (Again, this is not a stringent test; it should be easy for even a little 6 cm refractor.)

Some geographic settings almost guarantee turbulent air. The leeward slope of a steep ridge is often subject to a cascade of disturbed air, whereas the windward slope may experience a smooth laminar flow that is conducive to good seeing. Locations where the evenings are characterized by a very extreme drop in temperature after sunset will give troubled seeing; small diurnal variation is better. A site that makes it necessary to observe over the roofs of heated buildings nearby is unfortunate; a large surround of open space is far better.

What sort of building?

Although the classic revolving dome is an attractive option, cost may dictate something more basic. The simple garden-shed style of observatory, which may be nothing more elaborate than one of the prefabricated sheet-metal units available from garden supply shops, modified to allow its roof to slide off, is a common choice. A rooftop observatory at the University of Victoria in British Columbia houses both a 30 cm Cassegrain telescope and a 25 cm Schmidt camera in a store-bought sliding-roof shed of this kind.

Two even more rudimentary options are suggested in Figure 6.14.

The first, which has been used by one of the present authors, involves only minor alteration of some existing structure such as a garage. My version of the hatch-roofed garage observatory required only a weekend's labour to produce the telescope's enclosure shown in Figure 6.14*a*. Ease of construction is this design's chief attraction. Its major drawback is that its overhead opening permits observation of only a segment of the sky – a pole-to-equator strip about 50 degrees wide. This limitation is somewhat mitigated, however, by the fact that the entire northern sky rotates past the southward-facing roof aperture, from season to season.

The second option (Figure 6.14*b*) features a permanent mounting for the telescope, but provides shelter for the instrument only when it is not in use. Variations on this simple design are widely used because of the low cost, and also because they let the observer work outdoors under an unobstructed sky. The structure is just sufficiently

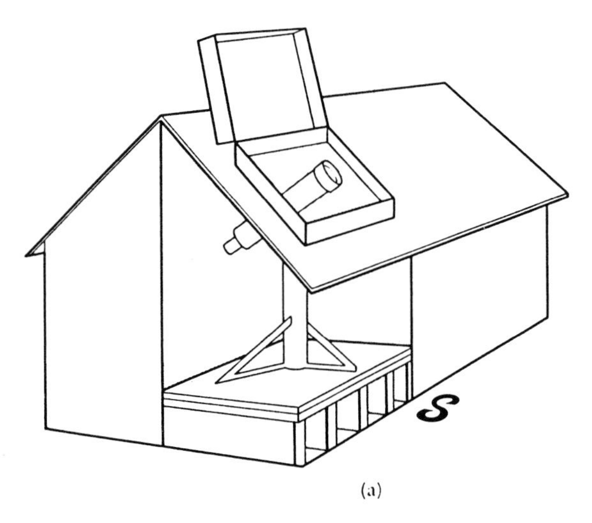

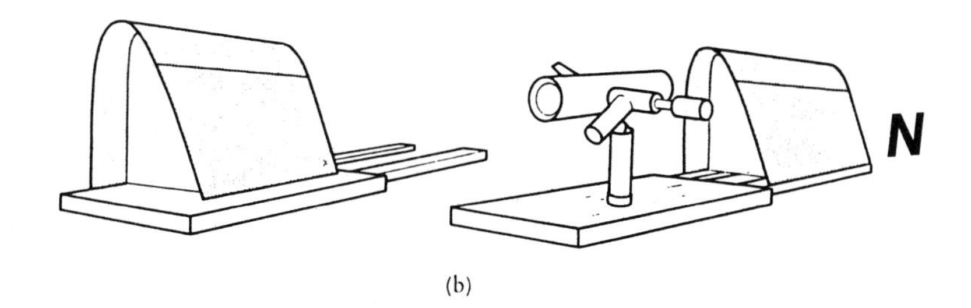

Figure 6.14. (a) Alteration of an existing structure (here a garage) to provide an observatory. *(b)* Permanent mounting and telescope shelter.

large to cover the telescope when the instrument has been swung down into its rest position. It is fitted with large-diameter casters that allow it to be wheeled completely away from the telescope on wooden rails. At its fully open position the little shed lies northward of the telescope, leaving an uncluttered view of the sky to the south, as well as eastward and westward.

The rotating dome

Compared with all the structures described above, the classic dome requires considerably more skill from its builder, and it will be much more costly.

Nevertheless, it is the style of building that most serious amateurs consider a proper observatory. In two ways its specialized form is highly advantageous: it offers the best possible shelter from wind, providing the observer with almost total enclosure while still giving access to the sky; also, its sky-access is unusually complete. The rotating dome can give access to virtually all of the visible sky, without obstructions or blind spots (Figure 6.15).

Essential elements of the typical amateur-built observatory are a

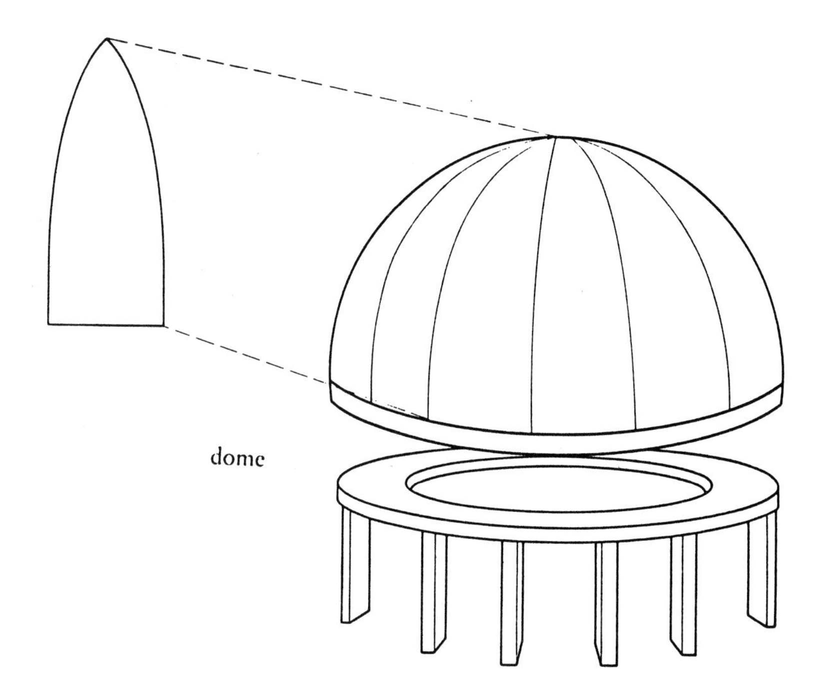

Figure 6.15. The rotating-dome observatory.

solid concrete base with cement pier for the telescope, a cylindrical lower wall framed with ordinary housing studs and sheathed with plywood and, rotating on a track atop this circular wall, the dome. The dome itself is usually a lightweight structure whose curved frames are sawn and glued plywood, and whose skin is formed of ten or more petal-shaped plywood segments.

A interesting variation on the usual two-part building (where the dome rotates on top of the cylindrical wall) is Canadian amateur George Ball's homebuilt observatory. He constructed the lower wall and dome as a single unit, the whole of which rotates on a circular track that lies on the concrete base, at ground level. This arrangement is structurally simple and strong, placing the weight on a track down on the building's foundation, rather than up on top of the wall. The fact that the entrance-doorway rotates with the entire building can prove rather disconcerting for the visitor; having entered from the south, one may be puzzled to find oneself exiting on the north side of the building – and wondering where one is!

Another variation on the standard hemispherical dome is the semi-cylindrical dome adopted by some builders because of its simple geometry. One of the authors (Jack Newton) has used this design for the building that houses his 50 cm reflector. Like the usual observatory, this structure has a cylindrical lower wall, topped by a track on which the dome rotates. The dome itself has a form that consists of two simple curves, a cylindrical segment having the same vertical axis as the lower wall, and an intersecting cylindrical surface whose axis is horizontal. (See Plate 140.) A shape of this kind, involving no compound curvature, is more quickly and easily built of plywood than the conventional hemisphere.

Plate 140. An Observatory by Jack Newton.

Support for the telescope

One feature that all really satisfactory observatories have in common is a pier-structure sufficiently rigid and massive to support the telescope without vibration.

Look back for a moment at Figure 6.14*a*, which shows the garage observatory. The floor is a heavily constructed wooden platform consisting of two layers of 18 mm plywood sheathing supported by closely spaced 5 cm × 20 cm framing underneath. One might suppose that this would be supremely rigid, or at least adequate to provide support for a 65 kg refractor. I must admit, however, that this floor is not a success. Even so heavily built a wooden structure still allows a slight vestige of flexure, much too small to be felt underfoot, but enough to be detected in the telescope at high magnification.

The floor of an observatory should be concrete, poured to as great a thickness as the builder can afford. The telescope's metal pier can be embedded permanently into the cement. Alternatively the pier itself can be of reinforced concrete, the floor and the pier being a monolithic unit.

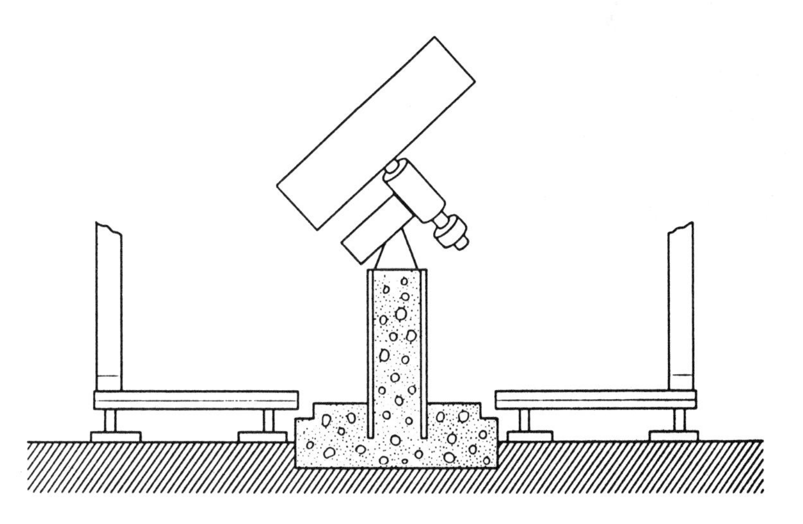

Figure 6.16. Interior cross-section of rotating-dome observatory showing concrete pier for telescope mounting.

If a large volume of concrete proves too costly for the amateur's budget, a satisfactory compromise is the arrangement shown in Figure 6.16. The key feature of this widely used foundation structure is the complete separation of the wooden floor from the concrete telescope base. The surface on which the observer stands at no point touches the pier or its concrete floor.

An advanced design

Typical of the most elaborate homebuilt observatories is a superb small aluminium domed building designed by landscape architect

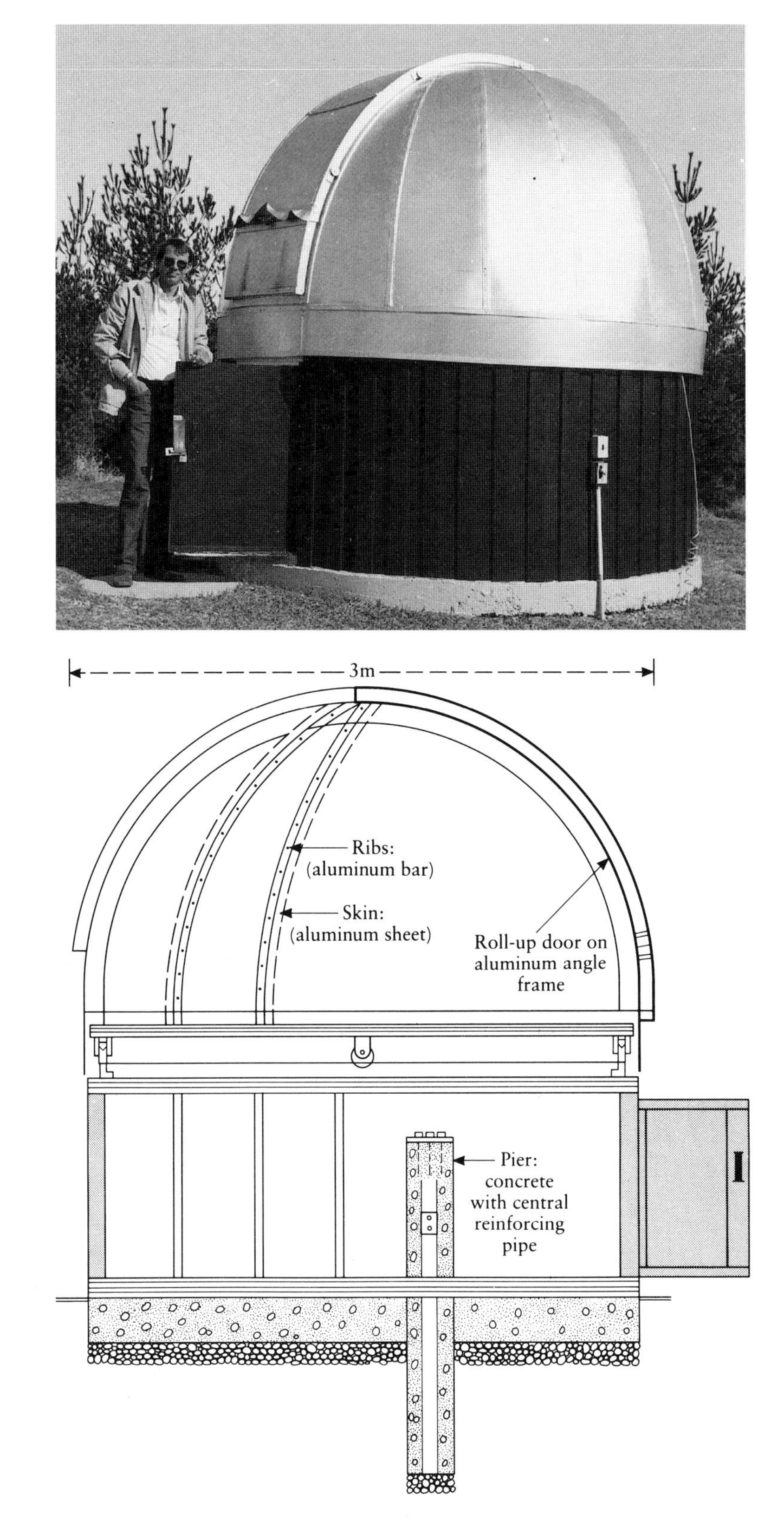

Plate 141. Observatory by John S. Hicks, Keswick, Ontario.

Figure 6.17. Aluminium-domed amateur observatory by John S. Hicks (plans available from the designer; see text).

John S. Hicks. The 3-metre-diameter observatory has a full-diameter base of concrete 20 cm in thickness, with a wooden floor separated from the chill of the concrete by an insulating layer of styrofoam. The telescope pier itself is sunk to a depth of about 1.25 m into the ground, for great stability. The rotating dome, in which aluminium is used for both frames and sheathing, is lighter than the more common plywood structure.

Drawings and photos of this interesting home observatory (Figure 6.17, Plate 141) are reproduced here with kind permission of the designer. Full construction plans for the domed observatory, and for a highly sophisticated roll-off-roof alternative, can be purchased from John Hicks by writing to P.O. Box 75, Keswick, Ontario, Canada L4P 3E1.

6.6 The Jack Newton Observatory

My 64 cm Newtonian is mounted in a new home observatory in Sooke, about 20 km southwest of Victoria, British Columbia, Canada. It sits atop the southern slope of Mount Matheson, at an elevation of approximately 310 metres. From our perch we enjoy glorious unhindered views of many miles of unspoiled beauty, extending from our own $7\frac{1}{2}$ forested acres to across the nearby Straits of Juan de Fuca and the town of Port Angeles, Washington, USA.

Because of its rocky location, our home is built on blasted bedrock (into which it is cemented as an earthquake precaution). While it encompasses five levels, all above ground, the height of the house is two stories. The roof-top observatory is readily accessible through a second level office with its own stairway leading up into the dome.

Construction of the telescope pier started just as soon as the blasted rock on the site had been levelled and packed. First a special 1.8 × 1.8 m footing was poured, with reinforcing rods protruding out of it. Construction of the first few feet of the 1 × 1 m concrete block square pier then commenced. Next, steel mesh was placed around the new base and the first floor of the house poured over the mesh. As the pier reached the height of the house floor at each level, concrete was poured into the pier. The pier, (which strongly resembles a chimney and is often mistaken for one), remains the

Plate 142. A front view of the Jack Newton Observatory.

same size as it continues upward through to the second floor, but then narrows down to a round 35 cm concrete cylinder (sona tube was used as the form) for the final 1 m to the telescope base. Its finished weight is around 13 000 kg and it is required to support a telescope weight of approximately 450 kg.

The pier is completely free standing, and does not actually touch the floors or ceilings anywhere in the house. This is very important to ensure that no vibration travels from the traffic in and through the house to the telescope itself. (At our house-warming party for the tradesmen who worked on the project, we opened the telescope for viewing. While 55 people were in the house partying at one time, no vibration was noticed in the scope at all.)

Plate 143. Rear view of the Jack Newton Observatory.

The floor and stairway leading to the dome are insulated to keep the telescope at outside ambient temperature. I turn on an exhaust fan to gently move air through the dome about an hour prior to doing any observing.

The telescope mount is built around a 43 cm Mathis worm gear. The German equatorial design best suits my needs. The right ascension housing was constructed from heavy-walled 20 cm pipe welded into 15 mm steel positioning plates. The right ascension shaft itself is 15 cm pipe mounted through a bearing at each end; a larger clutch was added to this assembly incorporating a four-screw pressure plate. The declination housing is 15 cm in diameter with a 10 cm pipe shaft mounted in bearings at each end. The counterweight shaft is just an extension of the declination rod to secure the 205 km of counterweight. The top end of the declination shaft is attached to a 30 cm × 90 cm × 13 mm steel plate. This plate has two larger rings welded to it to support the telescope tube. A 46 cm tangent arm declination clutch is attached around the declination housing and worm-driven with a reversible synchronous motor. A cradle was constructed of 6 mm × 50 mm wide steel bands reinforced with 2.5 cm angle iron. A 100 × 30 cm slab (25 kg) of 13 mm steel forms the base of the cradle. The mirror, telescope tube and cradle weigh in at over 135 kg. The finished mount is equipped with optical encoders on both RA and Dec. for a CAT (computer assisted telescope) computer.

The observatory dome, which I designed and built, is 4.9 m in diameter. Because I was fabricating the dome in a workshop some distance from the building site, I needed to construct it in such a way that it could be moved by truck down a steep, narrow, winding driveway, and then along a rough highway for about 10 km. With this in mind, I built the dome in halves, and then completed its assembly outdoors after it had been delivered to our cul de sac.

I used 18 mm plywood sliced into 20 cm widths and placed these to form a 4.9 m circle, laminated three layers thick. The ring was glued and screwed together for strength. The two main arches were built using the same diameter as for the base of the dome and placed 1.2 m apart. I then laminated 4.9 m × 6 cm × 9 mm cedar lath to form a 10 cm thickness and bent these into position using the base ring as the form. I needed 22 of these laminated strips, which were also glued and screwed for strength. The next step was to form the front of the dome. I did this by placing the lathe strips in 60 cm intervals from the outside of the base ring to the arches. I then covered the form with 3 mm-thick mahogany door skins, available in 1.2 × 2.4 m sheets. Each section was cut, bent over the rib structure, glued and then nailed in place. When all the sections were in place, the entire dome was fibreglassed.

Once the dome had been moved to the side yard, I connected the halves of the base and bent electrical conduit to form a ring, which I attached to the bottom of the dome. I then readied the dome, telescope tube and telescope mount to be lifted by crane onto the house just before the roof trusses were lifted into place.

A 1 m pony wall on the roof of the house had been readied to receive the dome by the installation of 13 cm grooved v-wheels pointing upward every 30 cm along the top of the wall. When the dome was lifted into position, and lowered onto the wheels, the conduit became the track on which the wheels ride.

Finally, the dome in place on the roof, I attached the covering for the slit, which has a 1.2 × 0.6 m section at the bottom that flips out on hinges. The upper 3 m section slides over the top.

The optics for my 25 in. (64 cm) f5 telescope were produced by Galaxy Optics in Colorado. The mirror cell is pan style, with its base constructed from 12 mm aluminium plate and a 53 mm band screwed around the base to support the sides of the mirror. The mirror is 100% floated on bubble pack (the packing material with air bubbles that everyone loves to pop with their fingers!). The mirror is secured with six claws, which do not touch the surface of the mirror. The 3 m long tube is constructed from 3 mm aluminium, rolled in three sections and welded together. I had the barrel fabricated in a local shipyard.

I use an 11 cm minor axis diagonal and homemade off-axis guider. The simple guider is constructed using a 9 mm aluminium plate that rotates in a ring mounted on the side of the telescope where the focuser would be positioned. This forms a photographic platform and permits quick changes in a variety of equipment. The guider has two prisms which are mounted on one side of the 53 mm focusing tube. The prisms will rotate through the field and the whole plate will rotate as well. This is coupled with a 3× Barlow and a 12 mm eyepiece, which produces a magnification of 800× for guiding. I can pick up virtually any star in the field to guide on, and usually have a choice from a half dozen suitable guide stars.

The telescope is equipped with an ST-6 model CCD camera for deep-sky photography. It was produced in California USA by the Santa Barbara Instrument Group. I use an IBM 386–40 computer to image and guide the telescope. My warm office is in the level directly below the observatory, and from here I can manually or auto guide. I use a 486-level personal computer to process the images in my office.

6.7 Some amateurs' projects

This final chapter will be a miscellany, but its contents are linked by a common theme. Here follow examples of several typical building projects that amateur astronomers have completed, for an intriguing variety of purposes.

It is fun to look at other people's ideas. Sharing the results of one's basement-workshop labours is a major part of the joy of amateur astronomy. Valuable hints are picked up in this way, both in terms of ingenious ideas to be copied and in terms of mistakes that may perhaps be avoided. In the final analysis, every amateur builder likes to do things in his or her own way.

Stationwagon telescope

To Leo Van der Byl of Victoria, British Columbia, the idea of transporting a 46 cm telescope in the back of a family car is not far-fetched. His giant Schmidt–Cassegrain actually folds out of a built-in emplacement in the back of his car.

The fork mounting slides out of the back of the stationwagon on tracks bolted to the bed of the car. When the telescope clears the rear hatchway, a pier is placed under the mounting and the telescope is jacked free, permitting the car to be driven away. A hand-crank adjusts the angle of the fork for polar alignment. An especially intriguing feature of Leo's telescope is the way in which fast slewing in RA is managed. A separate large gear is attached to the right-ascension shaft at the base of the fork. A motor mounted on the fork allows the fork to 'walk' around this gear in either direction. This means that the main RA worm gear always travels in one direction at sidereal speed.

The telescope's builder is currently designing a computer that will use stepping motors to control both RA and Dec. to seek any spot in the sky merely by punching in a specific object's co-ordinates; the telescope will do the rest.

Van der Byl's telescope is optically interesting. It employs a 46 cm,

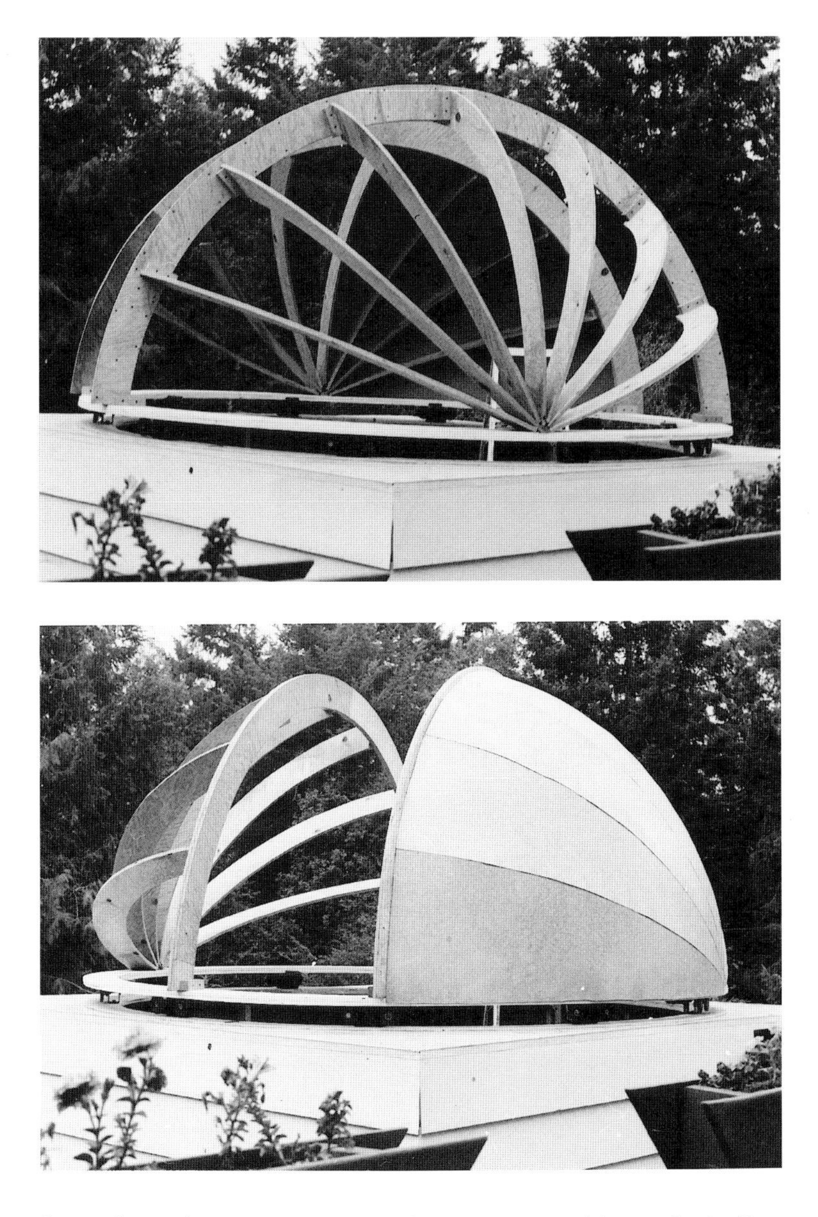

Plate 144. Paul Thomas' sundeck observatory.

f2.5 spherical primary mirror and a corrector, yielding a final effective Schmidt–Cassegrain focal ratio of f10. A remote controlled camera for the f2.5 focus will soon complete this ingenious system.

Homebuilt lensless Schmidt

The world's great observatories depend heavily on their large Schmidt cameras for discovery work and for wide-field sky surveys. The optical complexity of this system's superfast primary mirror and high-precision corrector has discouraged most amateurs from building these cameras for home use.

Plate 145. George Ball with his homebuilt lensless Schmidt camera.

George Ball's recently completed Schmidt is an example of how a skilled amateur optician can produce an effectively similar instrument without the labour and expense of the classic Schmidt. The lensless Schmidt uses a simple diaphragm placed at the primary mirror's radius of curvature, instead of a complex correcting lens.

Ball's Schmidt has a 24 cm primary mirror and a 15 cm diameter lightstop, or diaphragm, to produce an optical system that is in effect a 15 cm f5. Special features of this unit include homecast curved filmholders and a transfer lens that permits the user to look visually through the Schmidt camera for targeting and focusing. In spite of the optical simplicity, this instrument has turned out to be excitingly fine, yielding sharp star images to the edge of its wide field.

The new Schmidt camera shares a mounting with a 15 cm visual Schmidt–Cassegrain telescope and an aerial camera. When George Ball completes the 25 cm and 30 cm Schmidt telescopes that he is now building simultaneously, that heavily burdened mounting is going to become distinctly cluttered!

The Evans Van der Byl mobile telescope

Following is a description of a project that solved an astronomical society's problem of where to site a telescope for maximum use in the widest possible variety of members' needs.

For many years the Royal Astronomical Society of Canada's Victoria Centre had held funds that were bequeathed by Robert Evans to be used for a centre telescope. Among the difficulties to be resolved were the selection of a site accessible to all members, and the problem of vandalism that has plagued club observatories

Plate 146. The 50 cm Evans Van der Byl mobile telescope (its builder here shows the new instrument to the late Dr Helen Sawyer Hogg. Dr Hogg, who contributed the Foreword to this book, died while the *Guide* was in preparation).

situated at remote locations where nobody is permanently in residence.

One spring when Leo Van der Byl and I were returning from the Riverside Telescope Makers' Conference in California, Leo announced that he had been lucky enough to acquire an excellent 50 cm mirror blank. Furthermore, he had a suggestion regarding its use: why not build a large mobile telescope for the Victoria Centre? When he arrived home, Leo drew up plans and then proceeded

Plate 147. Portable 46 cm Schmidt–Cassegrain telescope, mounted on a stationwagon, designed and constructed by Leo Van der Byl.

almost singlehandedly to create the largest mobile Cassegrain telescope in Canada.

The telescope was built from the ground up. The trailer with its tandem wheels was welded together in a garage. Insulated walls were built, and piano-hinged to fold out; these walls were sheathed with aluminium siding. The roof was cast in fibreglass mould and heavily insulated. The roof slides off the trailer onto the towing vehicle. Three legs lift the telescope up from the trailer floor.

The large fork mount contains a Nasmyth focus through its declination axis, at f12. The primary mirror, which is f4 at its Newtonian focus, was ground and polished by Van der Byl with some assistance from George Ball.

The main drive gear is a 46 cm worm gear produced by Thomas Mathis of California.

The trailer houses not only the telescope, but also a computer centre. Once the telescope has arrived at an observing site and has been aligned, it can be made to point at a specific celestial target by simply entering the object's co-ordinates. The telescope is also linked to a TV camera and monitors, on some occasions, for public demonstrations.

A blink comparator

Can a backyard astronomer hope to find the dim glow of a comet or the pinprick of light from a tiny asteroid amid the blizzard of stars in the telescopic sky? Surely it is too much to expect that an amateur, with inexpensive and limited equipment, may detect the arrival of such an object before it is noted by the world's great observatories.

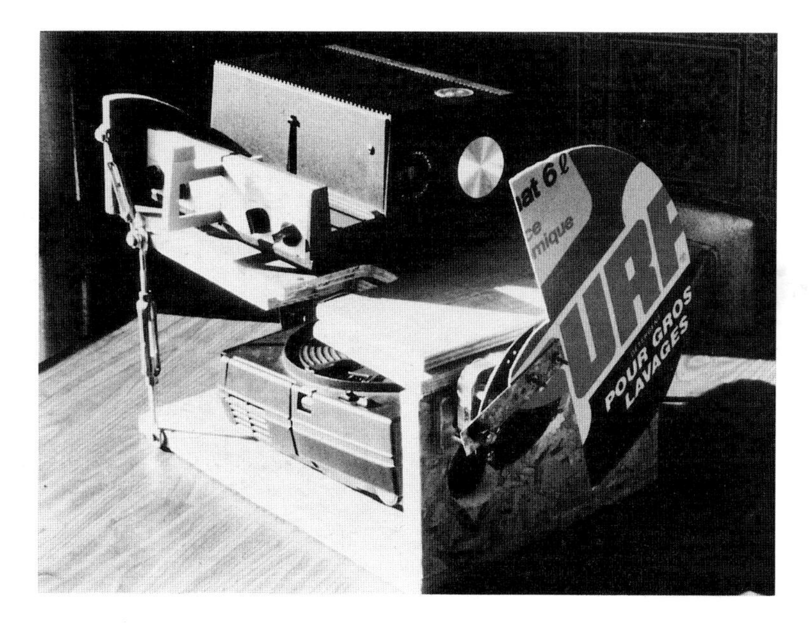

Plate 148. A blink comparator.

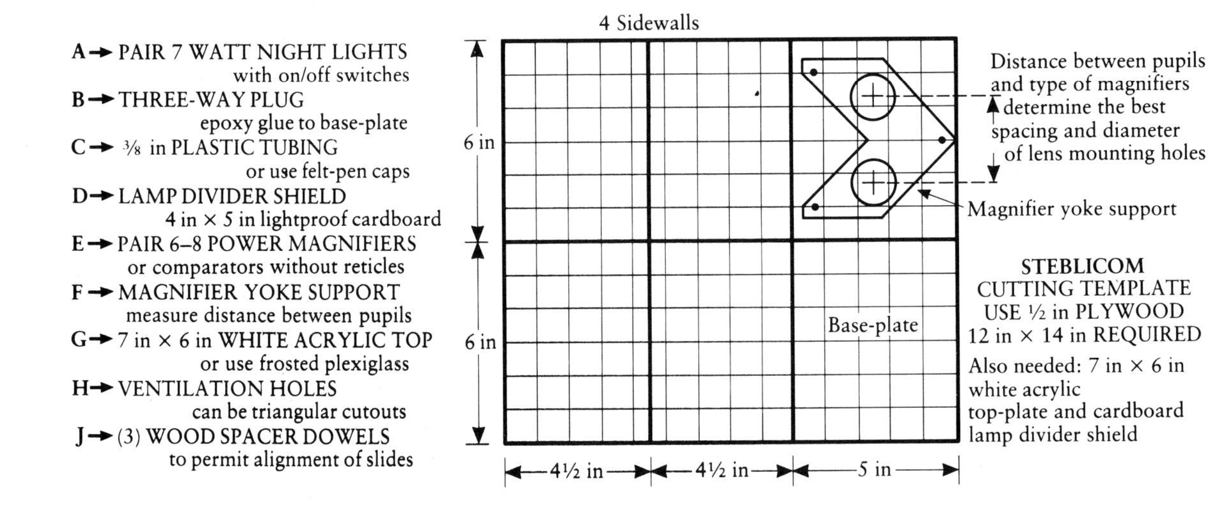

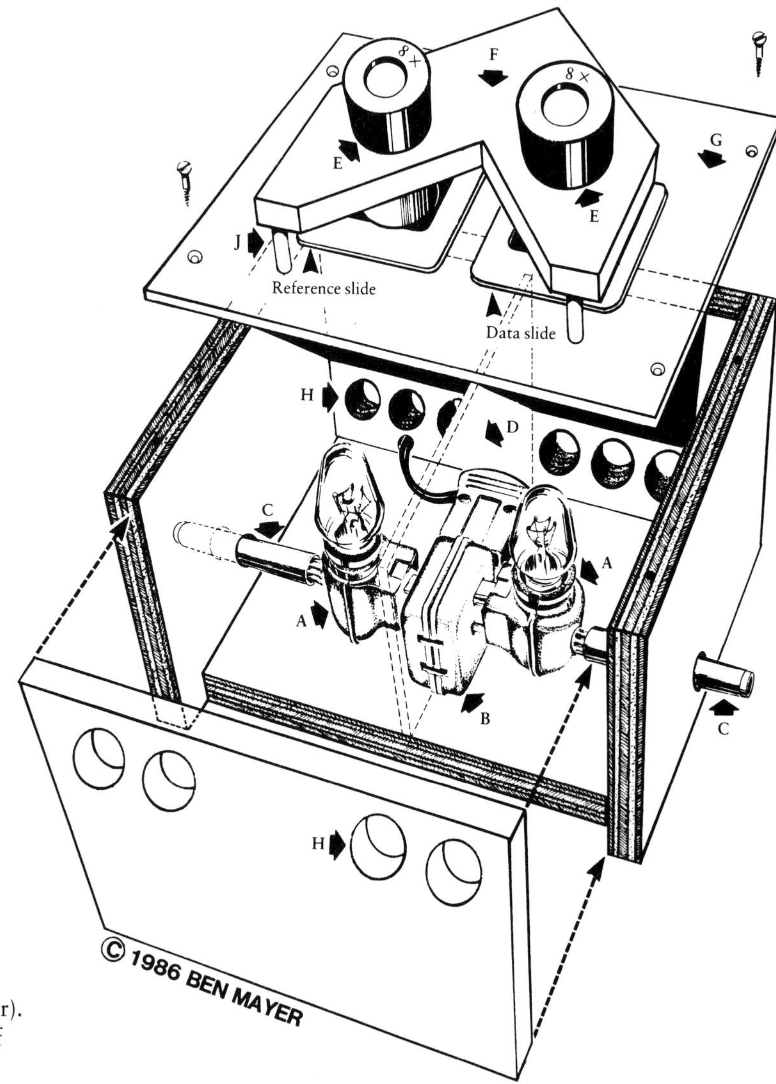

Figure 6.18. The Ben Mayer Steblicom (stereo blink comparator). Reproduced by kind permission of Ben Mayer.

The professional approach is to 'blink' pairs of photographs taken on separate occasions. The *blink comparator*, by rapidly alternating between the views provided by two photographs, displays an animated impression of the star-field. As the separate photographs are illuminated, any point of light that is absent on one of the slides but present on the other will appear to flash on and off.

The noted California amateur astronomer Ben Mayer has designed a highly attractive device of this kind for construction in the home workshop. His Steblicom ('Stereo Blink Comparator') is shown in Figure 6.18, which Ben Mayer very graciously contributed to this book.*

The basis of this little unit is a pair of naked-lights, each of which can be manually switched on and off, and a pair of binocular eyepieces, each mounted above one of these small lamps.

The matched slides that are to be compared are positioned on top of a translucent plastic plate under the pair of eyepieces. Blinking is done by merely switching the lights on and off, alternately, by hand (a small plastic knob is held in each of the observers' hands). With his or her eyes placed at the binocular eyepieces, the observer sees one slide with the left eye and the other with the right eye in rapid succession. The system is fine testimony to the fact that elegance can be derived from simplicity!

The construction details included in Ben Mayer's drawings should be sufficient to allow our readers to build Steblicom units for themselves.

* For more details of Mayer's innovative work with blink comparators, see Liller, William and Mayer, Ben, *The Cambridge Astronomy Guide*, Cambridge University Press, 1986.

Appendices

Appendix I Sky nomenclature

Traditionally, the brightest stars in each constellation have been designated by letters of the Greek alphabet. Usually (but with many notable exceptions) these letters are assigned in order of brightness, a constellation's brightest star being α (Alpha), its second being β (Beta), etc.

This style of nomenclature is used in all starcharts and observers' catalogues – and by astronomers in everyday conversation. Thus every serious observer will wish to become familiar with the appearances and sounds of the letters of the Greek alphabet, which is presented in the following table.

The Greek alphabet

A, α Alpha	I, ι Iota	P, ρ Rho
B, β Beta	K, κ Kappa	Σ, σ Sigma
Γ, γ Gamma	Λ, λ Lambda	T, τ Tau
Δ, δ Delta	M, μ Mu	Υ, υ Upsilon
E, ϵ Epsilon	N, ω Nu	Φ, ϕ Phi
Z, ζ Zeta	Ξ, ξ Xi	X, χ Chi
H, η Eta	O, o Omicron	Ψ, ψ Psi
Θ, θ, ϑ Theta	Π, π Pi	Ω, ω Omega

When a specific star is named, it is usually designated by its assigned letter plus the name of its constellation *in the genitive case*. The principal star in Orion, for instance, is α *Orionis*, and the second star in Centaurus is β *Centauri*. What is this 'genitive case' of a constellation name? It is simply the old Latin form that means 'of' or 'belonging to' the constellation. τ (Tau) Ceti means 'Tau, belonging to Cetus'.

In names that end with *-us*, that final syllable changes to *-i* in the genitive case. The ending *-ius* changes to *-ii*, and *-a* becomes -ae. The large number of constellations whose names end with a consonant (for example, Orion) have *-is* as their genitive ending. A few genitive endings that appear to fit none of these main categories are plural forms, where the constellation name itself is plural (for example Pisces, which alters to Piscium).

Perhaps in our era of 'small Latin and less Greek' these genitive forms have become an obsolete nicety; yet their correct usage continues to be the mark of the scholarly amateur (and professional) astronomer. The following list includes names plus genitive forms of 32 major constellations. In each case a stress-mark has been placed on the appropriate syllable, as an aid to pronunciation.

Constellation	Genitive form	Meaning
Andrómeda	Andrómedae	'Andromeda', the daughter of Cassiopeia
Aquárius	Aquárii	The Water-bearer
Aquíla	Aquílae	The Eagle
Áries	Ariétis	The Ram
Auríga	Aurígae	The Charioteer
Boötes	Boötis	The Herdsman
Cáncer	Cáncri	The Crab
Cánes Venátici	Cánum Venaticórum	The Hunting Dogs
Cánis Májor	Cánis Majóris	The Great Dog
Carína	Carínae	The Keel
Cassiopéia	Cassiopéiae	The Queen 'Cassiopeia'
Centaúrus	Centaúri	The Centaur
Cépheus	Céphei	The King 'Cepheus'
Cýgnus	Cýgni	The Swan
Gémini	Geminórum	The Twins
Hércules	Hérculis	'Hercules', ancient Greek hero
Léo	Leónis	The Lion
Líbra	Líbrae	The Scale (balance)
Lýra	Lýrae	The Lyre
Monóceros	Monocerótis	The Unicorn
Ophiúchus	Ophiúchi	The Serpent-bearer
Oríon	Oriónis	The Hunter
Pégasus	Pegási	The Winged Horse
Pérseus	Pérsei	'Perseus', ancient Greek hero
Písces	Píscium	The Fishes
Sagítta	Sagíttae	The Arrow
Sagittárius	Sagittárii	The Archer
Scórpius	Scórpii	The Scorpion
Sérpens	Serpéntis	The Serpent
Taúrus	Taúri	The Bull
Úrsa Májor	Úrsae Majóris	The Great Bear
Vírgo	Vírginis	The Maiden

Appendix II Time and timekeeping

Readers who are not familiar with advanced amateur work may be surprised at the wide range of astronomical activities that are time-dependent. Not only eclipse events, occultations of stars and movements of planetary satellites, but even the precise shapes of markings on the planets are phenomena whose measurement requires techniques that involve the clock.

Universal Time

For purposes of worldwide uniformity, all information provided in astronomical tables, yearbooks and periodicals and also all reported observations are recorded in *Universal Time* (UT). Unlike *civil time*, which varies from one time zone to the next, UT is independent of the observer's geographic location, and is equivalent to the time at longitude 0°. Thus, to convert the time of an event from UT to local civil time, we must subtract one hour for each time zone by which we are removed (westward) from the zero-degree or *Greenwich* zone. Because UT is expressed in the 24-hour system, the civil 1.00 pm is UT 13.00, 2.30 pm is 14.30 and so forth.

The process of converting time will often result in a situation that confuses some users of astronomical tables (including editors of newspapers, etc., who attempt to record the local data of a phenomenon such as the full Moon). An event that occurs at one given UT date may, in one's own time zone, occur at a civil time that falls in the previous calendar day. For example: the full Moon of February 1986 occurred at 00.31 on the 26th of the month. At the author's home city in the Pacific coastal time zone, however, eight hours had to be subtracted from that figure to give the local time of the event. That placed the full Moon, for us, at 16.31 (4.31 pm) on February 25.

The timing of phenomena

If a really useful observation of an event such as the occultation of a star by an asteroid is to be made, the instant of occurrence must be timed with the highest degree of accuracy that simple amateur methods can achieve. Although even a very roughly clocked observation can, in some circumstances, have a scientific value, most timings should be expressed accurately to a fraction of a second.

How can such a timing be achieved?

A method that is used almost universally among amateurs is as follows: the observer first ensures that he or she has all the necessary equipment on hand during the observation – a shortwave radio, a tape recorder and some instrument by which a sharp, brief signal can be made. Useful for the latter purpose is the type of digital stopwatch that emits a crisp beep when activated or deactivated.

The radio is tuned to one of the shortwave stations that continuously broadcast a coded time signal (WWV at 2.5, 5, 10, 15, 20 MHz is available throughout most of North America; MSF at 60 and 2500 KHz is a British service of this kind). Throughout the period in which the celestial event is expected to occur, the astronomer plays the radio time signal and runs the tape recorder. At the instant when the event begins, the stopwatch sound-signal (or other signalling device) is beeped. At the end of the event, a second beep is emitted. On the tape, when it is replayed, one has a permanent record of the timing, which can be assessed at one's leisure. The tape will have picked up the two signals *superimposed on the coded time signal*. Thus, from the time signal, one can form a very precise estimate of the time at which the event began, and of its duration.

Appendix III Your telescope's health: care and adjustment

Few amateur astronomers willingly abuse their telescopes. One has after all paid a high price, either in cash or in labour (if the instrument is homebuilt); one's instinct is to handle the telescope gently.

Ordinary careful usage includes such precautions as guarding against excessive accumulations of dust and moisture, and against jolts or traumas that can disturb the optical alignment. It is important to refrain from touching optical surfaces with the hands. Surprisingly, a mere touch can be one of the most damaging things you can do to a fine lens, because the sophisticated chemical coatings on modern optics are susceptible to permanent etching by the acids left behind in your fingerprints.

Dust should be removed from mirrors and lenses with soft lens-tissue or a light jet from a compressed-air can designed and sold for this purpose. Optics should be cleaned as seldom as possible; a few specks of dust will harm the lens far less than a scratch that you may accidentally cause during cleaning.

Recollimation of the optics is a chore that should never have to be undertaken if your telescope is a refractor or a catadioptric of the Maksutov design. These types are noted for their relative permanence of adjustment, even if handled roughly. Newtonian reflectors and Schmidt–Cassegrain telescopes may occasionally become misaligned, especially after being transported in cars, etc. This does not represent a weakness in these instruments, for they are designed to be readjusted with great ease, as follows.

Newtonian adjustments

In daylight, look into your Newtonian reflector through its eyepiece mount, with no eyepiece in place. Instead of the eyepiece, you should have a cardboard stopper in the focusing mount, pierced at its centre with a small circular hole about the size of an eyepiece lens. As you peer through this diaphragm, the correctly collimated telescope should show the outline of its flat secondary mirror perfectly centred

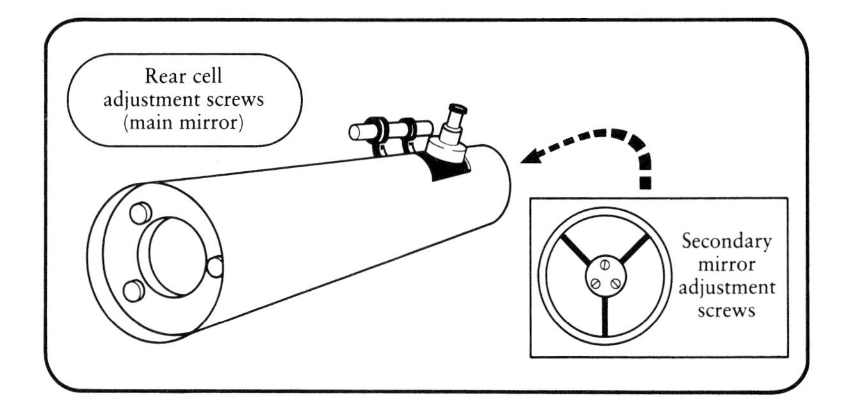

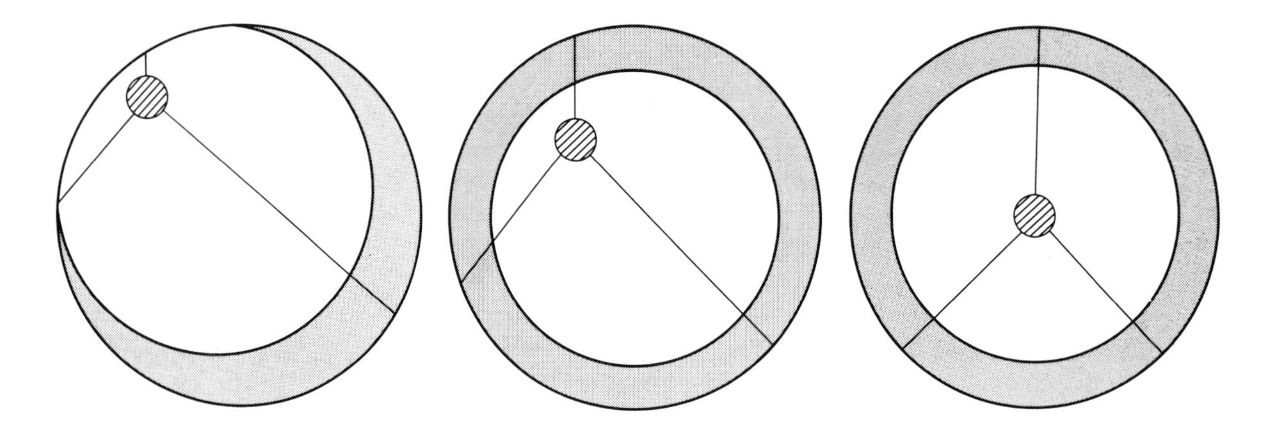

Figure AIII.1. Adjustment points on the Newtonian telescope. *(a)* Drawtube view of telescopic interior: both mirrors out of alignment. *(b)* First adjustment: square the diagonal (secondary) mirror by gentle loosening/tightening of its three adjustment screws. Correct alignment of the secondary will bring image of the primary mirror to central position within drawtube. *(c)* Second adjustment: square the main mirror by careful movement of its three adjustment screws. Correct alignment of the primary mirror will bring the small reflected image of the secondary to the central position.

within the eyepiece mount, and the outline of the primary mirror perfectly centred within the circular perimeter of the flat. If misalignment is observed, it can be corrected by following the few simple adjustment-steps illustrated in Figure AIII.1.

If the telescope, when focused on a star in the night sky, still shows a slight flaring of the stellar image in one direction, a final reduction of this condition can be achieved by trial-and-error adjustments of the three main mirror screws. By checking the appearance of the star image with a high-power eyepiece, you will be able to see the effects of exceedingly minute alterations in the settings of the adjustment screws.

Schmidt–Cassegrain adjustments

The popular commercially made Schmidt–Cassegrain telescopes are much less prone to misalignment than the Newtonian type. If collimation is required, it will be only the secondary mirror that needs adjustment; the primary is permanently fixed in alignment.

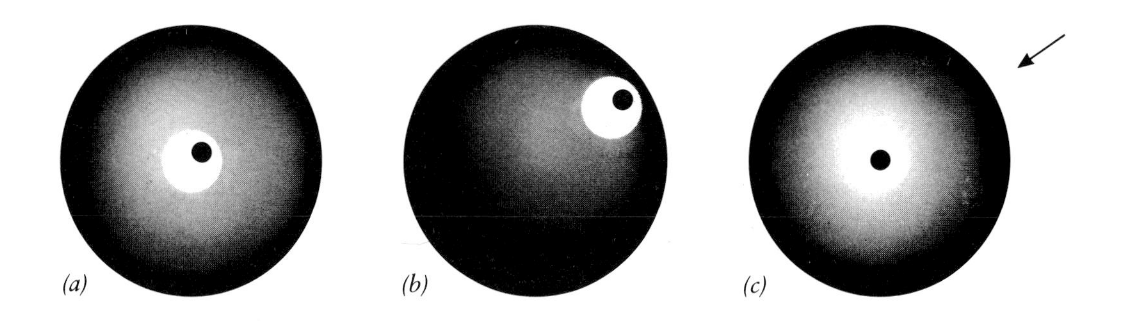

Figure AIII.2. *(a)* View of unfocused star-image with shadow of secondary mirror not centred within circle of light. *(b)* Move the telescope to shift the image towards field-edge in the direction in which the shadow is off-centre. *(c)* Recentre the image by tightening secondary mirror alignment screw at position indicated (and by loosening the two screws opposite). The shadow should now be at centre of light circle. If not, repeat steps *(b)* and *(c)*.

To check the Schmidt–Cassegrain's collimation, centre a first-magnitude star in the telescopic field, with a high power eyepiece in place. Put the telescope quite radically out of focus, so that the stellar image becomes a large circle of light in which the shadow of the secondary mirror appears as a black obstruction. If this shadow is not precisely centred within the light circle, the secondary mirror needs adjustment, as illustrated in Figure AIII.2.

As in the case of the Newtonian reflector, final superfine collimation can be done by focusing the star-image, using high magnification. If the stellar disc is not precisely centred within its surrounding diffraction ring, a very cautious touch of the secondary mirror's adjustment screws will correct the situation.

Appendix IV Star colours and spectral classes

'Oh, Be A Fine Girl; Kiss Me!'

Unlikely though it may seem, this brash remark has relevance to material in our foregoing chapters. At several points we have made reference to 'classes' of stars. M-Class dwarfs were mentioned, and we have talked about yellow G-Class suns. Certain bright young clusters were said to have been populated by superhot white B-Class giants.

What are these classes of stars?

The *spectral classes* represent a scheme of classification in which stars are grouped together by certain characteristics of their spectra. Each class has its own typical colour (sometimes quite noticeable even to the eye) and, in effect, these classes form a *sequence of descending temperature*. This is where the flirtatious invitation of our opening sentence has its use. The principal spectral classes, in descending order of temperature, are designated by the initial letters of those seven words – O, B, A, F, G, K, M.

The temperature–colour relationship is not something merely of dry academic interest. Every amateur astronomer should have at least a nodding acquaintance with this concept, for it adds an important dimension to his or her appreciation of the stars observed in the night sky. Looking at the steely blue-white glare of *Rigel*, for

Table AIV.1.

Spectral type	Temperature (surface K)	Visual colour	Examples
O	70 000	Blue-white	δ *Orionis*, λ *Orionis*
B	30 000	Blue-white	*Rigel* (β *Ori.*), *Regulus* (α *Leo.*)
A	15 000	White	*Castor* (α *Gem.*), *Vega* (α *Lyr.*)
F	9 000	Yellow-white	*Polaris* (α *UMi A*), γ *Virginis*
G	6 500	Yellow	Sun, α *Centauri*
K	5 000	Orange	*Arcturus* (α *Boo.*), *Pollux* (β *Gem.*)
M	3 500	Red	*Antares* (α *Sco.*), *Barnard's Star* (*Oph.*)

instance, one recognizes the typical aspect of a very hot star. In the dull red glow of an object like *Barnard's Star*, on the other hand, there is evidence of something near the opposite end of the temperature scale – a much cooler body.

Table AIV.1 summarizes features of the seven major classes. Temperatures are rough median values, and colours are the hues actually seen visually through a telescope, if stars are bright enough to show colour at all to the eye.

Appendix V Societies, clubs and activities

One of the most satisfying aspects of amateur astronomy is the communication of enthusiasm and knowledge, and especially the sharing of these with beginners. Such is the popularity of astronomy nowadays that almost every city has its own local clubs and activities that bring amateurs together. Your public library will have information on these groups. Following is a selection of the major national societies and a few of the best known annual events.

Addresses

American Association of Variable Star Observers
25 Birch St., Cambridge, MA 02138, USA.
American Astronomical Society
2000 Florida Ave. NW, Suite 300, Washington, DC 20009, USA.
Association of Lunar and Planetary Observers
PO Box 143, Heber Springs, AR 72543, USA.
Association of Planetary and Lunar Observers
New Mexico State University, Las Cruces, NM, USA.
Astronomical Society of the Pacific
390 Ashton Ave., San Francisco, CA 94112, USA.
British Astronomical Association
Burlington House, Piccadilly, London W1V 9AG, UK. (Meteor Section: 'The Harepath', Mile End Lane, Apuldam, near Chichester, West Sussex, PO20 7DZ, UK.
Central Bureau of Astronomical Telegrams
60 Garden St., Cambridge, MA 02138, USA (Electronic mail: MARSDEN CFA (.SPAN or .BITNET or .HARVARD.EDU)
International Occultation Timing Association
Lowell Observatory, PO Box 1269, Flagstaff, AZ 86002, USA.
Meteor Centre
Hertzberg Institute of Astrophysics, Ottawa K1A 0R6, Canada.
Royal Astronomical Society of Canada
136 Dupont Street, Toronto, ONT, Canada M5R 1V2.

Scientific Event Alert Network
Mail Stop 129, Natural History Bldg, Smithsonian Institute, Washington, DC 20560, USA.
Webb Society
96 Marmion Road, Southsea, Hampshire PO5 2BB, UK.
Webb Society of North America
1440 S. Marmora Ave., Tucson, AZ 85713–1015, USA.

Events

BAA Annual Exhibition Meeting
An amateur convention; held each year in May. Information from British Astronomical Association, Burlington House, Piccadilly, London W1V 9AG.
British National Astronomy Week
Held periodically, in Autumn: talks, demonstrations and telescopic observing. Information from NAW Co-ordinator, Royal Astronomical Society, Burlington House, Piccadilly, London W1V 0NL, UK.
Mount Kobau Star Party
Annual August gathering of amateur observers at one of the finest dark-sky sites in North America. Write to Mr Peter Kuzel, 4100 25th Ave., Vernon, BC, Canada V1T 1P4.
Riverside Telescope Makers' Conference
Telescope makers' competition and observing sessions, held annually in May. Contact Fox & Stephens, 9045 Haven Ave., Suite 109, Rancho Cucamonga, CA 91730, USA.
Royal Astronomical Society of Canada Annual Assembly
Joint amateur and professional gathering with talks, demonstrations and observing. Enquire through RASC National Headquarters, 136 Dupont Street, Toronto, ONT, Canada, M5R 1V2.
Southern Cross Winter Star Party
Annual gathering in January that brings observers from all regions to southern Florida for fine dark-sky astronomy while many other localities are buried in snow. Information from Bob or Sharon Grant, 5401 SW 110th Ave., Miami, FL 33165, USA.
Stellafane
The oldest annual gathering for display and use of amateur telescopes, held annually in August on Breezy hill in Vermont. For details write to Stellafane, 60 Victoria Road, Sudbury, MA 01776, USA.
The Texas Star Party
A huge conference in May, with lectures and excellent deep-sky observing. Held each year at 'The Prude Ranch'. Contact TSP Registrar, 1326 Mistywood Lane, Allen, TX 75002, USA.

Astronomy Day

Throughout North America this special day has been celebrated each year, usually in May, for many years. It is a co-ordinated sequence of events aimed largely at bringing amateur astronomy before the public eye through star parties, talks, and demonstrations arranged simultaneously across the continent by local clubs and groups. Contact your local society or one of the national groups listed above for details of the planned events of the current year.

Epilogue

The authors hope that this book has offered insights that will prove useful. We live in an age that demands 'usefulness', to a degree unsurpassed by any previous era. In keeping with this spirit, it is fashionable for books on astronomy to stress the practical contribution an amateur can make to science.

Let us conclude, however, on a different note.

For many amateur observers (even the deeply 'serious' ones) the extraction of pragmatic data from the night sky is not the prime motivation. It is true that our telescopes transmit information. Yet often we stand outdoors under a cold, dark sky in response to a simpler yearning. Like most of our basic drives, this one is hard to define. It has something to do with the nocturnal darkness itself, and the hidden splendour that is revealed in that darkness. It has much to do with the silence of the night, in which the perceptive stargazer catches an echo of the Universe's own mysterious silence.

The point is this: your patient, systematic study of variable stars, or of asteroidal occultations or the vagaries of solar behaviour may contribute something of value to science. Perhaps on a night of rare good luck you will even discover a nova. But don't let your delight in astronomy depend solely on such achievements. Ultimately, achievement may not be the source of the amateur's joy.

The sheer loveliness of the Cosmos seems in itself to convey a message. Its precise meaning is subtle; a lifetime of careful listening and watching may only barely suffice to capture some hint of what the stars have to offer the individual observer. Look up into the sparkling vault of the night sky with the uncluttered mind of a child. With this approach you may be fortunate enough to discover not only the science, but also the poetry of the wheeling galaxies and glowing pools of nebulosity.

At the end of an observing lifetime, perhaps many of us will recognize this quiet joy as the real prize.

Bibliography

Much of the best reading that we have enjoyed in astronomy, telescope-making and related fields was initially discovered through bibliographies in other books. We hope that the following list of titles will prove to be a similar springboard to our readers. Some of the books are classics that no amateur astronomer should miss. Others are less well-known works that proved useful to us in preparing *The Guide to Amateur Astronomy.*

Beatty, J. Kelly & Chaikin, Andrew, eds., *The New Solar System*, 3rd edition. Cambridge & New York, Cambridge University Press, 1990. (The finest compendium of planetary data and photos that we have seen.)

Berry, Richard, *Build Your own Telescope*. New York, Scribner, 1985. (Superb, simple plans for five amateur telescopes.)

Berry, Richard, *Choosing and Using a CCD Camera*. Richmond VA, Willmann-Bell, 1992.

Bishop, Roy L., ed., *The Observer's Handbook* (annual). Toronto, Royal Astronomical Society of Canada.

Bova, Ben, *Mars*. New York, Bantam, 1992. (A novel that dramatically portrays our post-Viking knowledge of the red planet).

Brandt, John C. and Chapman, Robert D., *Rendezvouz in Space: The Science of Comets*. New York, W.H. Freeman, 1992.

Burnham, Robert, Jr., *Burnham's Celestial Handbook*. New York, Dover, 1978. (This continues to be the standard descriptive catalogue – an essential item in every amateur's personal library.)

Carlotto, Mark, *The Martian Enigmas: A Closer Look*. Berkeley CA, North Atlantic Books, 1991. (Instructive for its description of image-enhancement techniques associated with space-probe planetology.)

Couteau, Paul, *Observing Visual Double Stars*, translated by Alan Batten. Cambridge MA, MIT Press, 1981.

Dickinson, T. & Dyer, A., *The Backyard Astronomer's Guide*. Camden East, ONT, Camden House, 1991. (Unusually thorough guide to telescopes and accessories available to the amateur astronomer.)

Dickinson, Terence, *Nightwatch*, 2nd edition. Camden East ONT, Camden House, 1991. (The most excellent introduction for the beginning observer that we have encountered.)

Giovanelli, Ronald G., *Secrets of the Sun*. Cambridge & New York, Cambridge University Press, 1984.

Hawkins, Gerald, *Mindsteps to the Cosmos*. New York, Harper & Row, 1983.

Henden, A. & Kaitchuk, R., *Astronomical Photometry.* New York, Van Nostrand Reinhold, 1982; reprinted, with corrections, Richmond VA, Willmann-Bell, 1990.

Jones, Kenneth Glyn, ed., *Webb Society Deep-Sky Observer's Handbook* (8 vols). Hillside NJ, Enslow, 1979–90.

Kronk, Gary W., *Comets: A Descriptive Catalog.* Hillside NJ, Enslow, 1984.

Levy, David H., *The Sky: A User's Guide.* Cambridge & New York, Cambridge University Press, 1991. (An outstanding introduction to the night sky, notable for its warm anecdotal dimension.)

Liller, William & Mayer, Ben, *The Cambridge Astronomy Guide.* Cambridge, Cambridge University Press, 1986.

Matloff, Gregory, *The Urban Astronomer.* New York, Wiley, 1991. (An optimistic view of what can still be done within the light-polluted environment of the city.)

Muirden, James, *How to Use an Astronomical Telescope.* New York, Simon & Schuster, 1984.

Parker, Don, *et al.*, *Introduction to Observing and Photographing the Solar System.* Richmond VA, Willmann-Bell, 1988.

Peltier, Leslie C., *Starlight Nights: The Adventures of a Star-Gazer.* New York, Harper & Row, 1965. (An inspirational classic for the amateur.)

Price, Fred, *The Moon Observer's Handbook.* Cambridge & New York, Cambridge University Press, 1988.

Ridpath, Ian, ed., *Norton's 2000.0 Star Atlas and Reference Handbook*, 18th edition. New York, Wiley, 1989. (Choose this atlas if you are just entering serious amateur astronomy.)

Schaaf, Fred, *Seeing the Night Sky.* New York, Wiley, 1992. (Unusual, project-organized approach to deep-sky observation.)

Texereau, Jean, *How to Make a Telescope*, 2nd edn. Richmond, VA, Willmann-Bell, 1984.

Tirion, Wil, *Cambridge Star Atlas 2000.0.* Cambridge & New York, Cambridge University Press, 1991. (A friendly introductory level atlas by the cartographer of the more advanced items that follow...)

Tirion, Wil, *Sky Atlas 2000.0.* Cambridge MA, Sky Publishing, 1981. (Superb atlas of deep-sky objects; a standard tool for all amateurs.)

Tirion, Wil, *et al.*, *Uranometria 2000.0.* Richmond VA, Willmann-Bell, 1987. (A very advanced star atlas; not for beginners.)

Vehrenberg, Hans, *Atlas of Deep-Sky Splendours.* Cambridge MA, Sky Publishing, 1983. (A famous photographic catalogue of the night sky; excellent!)

Webb, T.W., *Celestial Objects for Common Telescopes*, edited and revised by Margaret W. Mayall. New York, Dover, 1962. (Still substantially similar to the original 1859 edition, this charming old handbook is widely read for its unique embodiment of the amateur spirit in astronomy.)

Index

Page numbers in **bold** indicate photographic plates.